The Cayo Santiago Macaques

SUNY SERIES IN PRIMATOLOGY

EMIL MENZEL and RANDALL SUSMAN, editors

Edited by
RICHARD G. RAWLINS and MATT J. KESSLER

The Cayo Santiago Macaques

HISTORY, BEHAVIOR and BIOLOGY

State University of New York Press
ALBANY

Published by
State University of New York Press, Albany

Printed in the United States of America

For information, address State University of New York Press, State University Plaza, Albany, N.Y., 12246

Library of Congress Cataloging-in-Publication Data

The Cayo Santiago macaques.

(SUNY series in primatology)
Includes index.
1. Rhesus monkey. 2. Rhesus monkey—Behavior.
3. Mammals—Behavior. 4. Mammals—Puerto Rico—Santiago Cay—Behavior. 5. Santiago Cay (P.R.) I. Rawlins, Richard G., 1950- . II. Kessler, Matt J., 1947- III. Series.
QL737.P93C39 1986 599.8'2 86-19616
ISBN 0-88706-135-4
ISBN 0-88706-136-2 (pbk.)

We dedicate this volume
to our
parents

Contents

Contributors

CAROL M. BERMAN, Department of Anthropology, State University of New York, Buffalo, New York

LASLO Z. BITO, Research Division, Department of Ophthalmology, College of Physicians and Surgeons of Columbia University, New York, New York

BERNARD CHAPAIS, Département d'Anthropologie, Université de Montréal, Montréal, Québec, Canada

JOHN D. COLVIN, Department of Psychology, University of Bristol, Bristol, England

C. JEAN DEROUSSEAU, Department of Anthropology, New York University, New York, New York

CHRISTINE R. DUGGLEBY, Department of Anthropology, State University of New York, Buffalo, New York

HAROLD GOUZOULES, The Rockefeller University Field Research Center for Ecology and Ethology, Millbrook, New York; Yerkes Regional Primate Research Center Field Station, Lawrenceville, Georgia; Department of Psychology, Emory University, Atlanta, Georgia

SARAH GOUZOULES, The Rockefeller University Field Research Center for Ecology and Ethology, Millbrook, New York; Yerkes Regional Primate Research Center Field Station, Lawrenceville, Georgia

PHILIP A. HASELEY, Department of Anthropology, State University of New York, Buffalo, New York

DAVID A. HILL, Sub-Department of Animal Behaviour, University of Cambridge, Madingley, Cambridge, England

RICHARD M. JACOBS, Nutrient Toxicity Section, Food and Drug Administration, Washington, D.C.

PAUL L. KAUFMAN, Department of Ophthalmology, Medical School, University of Wisconsin-Madison, Madison, Wisconsin

MATT J. KESSLER, Caribbean Primate Research Center, University of Puerto Rico, School of Medicine, San Juan, Puerto Rico

ANN O. LEE JONES, Nutrient Toxicity Section, Food and Drug Administration, Washington, D.C.

PETER MARLER, The Rockefeller University Field Station for Ecology and Ethology, Millbrook, New York

BERNADETTE M. MARRIOTT, Psychology Department, Goucher College, Towson, Maryland; Caribbean Primate Research Center, University of Puerto Rico, School of Medicine, San Juan, Puerto Rico.

CAROL A. McMILLAN, Wenatchee Valley College North, Omak, Washington

RICHARD G. RAWLINS, Department of Obstetrics and Gynecology, Rush-Presbyterian-St. Luke's Medical Center, Chicago, Illinois

J. CECIL SMITH, JR., United States Department of Agriculture Beltsville Human Nutrition Research Center, Beltsville, Maryland

CATHERINE E. SCANLON, Animal Behaviour Research Group, The Open University, Milton Keynes, England

JEAN E. TURNQUIST, Department of Anatomy and Caribbean Primate Research Center, University of Puerto Rico, School of Medicine, San Juan, Puerto Rico

Foreword: The Cayo Santiago Macques

PETER MARLER

Strides in behavioral primatology over the past 20 years have been dramatic, so much so that it is difficult even for those who were in at the start to recall how primitive our knowledge was about the natural behavior of nonhuman primates when C. Ray Carpenter conceived of the Cayo Santiago project in the mid-1930s. This is not, of course, to underestimate the enormous impact of the earlier work and inspiration of Carpenter's mentor, Robert Yerkes, nor the significance of preliminary efforts at fieldwork by such pioneers as Bingham and Nissen. However, Yerkes placed primary emphasis on laboratory studies of behavior rather than on field research, with a focus on experimentation rather than description. Under his leadership, we had learned a great deal by 1940 about the learning abilities of monkeys and apes. It is still an inspiration to browse through the works of Yerkes and his many associates, such as the papers of Keith and Claire Hayes on the home-reared chimpanzee Viki. They reveal a fascinating array of remarkable talents. Yet, paradoxically, there was at that time no conception of how to place these talents in the natural behavior of chimpanzees or, for that matter, of any other nonhuman primate, if indeed there was a place at all. It was tempting to wonder whether the abilities were, in a sense, 'unnatural' and perhaps inculcated by the very intimacy of the mutually interactive, animal-human relationships that characterized so much of the early work on primate behavior.

Carpenter was a pioneer in several senses. He realized the urgent necessity for placing studies of primate behavior in an ethological perspective much broader than was typical of the Yerkes era. Perhaps even more importantly, he was able to call on his own experience in the field with gibbons and howlers to convince the younger generation that research on the natural behavior of primates was a practical, realistic undertaking capable of yielding rigorous and reproducible data, and not just anecdotes. Of all of Carpenter's many enterprises, founding the Cayo Santiago colony of rhesus macaques had perhaps the most enduring yield of all.

Unusual foresight and forebearance were displayed by those who, once the Cayo Santiago colony of rhesus macaques was established, nurtured it through a period of ten years or so in which the colony stabilized and established something close to normal social organization, with several troops. The scientific yields were not yet emerging, and in any case, they must have been difficult for all but the most inspired of administrators to appreciate. Only gradually did its true potential come to be widely appreciated.

Cayo Santiago offered two unique advantages. On the one hand it was infinitely more accessible than field research sites in Asia and Africa, and the animals were already habituated to the presence of observers. Perhaps the most valuable single asset, however, not available at any other study site except for the macaques studied by the Primate Research Institute of Japan, was the potential for obtaining long-term genealogical information. The annual practice of tattooing the yearlings, initiated by Stuart Altmann, and the maintenance of meticulous records of births, deaths, and maternal parentage eventually provided the basis for unique insights into primate social organization and its development that helped to bring behavioral primatology to full maturity. The foundations were laid for the work of such investigators as Conaway, Kaufmann, and Koford on population dynamics and reproductive behavior. They permitted Altmann not only to broach an entire spectrum of themes in the study of primate sociobiology, but also to bring to research on communication a degree of sophistication that was years ahead of its time. One of the many roots of modern sociobiology is to be found in the work he did as a graduate student from Harvard University, under Ed Wilson's direction, on the social behavior of the Cayo Santiago rhesus monkeys.

Nowhere is the value of the cumulative record of genealogies and group organization more evident than in the work of Donald Sade and his associates. I can still recall the excitement engendered by the first revelations about the dominance organization of social groups, the regularity of the progression in rank as a young monkey matures, and above all, the impact that a mother's rank has on that of her children, especially her daughters. The discovery of the apparent taboo on incestuous matings between sons and mothers struck an enduring chord among physical anthropologists that reverberates to this day, made more fascinating by Elizabeth Missakian's subsequent observation that such matings do sometimes take place, but in secret. Sade's research program, encompassing a unique blend of viewpoints from anthropology, ecology, ethology, and psychology, demonstrated in concrete fashion the revolutionary insights that flow from well-designed, long-term studies on populations of nonhuman primates subjected to a minimum of disturbance. Tensions were never far below the surface between investigators who saw long-term behavioral studies as the unique scientific niche for the Cayo Santiago colony, and more experimentally oriented researchers at the Laboratory of Perinatal Physiology who wanted to use the colony to study the effects of physiological or surgical perturbations, or simply as a source of subjects for laboratory research. Sade and those who succeeded him, especially Richard Rawlins, dissipated those tensions by showing unequivocally that the scientific potential for descriptive studies was dramatic and far-reaching.

The annual capture of monkeys also provided opportunities for blood sampling that has sustained a long series of important studies by John Buettner-Janusch and Christine Duggleby on the population genetics of the Cayo Santiago colony. So another frontier was established in our understanding of the relationship between social organization and genetic structure in vertebrate populations. It is in the behavioral domain, however, that research on the Cayo Santiago monkeys has had the greatest impact. This germinal influence is evident in all modern work on the social behavior of nonhuman primates, as is beautifully portrayed in Robert Hinde's recent synthesis in *Primate Social Relationships,* built on research that he and his colleagues have done over the past two decades, with key contributions from Cayo Santiago studies.

The emphasis on social behavior is also evident in the present volume, derived from a symposium organized by Richard Rawlins and Matt Kessler. They themselves have done more than anyone in recent years to keep traditions established by Carpenter alive and productive, committing both minds and hearts to a task that has not always been easy. A generation of students owes a special debt to Rawlins for making his encyclopedic knowledge of primates so freely available, and for the thoughtful counsel and support, always selfless and good humored, that made much of the research reported in this volume possible. The book presents an overview of current and recent research pursued on Cayo Santiago, stressing behavioral and demographic studies, and including enough topics in other domains, such as physical anthropology, to provide both a good sense of the interdisciplinary nature of the research endeavor and the potential of the Cayo Santiago rhesus colony as a biological resource. It serves to point new investigators toward future topics in vertebrate ethology and sociobiology for which the colony animals are ideal subjects. I hope that graduate and postdoctoral students in particular will be inspired to pick up the challenge. The opportunities for imaginative, innovative exploration of such elusive themes as the nature of social intelligence are almost limitless, working conditions are excellent, and supporting services are outstanding. The University of Puerto Rico is firm in its support of the basic concept of continuity of the Cayo Santiago colony, and in its commitment to sustain long-term studies of the kind that have yielded so many revelations in the past. We are only just beginning to sense some of the deeper complexities that characterize the natural behavior of nonhuman primates, especially of a social nature. I believe that the Cayo Santiago monkeys are destined to play a dominant role in the next generation of discoveries about the biology of primate behavior. This book serves both as a review of completed studies, and as an indication of new directions that the future holds in store.

Introduction

This volume is based on a special symposium entitled "Recent Research on Cayo Santiago," held to commemorate the 45th anniversary of the founding of this first American colony of free-ranging rhesus monkeys. The papers were delivered at the 1983 Annual Meeting of the American Society of Primatologists, hosted by Michigan State University in East Lansing, Michigan. Participants included investigators from the continental United States, the Commonwealth of Puerto Rico, Canada, and Great Britain. The book includes original presentations from the symposium, a detailed history of the colony, and a complete bibliography of nearly 300 scientific publications, based on work at Cayo Santiago from 1938 through 1985.

As organizers of the symposium, we sought to bring together some of the recent projects at Cayo Santiago, and to stimulate the interest of new investigators in future research at the facility. The colony, administered by the Caribbean Primate Research Center of the University of Puerto Rico School of Medicine, provides an international resource for field studies of macaque social behavior and offers a rich testing ground for behavioral hypotheses. In addition, this accessible and relatively homogeneous island population supports noninvasive research on the biology of the rhesus monkey in such diverse fields as animal husbandry, anthropology, genetics, gerontology, hematology, ophthalmology, parasitology, population ecology, and virology. An extensive collection of over 600 skeletons from Cayo Santiago monkeys of known age, sex, and maternal genealogy supplements the living animal resource for detailed anthropometric studies and osteological research.

The chapters in this book represent examples of both short and long-term research conducted on the island over the past several years. Some are reviews, intended to give the reader a synopsis of complex longitudinal studies of behavior, genetics, and population dynamics. Others document the results of opportunistic studies of behavior or biological surveys. The papers cover a broad range of topics, but all share a common dependence on the detailed knowledge of individual life histories, genealogical affiliation, and troop membership of the animals. Without accurate information of this type and depth, few, if any, of these projects could have been undertaken.

As the title indicates, the volume is organized around three subject areas; the history, behavior, and biology of the Cayo Santiago macaques. The first two chapters chart the historical and demographic evolution of the colony. The historical review is intended to document the key events in Cayo Santiago's past and to recognize the pivotal roles that a few dedicated individuals have played in perpetuating the colony and in developing the facility into a unique biological resource. The concept of a renewable primate resource which would support both biomedical and behavioral research is the legacy of Clarence R. Carpenter and his colleagues, who foresaw the need for such facilities in the 1930's. After 48 years, most of the problems which justified founding the Cayo Santiago facility still exist. The supply of healthy macaques for biomedical research has been limited in recent years by political embargoes and the costs for primate research have risen sharply, forcing large scale breeding of nonhuman primates on an in-house basis to meet research needs. Political events in countries where nonhuman primates live in the natural habitat have made field studies of behavior both dangerous and logistically difficult. Most important, few feral primate populations have life history data available for more than a single generation of animals. Despite its wealth of biological resources, the history of the facility is rife with periods of little or no support, and it has been difficult to maintain the animals and to sustain interest in primate research over the years at Cayo Santiago.

The dynamics of population growth at Cayo Santiago have been intimately linked to the history of the colony and its varied use over the years. From the 1930s and into the 1960s, animals were fed sporadically and routinely removed from the island for experimental use. The population was held in check by the random removals, but systematic studies of demography were handicapped by a significant bias due to the manipulation of the animals. The program of management stabilized in the late 1950s with regular provisioning and the introduction of a complete census of the population, but monkeys continued to be culled as needed. Early demographic profiles of the population were limited under these conditions since many of the animals were of unknown age and several of the component troops of the colony had been severely reduced by the trapping and removal of monkeys for biomedical work. With the establishment of the Caribbean Primate Research Center in 1970, the policy governing management of the colony changed to emphasize behavioral studies and manipulation was limited to a single annual trapping of the colony to tattoo the new birth crop for identification and to collect blood samples for genetic studies. In time, animals of unknown age died off and over the next 14 years a new body of demographic data on a totally known and minimally manipulated population of rhesus monkeys was obtained. Chapter 2 updates the long-term studies of population growth at Cayo Santiago and documents the recent changes which have occurred in the colony. The management protocol for the colony has had a tremendous impact on the numbers of animals maintained on the island and on the cost of operation. Projections of population growth derived from the data obtained under this protocol have led to more effective colony management. More importantly, the stable population has opened up research opportunities in behavior and biology which cannot be duplicated at any other location.

The behavioral section of the volume begins with two very different investigations of infant social development and a detailed examination of the role of vocal communication as a tool for investigating macaque social relationships. We start with these topics because they delineate, through independent lines of investigation, the importance of not only the maternal role in defining the social network within which a young animal interacts, but also that of the extended family or matriline as a source of care and protection.

In Chapter 3, Carol Berman reviews her findings from 10 years of research on the role of mother-infant relationships in determining the social development of rhesus monkeys. Unlike the earlier studies of Harlow or Hinde, which examined the behavioral ontogeny of infants under aberrant conditions induced by temporary or permanent removal of an infant's mother, Berman's work at Cayo Santiago examines infant development within a normal social context where both the mother and the members of the maternal genealogy are present as a component of the entire social group. Both the maternal influence and kinship emerge as important factors in organizing the behavior of young monkeys. Throughout early development, infant social interactions are mediated by the mother and influenced by her interactions with other troop members. Although the infant becomes increasingly independent with age, it continues to rely on maternal intervention for protection and the early agonistic experience of the monkey is directly correlated with the dominance rank of its mother; infants of high-ranking females are threatened less than those of subordinate animals and high-ranking infants receive more aid from unrelated animals than lower ranking monkeys. Thus both maternal and genealogical dominance rank strongly affect the behavioral environment of the infant and the impact of these early interactions is long lasting.

The strength of the maternal bond and the behavioral plasticity of the infant rhesus monkey is remarkably demonstrated in Catherine Scanlon's account of the development of a handicapped infant in

Chapter 4. A congenitally blind animal was born into the social group in which she was observing the role of play in the development of social skills among infants. The birth provided a natural experiment for comparison of the behavioral ontogeny of a handicapped animal with that of 20 normal infants of the same age and sex already under observation. The mother and kin of the blind monkey compensated for the infant's deficit by increased protective behavior and additional care, often at the expense of other offspring. The infant, however, showed no signs of behavioral abnormality and maintained frequencies of contact with its mother—feeding, playing, and moving about the forest canopy—which were virtually identical to the sighted animals. At least within the first year of life, the support of the extended family was sufficient to offset what was clearly a significant physical handicap for an infant monkey and permitted it to successfully negotiate both its social and physical environment.

The important function of the kin group as a source of care and protection for members of a matriline extends well beyond the early years of life into adulthood and is most clearly observed when animals are involved in agonistic encounters. The pattern and distribution of aid given to fighting animals follows lines of genealogical affiliation. In Chapter 5, Harold and Sarah Gouzoules have combined their efforts with Peter Marler to study vocal communication as a means of investigating social relationships in the rhesus troop. Their first findings revealed that in agonistic encounters the types of screams given vary according to the identity of a monkey's opponent and the severity of the fight. For example, screams given when a fight occurs between related animals differ from those emitted by unrelated monkeys. The results challenged the notion that primate vocalizations reflected only an animal's emotional state or motivation. Instead, the screams were representational and revealed information about the abstract social relationships maintained within the troop. To demonstrate a representational function, the calls of known individuals were recorded in a variety of agonistic contexts, then replayed from hidden speakers to the troop when the original caller was out of sight. As seen in the infant development studies, the response of the monkeys to the recorded screams varied according to genealogical affiliation. Kinship proved to be important in determining both the type and degree of behavioral response observed. The replay of screams from immature monkeys consistently aroused both the animal's mother as well as related females to a greater degree than that seen when the same calls were played to unrelated monkeys. Even more interesting, the results showed that vocal recognition in the absence of physical visual cues extended well beyond immediate relatives in a complex partitioning of the social group. These animals are clearly capable of differentiating the abstract relational attributes conveyed in vocal cues alone.

Rhesus females enjoy the security of matrilineal protection throughout their lives since they remain with their natal troop, but males disperse from the group at adolescence, exposing them to a socially hostile environment. The young males must adapt by attempting to enter a new social group, and in the process, are exposed to the xenophobic response of members of the new troop. This often results in the death of the emigrating male when wounds received in fights become septic. Considering the risk that dispersal presents to the animal in contrast to the obvious social support it would retain if it did not emigrate, investigators have attempted to identify what compels the young males to leave the troop. On a population level, dispersal clearly contributes to gene flow, but the benefits to individuals are not obvious. The next four chapters move beyond the early stages of development in the life cycle to the complex behavioral patterns associated with adolescence and adult mating behavior, including male dispersal, spatial relationships among adult males, affiliative behavior of adult males and females, and an examination of lineage-specific mating behavior. Each investigation reveals much about the behavioral interactions between animals patterning the social structure of the troop.

In Chapter 6, John Colvin examines the proximate causes of male emigration at puberty. There is considerable variation in the age at which the males depart from the natal troop, but most are gone by the age of 6 years. Some males remain with the natal troop past this age, and social dominance is shown to be a key factor in the delay of emigration; subordinate animals routinely disperse early, but high-ranking males may not. The data indicate it is the male's social situation within his peer group that most strongly influences the timing of emigration, rather than his social relationships with either siblings or adult males in the natal troop. Social interaction between the animal and his mother, or other adult females in the troop, exerted some influence on the timing of emigration, but the relative dominance rank of the young male within his peer group was the best predictor of the age at which dispersal occurs. The troop to which the male emigrated was most frequently one previously entered by his older brothers.

Once an adolescent male has dispersed from the natal troop and successfully integrated into a new social group, often over a period of several months, its social relationships change radically from those maintained prior to puberty. The male competes for resources and social dominance without the support of its matriline. The affiliative relations which evolve between adult animals provide critical information about the social structure within the troop. Since Zuckerman's early descriptive reports on baboons, there has always been an underlying assumption that affiliative interactions between adult male and female monkeys relate solely to reproduction, but spatial proximity and affiliative behaviors persist among the animals outside the discrete period of seasonal breeding which characterizes most primate populations. In Chapters 7 and 8, David Hill and Bernard Chapais examine seasonal differences in the spatial relationships of adult males at Cayo Santiago and attempt to determine why adults of both sexes continue to affiliate outside of the mating season.

In Hill's study, adult males were physically closer to females more frequently during the mating season than in the following birth season, and high-ranking males were close to more females more often than lower ranking males. Persistent proximity relations were observed throughout the annual reproductive cycle, including anestrous, but were more often seen between high-ranking males and females. Females maintained proximity relations with males during the birth season, but both sexes did so during the mating season. Of the relationships which continued throughout the annual cycle, regardless of reproductive season, all were maintained by females. The persistence of male-female relationships beyond the mating season suggests that sexual attraction alone is not responsible for the continued association of adult males with the social group.

Chapais used data from birth season interactions among the adult females and central males to investigate why males associate on a long-term basis with females who are not sexually receptive. His results showed that only the highest ranking males were attractive to the adult females and that the males, in turn, were attracted to the highest ranking females. The females were aided by these males during agonistic encounters against lower ranking animals. The results could not be interpreted in terms of a lengthy process of mate selection since concordance was not found between the pattern of male-female affiliative bonds during the birth season and that of sexual activity in the following mating season. Instead, the animals appear to benefit mutually from these relationships by increasing their competitive ability over lower ranking males. However, affiliation of an adult male with adult females during the birth season may facilitate maintenance of a central high-ranking position within the troop prior to the mating season and contribute to improved reproductive success within the social group.

In the final section dealing with behavior, Chapter 9, Carol McMillan investigates yet another strategy for reproductive success—lineage-specific mating. As a result of analysis of blood group genetics in the Cayo Santiago colony it has been found that genetic distances between the matrilines within a social group are not smaller than those between matrilines between social groups. It was expected that the dispersal of males from the natal group, coupled with random mating across the matrilines of a single troop, would decrease the genetic distance between genealogies within a troop, due to shared parental genes of the offspring. In other words, the genetic profile of the members of the matrilines within a group should be more similar to each other than to members of lineages in other troops. This was not the case and McMillan suggested that nonrandom mating by males amongst the matrilines might provide a possible explanation for the genetic data. A lineage-specific mating hypothesis was proposed and tested by direct observations totalling over 2,555 hours on a single social group. The hypothesis was only weakly supported by patterns of female mating. Roughly 17 percent confined their mating to males of one lineage. A moderate trend for lineage-specific mating was found for males, where 40 percent mated consistently with females of a preferred lineage, and related males chose females from the same matriline as consorts. Recall that in Chapter 6, Colvin found adolescent males dispersed into social groups previously entered by older brothers. Clearly intergroup transfer by related males may act to increase the genetic distances between matrilines within a social group. On a proximate level, it appears that individual familiarity plays an important role in regulating both the social and reproductive relationships of the animals.

Life history data also provides the detailed information needed for interpreting biological surveys of the population. Several different types of biological investigations are presented in this book, but all require knowledge of the age, sex, and matrilineal affiliation of the animals examined. The diversity of the studies is unified by this common dependence. Also, it is interesting to note that although each investigation represents an independent line of evidence concerning the biology of the population, the results of the studies are generally complementary and mutually reinforcing. For example, information on age-related changes in the locomotor behavior of the monkeys is consistent with the distribution of arthropathy and limb-joint mobility of the animals in the colony.

Rhesus monkeys are the most widely used nonhuman primate model for the study of human nutritional problems and in Chapter 10, Bernadette Marriott and colleagues use hair mineral analysis to study the relationship between hair mineral excretion and mineral nutrition state in a known primate population free from the bias of environmental contaminants or hair treatment. Because the macaque digestive system is similar to humans, such studies have clinical relevance as well. The working hypothesis is that hair mineral values should reflect excessive or deficient dietary intake of critical minerals and serve as an index of nutritive status. The results show that significant differences in mineral excretion exist between the sexes and age groups. Calcium, magnesium, and manganese were excreted in higher concentrations by males than females, but females had greater traces of phosphorus than did males. Pregnant females differed significantly from nonpregnant females. The overall concentrations of calcium, copper, iron, manganese, and zinc were lower for adults of both sexes than in juveniles. These findings show strong parallels with results of human nutritional studies and represent the first multielement survey of this type on rhesus monkeys. The results suggest the population is well exercised and in good health, providing yet another measure of the biological impact of the colony management protocol.

The next two chapters deal with age-related deficits in the visual and musculoskeletal systems of the Cayo Santiago monkeys. The studies are particularly interesting because they give clear evidence of the anatomical constraint of behavior in a free-ranging population of animals. The two surveys show close correspondence between the predictions of behavioral constraint based on hard tissue changes with age and observed patterns of behavior on the island. From the standpoint of functional morphology, such information is very valuable since anatomists have long suggested that hypotheses about the physical limitations of animal behavior, derived from the analysis of morphology, should be tested in the field. Again, it is the unique depth of life history information that permits these investigations to be carried out at Cayo Santiago. The difficulties of measuring age-related variation on a population scale and of determining age in unknown animals have hindered development of aging studies in long-lived primates. Both the large numbers of animals and the documented histories contribute to Cayo Santiago's utility in advancing this type of research.

While it comes as no surprise that advancing age results in increased physical problems for these animals, the close parallel between the aging process in the rhesus monkey and man is striking. The rhesus monkey lives to a maximum age of about 30 years and the biological deficits associated with aging in this species proceed at relatively the same rate as in humans. The findings are particularly important in view of the presently increased research interest and support for studies of biological aging. The discovery of a suitable and shorter lived animal model for the human aging process has great value in that both prophylaxis and therapy for serious human health problems can be attempted in the nonhuman primate and the efficacy of such protocols can be assessed in a much shorter period of time. From the perspective of understanding the evolutionary biology of the rhesus monkey and nonhuman primates in general, the data are invaluable since they document changes which occur in two critical biological systems—vision and locomotion—essential to primate adaptation.

In Chapter 11, Jean DeRousseau, Laszlo Bito, and Paul Kaufman examine living animals and skeletal materials at Cayo Santiago for signs of age-related impairment in vision and locomotion and compare the free-ranging Cayo Santiago macaques to other captive colonies of the same species. At Cayo Santiago, degenerative joint disease increases in frequency in animals of both sexes with advancing age at all major appendicular joints and along the spinal column. Surprisingly, age-related osteoporosis was not routinely observed. Loss of passive joint mobility occurred in the animals prior to development of osteoarthritis, and the reduction in mobility correlated with age related changes observed in the locomotor behavior of the animals. The monkeys were also examined for age-related eye disorders, including presbyopia, cataracts, glaucoma, and retinal degeneration. Potentially handicapping conditions such as cataracts or ocular hypertension were not found, but some evidence of the retinal pathology was discovered and loss of accomodation in the lens was observed with increasing age. The Cayo Santiago monkeys were found to be predominantly myopic.

In Chapter 12, Jean Turnquist continues the investigation of joint mobility by extending the survey to incorporate the entire life cycle. Mobility declines rapidly in early life as a function of epiphyseal closure and remains stable through early adulthood, but decreases rapidly with old age. Again, the maximum range of joint flexion and extension correlates with and is limited by developmental changes in the locomotor and postural habits of monkeys. The evidence would suggest that as the animals get older they are increasingly handicapped with respect to their ability to negotiate the environment, but the findings certainly point out the crucial importance of the social group to the survival of individual animals in that these anatomical deficits appear to be buffered by the system of support maintained by the social networks within the troop.

The final two chapters consider yet another parameter of the biology of the Cayo Santiago macaques—population genetics. It is this level that the most unique advantage of the colony becomes apparent. Because the history of the colony and the demographic changes within the population have been closely monitored over a time span equivalent to five or more generations, the long-term genetic consequences of both management policy and the social behavior of the animals can be tracked in a totally ascertained population of nonhuman primates. The combination of life history data with genetic monitoring over this length of time is unique and extremely important in that the colony offers perhaps the only location in the world where classical theories of population genetics and the distribution of genetic variation can be linked to patterns of behavior in nonhuman primates. In Chapter 13, Matt Kessler and his colleagues report on the occurrence of a rare phenotype in the colony, the golden macaque, and document the transmission of the genotype through four matrilines in the colony. Compared with previous birth seasons, the incidence of this hereditary anomaly increased eightfold, which might indicate an increased distribution of the autosomal recessive allele which controls pigmentation of the pelage, and/or increased mating among carriers of the gene in the colony.

Although the increased frequency of appearance of the golden macaques on the island might suggest the genetic load due to inbreeding was accelerating, an 11-year study of blood group genetics of the population, reviewed by Christine Duggleby and colleagues in Chapter 14, shows this is not the case. Although over 88 percent of the population alive at the beginning of the genetics study in 1972 were the descendents of 15 females living on the island in 1956, the majority of adult males at that time were from troops previously removed from the island and presumably unrelated to the 15 founder females. Over the years, the data show no evidence of inbreeding as measured by the loss of genetic variation and reduction of heterozygosity. Even the culling of the colony did not permanently restrict population variation. Male migration and outbreeding is apparently sufficient to prevent the deleterious effects of increased genetic load due to inbreeding. Of particular interest is the discovery that the socially important structures within the population, the troop, and its constituent genealogies, show genetic counterparts over multiple generations. Most exciting is the new evidence that there are clear differences in individual fertility and survivorship which can be related to specific genotypes represented within the population as a whole, and that for the first time, the genetic consequences of individual behavior and population dynamics can be examined. Perhaps it will now be possible to move beyond description to explanation of events in the life cycle of the individual, the social group, and the population as a whole.

We close this volume with a complete bibliography of publications from the Cayo Santiago facility. It is included in response to the innumerable requests made for such a document and in a sense, its greatest value is not only as a research resource for new investigators, but as a record of the history of research at this unusual facility. All of the varied interests of the researchers over the years are reflected there, as well as the underlying sources of financial support which carried the colony along. From 1938 to 1985, nearly 300 published articles have appeared in peer-reviewed journals and books.

We thank all of our colleagues who contributed to this book, and especially wish to extend our gratitude to Dr. W. Richard Dukelow, Chairman of the Local Arrangements Committee of the 1983 meeting of the American Society of Primatologists (ASP) for his help in making the symposium a success. Dr. David M. Taub, Chairman of the ASP Annual Meeting Program Committee also provided invaluable help. All of us owe a tremendous debt to the past and present staff members of the Caribbean Primate Research Center (CPRC) of the University of Puerto Rico School of Medicine, and especially to the loyal members of the Cayo Santiago 'crew', particularly Angel ("Guelo") Figueroa for providing continuity of the census during the professional lifetimes of several 'Scientists-in-Charge' of the colony; Emilio Tolentino and Hector Vasquez for their dedication in caring for the animals over many years; John Berard, Mary Knezevich, Sammy Martinez and Janis Gonzalez for technical help; and Edgar Davila for census information and preparation of specimens for the skeletal collection. Thanks are also due to the citizens of Punta Santiago for their support and friendship over the years. We would also like to acknowledge the support provided for Cayo Santiago from the Animal Resources

Branch, Division of Research Resources, National Institutes of Health, Bethesda, Maryland under contract RR-7-2115 and grant RR-01293 to the University of Puerto Rico. We want to thank Drs. William Goodwin, Leo A. Whitehair, and William T. London for their personal friendship and support of Cayo Santiago and the Caribbean Primate Research Center. Also, the continued efforts of the past directors of the Caribbean Primate Research Center, Drs. C.H. Conaway, W.T. Kerber, S.O. Ebbesson, G.W. Meier, and L.A. LeZotte, Jr. are sincerely appreciated by all who have worked in Puerto Rico. Special recognition is due Attorney Carlos Torres, Administrative Director of the CPRC, for his competent judgement, understanding, and strict fiscal management of the primate center since its inception in 1970. We extend our gratitude and respect to Dr. Norman Maldonado, former Chancellor of the Medical Sciences Campus, University of Puerto Rico, for his continued interest in Cayo Santiago and personal support. We wish to thank our editors, Dr. Emil Menzel, Dr. Randy Sussman, Kay Richardson, and Nancy Sharlet for their considerable help in producing this volume. Finally, we would like to thank Cindy Evans and Ava Gaa Kessler for their understanding and patience during the compilation of this volume.

Richard G. Rawlins, Ph.D.
Assistant Professor and Director Of Primate In Vitro Fertilization
Co-Director Human IVF Laboratory
Dept. of Obstetrics and Gynecology
Rush-Presbyterian-St. Luke's Medical Center
Chicago, IL 60612

Matt J. Kessler, D.V.M.
Director — Caribbean Primate Research Center
University of Puerto Rico School of Medicine
Sabana Seca, PR 00749

Illustration 1.1 Two members of the founding stock of the Cayo Santiago colony of rhesus monkeys.

CHAPTER ONE

The History of the Cayo Santiago Colony

RICHARD G. RAWLINS AND MATT J. KESSLER

Of all the South Asian primates, none is as well known as the rhesus monkey. Detailed information on its anatomy and physiology has long been available and numerous aspects of its behavior have been examined in the laboratory, but surprisingly little data relate this species to its native habitat. A great deal of our understanding about the organization of rhesus social behavior and the ontogenetic and evolutionary processes which underly it, has come from study of the Cayo Santiago colony. Cayo Santiago, a 15.2 hectare island which lies one kilometer off the southeast coast of Puerto Rico at 18° 09′ N 65° 44′ W, has been home for an introduced population of free-ranging rhesus monkeys *(Macaca mulatta)* since 1938. It is the oldest continuously maintained colony of its sort in the world, serving for over 47 years as an international research resource in primate biology.

One reason that a transplanted island population has been a favored source of information on *Macaca mulatta* is that Cayo Santiago frees investigators from restrictions imposed by problems in the field which limit long-term observations on groups of wild animals, and which often result in a superficial view of the complex behavioral interactions that comprise and sustain a social network. Researchers working with indigenous populations face conditions where terrain may be impassible, the observability of the animals low, an accurate estimate of population age structure and genealogical affiliation impossible to obtain because of poor animal identification, or the logistics of support and protection from political terrorists are too difficult. Cayo Santiago offers an effective compromise between laboratory and field conditions because its free-ranging animals live relatively undisturbed in a seminatural hatibat and are easily seen. More important, long-term histories of individual social and physical development as well as troop and population demography are available. Certainly some aspects of primate behavior and ecology cannot be studied with a provisioned colony such as Cayo Santiago, but the caveats are largely offset by the unique opportunity to follow the behavioral ontogeny and population dynamics of highly social species over many generations.

CARPENTER'S LEGACY

Cayo Santiago was founded as a primate facility just as modern primate studies began to emerge through the work of men like Yerkes, Zuckerman, Kohler, and others. As Yerke's student in the 1930s, C.R. Carpenter was one of the first behaviorists to actually observe primate behavior in the wild, both in Panama and Asia. Frustrations experienced when conducting fieldwork and an ever increasing demand for monkeys as laboratory specimens led Carpenter to formulate plans for an island population of both gibbons and rhesus macaques which could be observed over long periods of time and which would provide healthy animals for biomedical work.

Carpenter recounted his impressions of the early development of the idea for the Cayo Santiago colony in an address to the University of Puerto Rico School of Medicine in August 1959. Unlike other published accounts by Carpenter, the address provided a uniquely personal recollection of the efforts which led to development of the island colony:

> As far as I know, the beginning of the idea of a rhesus colony being founded on some island in the American tropics has quite a long history. I discussed the possibilities extensively with Dr. Tom Barber, of the Museum of Comparative Zoology at Harvard in the early 30's. Harold Coolidge was interested in the plan for many years. While I was working on Barro Colorado Island, we talked about stocking Orchid Island with spider monkeys. I went over to Coiba Island to observe the possibility of introducing chimpanzees, gibbons or some other interesting Old World primates into some island in the American Tropics. More directly, the original ideas for this colony came from the Asiatic Primate Expedition which Coolidge, Adolph Schultz and I carried out in 1937. During this expedition we collected, I believe, seven beautiful gibbon specimens. At that time I was associated with Bard College of Columbia University. As you may know, the School of Tropical Medicine was also under the auspices, somewhat indirectly, of the Columbia University, or had a type of cooperative arrangement which involved the College of Physicians and Surgeons. Dr. George Bachman was Director of the School of Tropical Medicine and when he heard about this colony of gibbons, he became very interested in it and in the possibilities of using gibbons for certain special kinds of medical experimentations. So, Dr. Bachman became intensively involved and I'm sure committed relatively large sums from his restricted budgets. He built a special animal house down here near the School [of Medicine in P.R.] for the gibbon colony and after I arrived back in New York from Sumatra, the seven gibbons were sent down here [to Puerto Rico].
>
> The story then shifts to Dr. Philip Smith and Dr. Earle Engle,

of the College of Physicians and Surgeons. They heard of the developments up to that point, and were interested particularly in endocrinology, sexual physiology, sexual behavior and anatomy. They, Dr. Bachman and I got together and began to play with the idea of establishing a joint free-ranging colony including gibbon and rhesus monkeys. The use for gibbons was not clear; it was not clear to me except from the behavioral studies point of view. I had become fascinated by them during the field studies in Thailand. I was also interested in gibbon vocalization and communication.

Smith and Engle wanted healthy animals for use in the laboratory and therefore they were most interested in the rhesus monkeys. We began to think of the possibilities of establishing a free-ranging or semi-free ranging breeding colony of rhesus monkeys. Dr. Bachman explored the possibilities of islands off Puerto Rico. I think he visited practically every small island around Puerto Rico and made an assessment of their suitability for a breeding colony. He selected Cayo Santiago as being, in his judgement, the best possible place.

Very few rhesus monkeys had been bred in the Western Hemisphere. Dr. Carl Hartman at Johns Hopkins had studied the reproductive cycle under captive conditions. We knew too little about what might be called primate husbandry, so one of the first objectives was to establish a successful breeding colony. Remember that this was 1938-39, and it was a critical time. Chamberlain and Hitler were having discussions and we were expecting war. We anticipated that the import of rhesus monkeys from India would be cut off completely and we hoped to have at least a seed bed of animal specimens for research in the Western Hemisphere. Since we had considerable work going on the rhesus monkey this made good sense and was one of the major objectives.

We decided to approach the Markle Foundation for funds: $60,000 for three years to do this job. . . a ridiculously small sum of money. The Foundation consulted with Dr. Hartman, who advised them that rhesus monkeys would *not* breed in Puerto Rico. This was at a very critical point when the Markle Foundation was trying to decide whether to give us the $60,000 or not. I telephoned a group of people and asked them to express their ignorance by telephone or telegraph to the Markle Foundation. The Markle Foundation finally gave us the grant, regardless of reservations. You see how ignorant we were in 1938 to think that rhesus monkeys would not breed in a tropical situation! The establishment, therefore, of a successfully producing free-ranging or semi-free ranging colony was the major objective. We hoped to do it for rhesus monkeys and for gibbons.

Another objective was to explore the suitability of animals for different kinds of experimentation. You will recall from the taxonomic classifications that there are something like 500 different species of primates. It is inconceivable that over 500 different species and subspecies and varieties of primates are all equally suitable for the great variety of experimentation planned or being done on primates. It is highly probable that there are certain problems that are extremely appropriate with certain types of primates, and there are other problems for which you should use still other kinds of primates. Here is another problem which is just as alive and important today as it was in 1938-39.

I was interested in the social behavior, the way a population organizes itself, the kinds of social behavior that occurs. Smith and Engle were interested in the old aging or senility problem in 1938. This reflected itself in the fact that I brought back 15 extremely old males which we trapped with great difficulty in central India. They were also interested, especially Engle, in the reproductive cycle. I think, as a matter of fact, that Dr. Smith just "went along with" Dr. Engle. I don't believe he was ever very enthusiastic about the project. He was cautious and conservative. A thousand dollars was a lot of money to him, and sixty thousand dollars was something that was fantastic to spend on such a venture as we were cooking up in Puerto Rico. There were many times, I'm sure, when he told Engle, "I told you so!"

Now that we had the grant, how did we launch the project? We had to have breeding stock. The right kind and numbers of breeding specimens could not be bought on the market. I had to go to India and collect them. I well remember the day when Dr. Smith and I went over to the bursar at Columbia University and got a check for $5,000. With this amount of funds, I set off around the world to collect 500 rhesus monkeys from several provinces of India. I went by way of Indo-China, Thailand, and Malaya, where I was to collect 20 or 25 gibbons to supplement the seven we already had.

This adventure illustrates a problem which one can only learn to solve by experience, and that is estimating accurately what kind of funds are necessary to achieve remote and complicated objectives. I have been the victim several times of attempting to accomplish more than could possibly be done with the money and other resources available. I went to India by way of Indo-China and collected 18 or 20 fine gibbon specimens, several siamangs and an orangutang from Sumatra, and assembled them at Penang, Malaya. Then I committed a very grave error.

There was a famous neurosurgeon at Johns Hopkins University whose nephew was stranded in Singapore. I permitted a "halo effect" to operate. I thought that surely a distinguished neurosurgeon's nephew could be trusted, so I put him in charge of this shipment of animals from Penang, one of the finest shipments of primates that ever left that port. The boy had jumped ship in Singapore and he was in trouble with the U.S. Counsel there. He could not get out of the city. I gave him a job and this solved his immediate problem. The only animals that reached this country were the orangutang and two or three of the small gibbons. All the rest died in the Red Sea from the lack of food, high temperature and poor care.

This was a great tragedy for me. I don't know how it could have been avoided. Dr. Bill Mann, the previous year, had the same experience in the Red Sea, even though he had a trained crew of special animal men from the National Zoo. That is a tricky region through which to ship animals, because they are on deck and if you get five or six days of still, hot weather on the Red Sea, they die from heat exposure.

I went on to Calcutta, India and began the business, a very nervy business I thought then, of trying to trap 500 animals of assorted ages including 100 females with infants. The composition of the shipment took into account the socionomic sex ratio of one male to five females. I collected the animals in India according to that ratio, assuming that this would give you a kind of balance in the population organization that would prevent the kind of fighting and killing that occurred in many colonies and most dramatically in the baboon colony in the London Zoo, the colony that Zuckerman studied to collect the material for his classical book *Social Behavior of Primates.* I collected 100 mothers with infants, 15 old males, over 200 young females which would reproduce in a year, 150 juveniles, and about 50 adult males.

It was known, at that time, that one of the great hazards of importing rhesus monkeys from India was the tuberculosis problem. They have tuberculosis in India because of the contact with people and I determined to test every animal selected for shipment. So I tested all of them for tuberculosis and eliminated those that showed a positive reaction.

There is another problem of major dimensions which relates to the animal dealers in Calcutta, in fact any port, New York or New Orleans. I formed the conviction after trading with dealers in Calcutta and studying the situation around the world, that animal traffic is a world-wide racket. It's a racketeering proposition with few exceptions. The risks are great and they make fantastic charges for animals that are bought for practically nothing in their local area. Shipping costs are also very high. I had to enter into an agreement with one of the powerful animal dealers in Calcutta and negotiate a unit price before any animals were trapped. The dealer in turn dealt with what we might call the animal unions, of which there are two. There is the Mohammedan union that collects the animals and the Hindu union that cares for them after they are collected. As you know monkeys are "sacred" animals in India. Therefore, the Hindus will not trap them because this is rough, cruel business. The Mohammedans will do this and there is an organized network of animal trappers out in central India. I had to deal with the chief man of that particular union through the animal dealer, a woman, Mrs. Chater.

When the animals were trapped, they were brought to an estate house which was rented and used for this particular shipment of animals. The animals were brought in, sorted, cared for, and tuberculosis tested. Special cages were required for the large males and females with infants. Then we were confronted with the critical business of arranging for shipment.

No shipping company wanted to transport the animals. You could tell them that you represented science and research, the College of Physicians and Surgeons and the Markle Foundation, but to no effect. This argument left them cold. I learned that the arrangements had to be made with the captain of a ship. The captain expected an amount of extra money over and above the shipping charges, which were already very high. I think I promised my captain $150. I never paid $50 of it because I ran short of funds. This was not all. The first officer comes in for his $50 and the deck steward expected another $50. I think they deserved the bonuses because 500 animals practically cover the whole deck of a large freighter. The nuisance to the officers of a ship is quite considerable. In addition to getting arrangements for shipping, there is the problem of arranging for food for the monkeys.

As far as I know, the standard method of feeding animals in transit was to use unhulled cooked rice. I thought I could do better than that, so I got the formula of the monkey biscuit that had been developed at Orange Park and tried to have this duplicated in Calcutta. I had 500 pounds of food made up into biscuits using, as closely as I could approximate, the ingredients that I could get around Calcutta and put the lot on shipboard along with a very large supply of unhulled rice, thank God. Three days out of Calcutta the biscuits molded and soured, and a couple of more days later I threw them overboard. By the time we reached Columbo, I had learned more than I knew about the food consumption of 500 rhesus monkeys, particularly the big males, and I was able to stock up with large quantities of fruits and vegetables there.

At Columbo, were were informed that instead of going

through the Mediterranean as scheduled, the Cunard Line wanted us to make a test run around South Africa. War was near. This was late September. All the animals were on the exposed deck. The ship was scheduled to first dock at Boston and then go down to New York from where we would tranship to Puerto Rico. So you can imagine something of the anxiety that developed with the responsibility for 500 animals caged on deck and with the uncertainties of weather around the Cape as well as near Boston and New York.

I worked 14 or 15 hours a day. I cleaned cages or fed animals all day long, exhausted, in rough weather or calm, then I went to sleep. This trip required 47 days. We escaped possible severe cold weather around the Cape of South Africa. However, the crew was dressed in heavy winter clothing for the first time in the history of that ship. We got into Boston and to New York in weather that did not fall below the tolerance levels of rhesus monkeys.

I didn't quite accomplish the job on $5000. I committed several gibbon specimens to the Brookfield Zoo in Chicago, for a thousand dollars. When I returned to New York, I found Smith and Engle in a panic about deficits that were occurring down here [in Puerto Rico] on construction work, planting of mahogany trees on the island, the tragic loss of the shipment from Penang, and the costs that I had incurred with the rhesus monkeys. They had sold some 50 of the adult females to Carl Hartman.

You can imagine how delighted I was to turn this shipment over to Mr. Michael Tomilin in San Juan and the caretakers here in the animal house, to Mr. Lamella, and Bachman or whoever else wanted to take care of them. Incidentally, one of the major errors of underestimation that we made in planning for the colony and locating it in Puerto Rico was the difficulty and cost of transportation. The cost of shipping materials to Puerto Rico, the inconvenience of it, etc., made it seem for me, at times, to be a tremendous error in locating a primate colony here. I began to think of other locations on the mainland of the United States or an island off the coast of the southern United States. There are compensating advantages perhaps, including climate and particularly the enthusiasm of support of Dr. Bachman and the facilities of the School of Tropical Medicine.

Another error that we made was to think that we could feed rhesus monkeys on the fruits and vegetables grown in Puerto Rico. This we attempted to do, but various kinds of deficiency diseases developed. The animals lost weight, and many animals died. The rhesus need more protein than is provided in the fruits and vegetables of a semitropical region. So within six months, we found ourselves trying to supplement the large amount of fruits and vegetables that we could buy in Humacao at reasonable prices, with fox chow from St. Louis, Missouri! In relation to the very inadequate budget we were operating on, this purchase of concentrates was a severe drain . . . buying this food stuff, shipping it down here, and trying to keep the colony alive.

I would like to suggest that in a project of this kind, there should be a clear cut division between the operational problems and the investigatory problems, and that these should be reflected in a budget. I think that the major problem that we ran into during the two years that I was primarily responsible for this rhesus colony was the fact that the housekeeping demands consumed most of my energy, all of Tomilin's energy, and all of the budget, so that there was little left for doing the research for which the project was planned and developed.

Now, having the colony set up with the rhesus and gibbon, I began to see some enormous possibilities for utilization of the resources as existed even in 1940-41. Study of parasites, study of disease, study of reproductive cycles, social organization and anthropological studies were a few of the possibilities. I began to dream about these unique resources and how they would be used by different people in the United States. I suppose it was an unrealistic dream to expect scientists from many different fields to cooperate, because they aren't usually trained that way. They are trained as specialists and they're trained by necessity to look out after the needs of their special projects. I am not sure how realistic such expectations are, even today. The animals were not being used here in the School of Tropical Medicine as they should have been. The gibbon colony was here for two years and it was never studied except in a minor way. The rhesus monkey colony was on the island, with enormous possibilities but no investigators. I understand this condition grew worse later on when the colony was simply maintained at a subsistence level.

The suspense created by doubting Thomases, like Hartman, was relieved after six or eight months by the birth of the first baby. You cannot imagine how welcome this baby was! I had brought back 100 females with babies, and it took them some time to get through the lactation period and to reproduce again. Predominantly the colony was then made up of your nonproducing females and it took time for them to mature. I don't think there was ever a baby, human or nonhuman, that was as much photographed as that infant rhesus monkey. Then the colony had started reproducing.

You might be interested in certain phases through which this colony passed. The first phase was one of organization of the groups, during which, many of the infants that were brought from India were killed. This is a very significant point, because it shows that the social organization of an animal population is protective for the rearing of the young. Group organization isn't just a superficial level of behavior that has no biological survival value. The females give them protection to a certain extent, but group organization was necessary for survival of the young. Incidentally, the animals were collected from seven different provinces in central India, and I expected that they might form groups in relation to the previous locations. This seemed not to be true; they were mixed up. There was a tremendous amount of fighting, killing, and a number of males were driven out to sea. I don't know how many were lost by drowning. This settling-in period resulted in the formation of five groups which I studied very intensely the second year. The results are reported in the Journal of Comparative Psychology and I also made a film showing what the behavior of these animals was like at the time (Carpenter, 1959).

Carpenter collected the original stock of rhesus monkeys from a 12-district area in the mountains, near Lucknow, India in September 1938 and shipped them 400 miles to Calcutta. They remained in the city from a few days to three weeks and were shipped on September 30, 1938 for the

Figure 1.1 Cayo Santiago as it appeared in 1938. Courtesy The Pennsylvania State University Libraries.

14,000 mile journey to the United States. The animals were transhipped and brought to Puerto Rico from New York on board the SS Coamo, which was later sunk by Axis submarines during World War II (Frontera, 1958).

While the animals were in transit, Bachman was at work preparing the island for the arrival of the primates. Cayo Santiago was owned by the Roig family of Humacao, and had been used principally as pasture for goats. Vegetation was sparse, consisting largely of grasses, scrub, and several coconut groves. With the lease of the island (later titled to the University of Puerto Rico by the Roig family) for the primate colony, Bachman obtained the services of the Civilian Conservation Corps to forest the island. In addition to coconut palms and mahogany trees, different kinds of tropical tubers such as yams and yautias, were planted. Eventually, over 4000 plants, including almonds, guavas, limes, mameys, and bananas were set out (Locke, 1938). Only the coconuts survived. The Corps also built the original pier on the western shore of the island and a large number of small rock caves covered with wire mesh were also reputed to have been built by them as shelters for the monkeys, which they apparently believed to be cave dwellers. These structures were never used by the animals, but still stand, covered with vegetation, as testament to the early ideas that were held. The Corps also excavated a number of wells on the island, but the water was brackish. Consequently a catchment and cistern storage system for rainwater was and still is used to supply the monkeys with water. The first catchment was constructed by the Civialian Conservation Corps on the Small Clay of the island in 1939 and was supplemented by the addition of a 10,000 gallon cistern built in 1941-42 by the National Youth Administration on the northern slope of the main part of the island, the Big Cay. The latter was fed from the gutters of the small house which also occupied the hill (Frontera, 1958). Today, the house remains in use as a laboratory, and the cistern continues to collect and disperse water.

While physical preparation of the island was being undertaken, the first Scientist-in-Charge of the colony was hired. In 1938, Mr. Michael I. Tomilin was appointed as primatologist in charge of Cayo Santiago and worked closely with Carpenter during and after the release of the monkeys to establish the population. Tomilin was a unique individual in his own right and today stories of his exploits, such as his daily swim from the island to the mainland and back, persist in the small village of Punta Santiago, the point of embarkation for the island. The factual information on Tomilin was drawn together by Frontera (1958):

> Probably the outstanding and, in a sense, the most controversial figure during these years was Mr. Michael I. Tomilin, primatologist in charge of the Sanitago colony. Born in

Figure 1.2 Michael I. Tomilin's home on Cayo Santiago.

Siberia, son of an orthodox priest, he was educated in Czarist Russia 'gymnasium' type of school and at the Tomsk Institute of Technology where he studied biology. During World War I, he worked in an armament factory and later fought on the side of the White Russions during the revolution. After four years, he reached China by way of Vladivostok and Japan. He worked his way to the western coast of the United States where he spent several years in lumber camps near Seattle, Washington. Later, he went to Stanford University in Palo Alto, California where he obtained his A.B. degree in 1930 and M.A. in psychology in 1932. He was probably well advanced towards his Ph.D. when he took a position at the Yerkes Primate Colony in Florida, sponsored by Yale University. In 1935, he was appointed primatologist at the Philadelphia Zoo, from where he came to Puerto Rico in 1938.

Tomilin and his wife, Eugenie, also a refugee from the Bolsheviks, arrived in Puerto Rico in September 1938, and took up residence on Cayo Santiago (Locke, 1938). The shipment of animals arrived on November 14, 1938 and in the second week of December 1938, the release of 409 rhesus macaques and 14 gibbons which Carpenter had collected was begun. Three pig-tail macaques (*Macaca nemestrina*) were also set out on the island and by January 1939, all of the animals had been set free on

Figure 1.3 Michael Tomilin feeding recently released monkeys. Courtesy of The Pennsylvania State University Libraries.

Cayo Santiago (Carpenter, 1942; Lebron and Otero, 1940; Frontera, 1958).

The establishment of the monkey colony was widely publicized by the local and international press. More than 40 articles were published announcing the event, including a detailed story in *The Economic Review* (1940) on Puerto Rico's "research investment." This proved to be a mixed blessing. As a result of an article in *Life* magazine (January 1939) illustrated with photos by Hansel Meith, great concern arose in the local community of Punta Santiago about the threat of disease from the monkey colony and its proposed research. The article had reported that the purpose for establishing the colony was the pursuit of research on poliomyelitis and leprosy. This alarmed the citizens of Playa de Humacao (Punta Santiago) and a delegation of heads of households convened under the leadership of Señor Luis Luhring. The group requested that the School of Tropical Medicine send a representative to meet with them and to explain the purpose of the research. Morayta (1969) described what happened.

The School of Tropical Medicine sent Dr. Pablo Morales Otero, Sr. Felix Lamela, and additional representatives from the Department of Education including Sr. Oscar Porrata Doria and the Superintendent of Schools, Sr. Issac Santiago. Also present at the meeting were Tomilin, the tax collector Sr. Cook and his wife, the president of the Central Committee of the Parents and Teachers Association of Humacao, Sr. Casimiro Olmiolo, and a representative of the Bull Insular Line, Mr. Brizzie. Luis Luhring, president of the concerned group of heads of households in the Playa opened the meeting and Doria, Otero, Lamela, and Santiago spoke on behalf of the colony. It was made clear that the use of colony was to be for raising healthy monkeys for experimental use elsewhere. Luhring put forth a motion to the assembly expressing thanks to the School of Tropical Medicine and the visitors for explaining the objectives and proposed that all opposition to continued work on Cayo Santiago be withdrawn. The motion carried unanimously. Ultimately, Bachman himself was forced to write a lengthy letter to the science editor of *Life,* Mr. Heiskell, in an attempt to correct inaccuracies published about the Cayo Santiago colony and to explain that the principal objective of the program was production of healthy animals, free from sickness and of known life histories, for use in psychological and biomedical investigation (Morayta, 1969).

In general, both the local and overseas press strongly supported development of the primate colony. The local papers soon stopped referring to the work at the facility as "monkey business" and the raising of primates for research work was soon heralded as an important new industry for the island of Puerto Rico (Morayta, 1969).

Tomilin and his wife were very popular with the local press and received large numbers of visitors on the island. On one occasion in 1942,

Tomilin presented a group of visitors from the naval base at Ensenada Honda (now Roosevelt Roads) with a female rhesus. The officer, Lt. Commander Ray, took the animal back to the base and gave the monkey to the wife of one of the other visitors in the group, Mrs. H. Gericke, who named it Heloise and took it home as a pet. The macaque took one look at the Gericke house cat and became terrified, so the animal was given to a Marine officer and was stationed at Marine quarters as a mascot. It became friendly with the camp dog and apparently was a regular patron of the U.S.O. shows at the Marine camp (Gericke, personal communication, 1983).

Tomilin also spent time trapping escaped monkeys as well. The New York Times reported that one of the Cayo Santiago monkeys left the island by swimming the distance, about 1 kilometer, from Cayo to the mainland in search of "a better environment," but Tomilin captured it and returned the monkey to the colony of (Morayata, 1969).

Tomilin went to considerable effort in trapping and testing the released animals for tuberculosis. About 3.0 percent of the animals released on the island died of tuberculosis within the first year and 2.2 percent tested positive when reexamined by Tomilin and his helper, Sr.

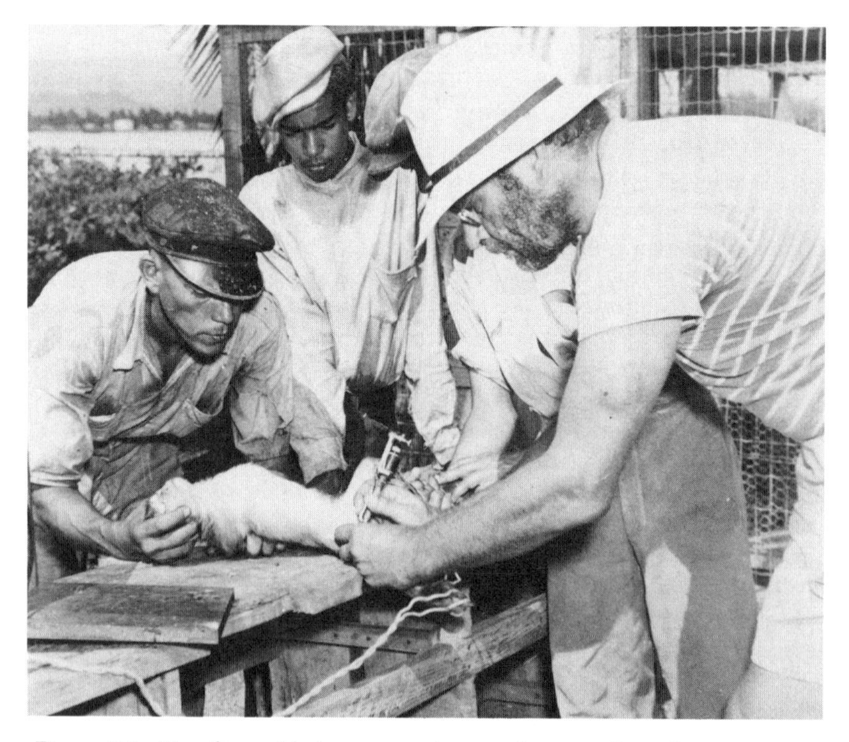

Figure 1.4 Tomilin and helpers tattooing monkeys on Cayo Santiago.

Louis Gonzalez in January and February of 1940 (Carpenter and Krakower, 1941; Frontera, 1958). By March 1, 1940 the colony contained about 350 monkeys of all ages (Carpenter, 1942). The gibbons did not fare so well. Because they repeatedly attacked observers they were kept caged after June 1940 and rarely released (Carpenter, 1972).

EARLY SCIENTIFIC WORK

By 1941, a healthy breeding colony of rhesus macaques was operating and a considerable amount of original research was underway. Between 1940 and 1942, there were 194 new infants born on the island (Carpenter, 1964). One of the first publications produced from research with the population appeared in 1940. Lebron and Otero (1940) had surveyed the throats of the newly arrived macaques for hemolytic streptococci in the weeks just prior to their release on the island and found over 45 percent carried Group A bacteria as well as Group C and G bacteria (36 percent and 18 percent respectively), but by 1940 none of the monkeys examined carried Group A streptococci. The findings laid the groundwork for the speculation that the observed low incidence of these same strains in humans living in Puerto Rico was the result of the unique streptococcal conditions on the island as opposed to those commonly reported for the temperate North American climate (Lebron and Otero, 1940).

Also of interest was an outbreak of diarrhea in the colony which led to the death of a number of animals. Cultures were made and *Shigella* was isolated. The epidemic was linked to inadequate diet and better food led to its disappearance. The outbreak drew the attention of Dr. James Watt, then of the School of Tropical Medicine and working in collaboration with the Puerto Rican Department of Health. Watt intended to study the naturally occurring *Shigella* infection in the monkeys, but the disease had disappeard from the population by the time Watt was ready to investigate its outbreak. Of greater importance, this early contact with the Cayo Santiago colony later led Watt to formulate the concept of a National Primate Center, which ultimately resulted in congressional support for the seven Regional Primate Centers which exist today (Windle, 1980).

Carpenter returned to Cayo Santiago with his wife in 1940 and began his study of sexual behavior in the rhesus monkey. The observations were made between February 29 and April 27, 1940 and resulted in a landmark paper published in 1942 (cf. Carpenter, 1942). Tomilin continued tuberculosis testing in the population and by September 1941, no additional cases were found and the island was considered free of the disease. Carpenter and Krakower (1941) subsequently reported on tuberculosis testing in the colony and made some important observa-

Figure 1.5 Rhesus monkeys on Cayo Santiago in 1939. Courtesy of The Pennsylvania State University Libraries.

tions on the use of intradermal tuberculosis tests in rhesus monkeys. Other studies during the period included work by Poindexter (1942) on intestinal parasites and the hematological studies of Suarez, Diaz-Rivera, and Hernandez-Morales (1942).

The success of the colony led Carpenter (1940) to advocate the establishment of several similar island colonies for primate husbandry off the southern coast of the United States. Only recently has Carpenter's timely advice been heeded, with the development of the Key Lois, Raccoon Key, and Morgan Island breeding colonies of rhesus macaques (Pucak et al., 1982; Cheslak and Taub, 1983). The concept of a free-ranging gibbon colony did not work out, even though the animals survived capture, transport, and release in a nonnative habitat. They were removed from Cayo Santiago in the spring of 1941 and sent to the continental United States. Six went to Pennsylvania State University and eight to zoological parks, thus ending the Puerto Rican gibbon colony which had been a major focus and source of interest in the founding of the facility (Carpenter, 1972).

TROUBLED YEARS FOR THE COLONY

By 1941 the Markle Foundation grant had ended and both financial support and scientific use of the population dwindled as World War II involved the United States. Between 1941 and 1942, 137 monkeys were shipped to research facilities in the United States. From 1942 to 1943 another 300 animals were removed for research stateside, and 100 more were taken off for use by the army in war-related research during 1943 and 1944. About 200 animals remained on Cayo Santiago. As a result of the acute shortages of provisions and support, the Tomilins left the island (Frontera, 1958), ending an era of excitement and development colored by unique individuals who are remembered fondly in Punta Santiago by all those who came into contact with them. The animals were subsequently cared for by Rafael Luis Nieva, who moved out to live on Cayo Santiago after the Tomilins left (Nicholas and Locke, 1951).

Compounding these problems was the closure of Columbia University's School of Tropical Medicine in Puerto Rico. Growing sentiment on the part of island residents for greater involvement in the operation of the school led to the resignation of Dr. Bachman, one of Cayo Santiago's founders, and eventually resulted in the withdrawal of Columbia's support for the entire facility (Windle, 1980). As a consequence, the Cayo Santiago colony was transferred to the College of Natural Science of the University of Puerto Rico in 1944 (Bailey, 1965; Windle, 1980). Apparently, little was done with the animals during this period and there is no information on its administrative history (Frontera, 1958).

Plagued by a lack of interest and funds, the College of Natural Science advertised in *Science* in July of 1947 (106:32-33) to find a buyer or sponsor for the colony. At the time, about 400 monkeys were living on the island (Frontera, 1958). A young Puerto Rican scientist studying at the University of Michigan, Jose Guillermo Frontera, believed that the primate colony was an invaluable asset to Puerto Rico, and wrote to the Dean of the College of Natural Science, Dr. Facundo Bueso, asking him to delay any further action until he (Frontera) returned to Puerto Rico. Bueso agreed and a year later, in the Fall of 1948, the College secured two years of financial support at $15,000 per annum for support of the colony from the National Institutes of Health after Dr. David Price and Mr. Ernest Allen visited the facility and recommended government investment in the maintenance and development of the colony. The grant was continued for a third year and separate funds to develop facilities for neuroanatomical studies of the macaques were also obtained (Frontera, 1958). Windle (1980) felt this aid probably saved the colony. Although undocumented, it is frequently noted among faculty members at the University of Puerto Rico (UPR) who knew Frontera, that it was Frontera's efforts that truly rescued the animals. He is reported to have used personal funds to secure food for the monkeys during times of poor support and to have delivered fruits and vegetables in person on several occasions to keep the monkeys from starving.

With the opening of the University of Puerto Rico's new School of Medicine in 1950, jurisdiction over the colony and responsibility for its support was transferred to it from the College of Natural Science. This did not alleviate the financial problem however, and the colony continued to be maintained by sporadic inputs of cash from unused portions of NIH and other research grants (Frontera, 1958). Operational funds were provided mainly by the School of Medicine and the colony continued to supply animals to the school and federal agencies for research (Bailey, 1965). Only about 225 macaques remained on the island at this time. The impoverished condition of the colony continued through 1955 and the potential of the research facility as envisioned by Carpenter was unrealized. Why Carpenter apparently had so little involvement with the facility he and Bachman created, except during the first year or so after its founding, is not known.

NEW SUPPORT FROM WASHINGTON

In the spring of 1955, Frontera, by then a professor of neuroanatomy at the School of Medicine, UPR visitied Dr. William Windle, Chief of the Laboratory of Neuroanatomical Sciences of the National Institute of Neurological Diseases and Blindness (NINDB) in Bethesda. Frontera, an NINDB grantee, raised the issue of the Cayo Santiago colony and its financial problems and suggested to Windle that the resource could be developed with stable support.

As a result of the interest stimulated by Frontera's visit, Windle visited Puerto Rico in August 1955, as a guest lecturer for the Veterans Administration Hospital in San Juan, and conducted a survey of the Cayo Santiago colony.

> On August 22, 1955, I went to San Juan where I was met by Drs. Max and Maria Ramirez de Arellano, who had studied under me at Northwestern University Medical School and were practicing neurosurgery and neurology in Santurce. Through them, I met people who could be helpful later in overcoming obstacles on the way to re-establishing programs of biomedical research in Puerto Rico. Much of our success was due to their efforts. Dean Harold Hinman arranged to have them take me to Cayo Santiago, and afterward he offered me use of the monkey colony if I would transfer my primate research from Bethesda to Puerto Rico.
>
> We drove to Playa Humacao and were taken out to the Cayo by Sr. Encarnacion (known in the Village as Don Maso). He opened a shed near the dock, took out some over-ripe fruit and a small quantity of dry corn, and scattered these on the

> ground for the monkeys. The animals rushed in, struggling with each other for the food. Many of them appeared to be healthy, but some had deep wounds and healed scars from fighting. We were told that although the amount of food was meager, the monkeys stayed on the small island and did not try to swim to the mainland as they had done when they were starving during the war years. There appeared to be about 150 monkeys, all except one stumptail being rhesus. A more accurate estimate in the winter of 1955-56 was 115. I went back to Bethesda on August 26 and reported my observations and conversations, including Dean Hinman's offer. I recommended that the NINDB provide support for the primate colony through grants to the Medical School, especially to Dr. Frontera and his associates in the Anatomy Department (Windle, 1978).

In a 1965 address to the Meeting of Scientific Counsellors of the Laboratory of Perinatal Physiology in San Juan, Dr. Pearce Bailey, Director of NINDB, explained the interest of the agency in the Cayo Santiago animals.

> The original NINDB plans for the establishment of a laboratory of perinatal physiology in Puerto Rico came about as a by-product of a larger collaborative program of perinatal research. The larger program envisaged the uniting of several important medical centers in the United States to prospectively collect data on perinatal factors affecting the developing nervous system in man from conception to adulthood. Using a master protocol in a multidisciplinary and multiinstitutional approach, these institutions aimed to establish reliable developmental sequences, develop better diagnostic methods—prenatal, natal, and postnatal—and search for new clues toward the prevention and treatment of cerebral palsy, mental retardation, and other neurological disorders of early life.
>
> As plans for the perinatal collaborative program developed, it seemed wise to parallel these clinico-pathological investigations of the larger program with animal studies. Prior to World War II, Drs. Windle and Becker at Northwestern [University] had studied the effects at birth of asphyxiation and resuscitation on the behavior and brain pathology of rats [later when the animal was sacrificed] when compared to that of their litter mates. So provocative had been these experiments that Windle and others believed that they should be extended to animals more comparable to man. The rhesus monkey seemed to be the animal of choice. The genital tract of the rhesus is very like a miniature human one. The sexual cycle of the rhesus corresponds to the human menstrual cycle

of 28 days and in addition, the rhesus extrudes but one ovum at a time, which when fertilized delivers but a single helpless offspring. Not too unlike the human infant.

Harlow and others had already demonstrated the unique adaptability of the rhesus to psychological and sensory tests and the monkey seemed to be an ideal animal for the standardization of neurological examinations and the application of such clinical laboratory tests as x-ray, EEG, air encephalography, arteriography, spinal fluid studies, and other tests. A free-ranging colony would serve as a reproductive reservoir from which some monkeys could be transferred to caged colonies for experimental purposes and also serve as an instrument for the study of primate ecology and behavior in a more or less natural habitat. A laboratory could probably be leased or constructed on the grounds of the School of Medicine of the University of Puerto Rico for testing the feasibility of inducing in primates counterparts of human mental retardation, kernicterus and other disabilities of early life. Monkey facilities were not available to us then at NIH and at that time India was clamping down on shipments of rhesus monkeys to the United States.

So when news came that the monkey colony on Cayo Santiago might be available, we wasted no time in going to Puerto Rico. The Appropriations Committee of the Congress had already given its blessing in principle as had the higher echelons at NIH (Bailey, 1965).

Several months later, Windle and Bailey returned to Puerto Rico for another evaluation and conferences with officials of the School of Medicine. Bailey described his first trip to the island colony as follows:

During the first visit, we went in a row boat, under our own steam, from the Playa de Humacao to the Cayo in a tricky current and under a blazing sun. The trip took about one hour. The colony appeared in a lamentable condition. There was gross evidence of malnutrition, cannibalism, and the island was infested with rats, which would beat the monkeys in a struggle for coconuts. The delivery of food from the shore was irregular and inadequate and there was evidence of water shortage. The Russian born primatologist Tomilin had long since departed and with him went all records, identifications and measurements of the monkeys. He refused to give them up, even when exposed to the persuasive talents of Dr. Gertrude van Wagenen, who knew him well (Bailey, 1965). (Note: Tomilin's records were eventually purchased by NINDB with the agreement that they be used in future research only and they were deposited in the office of the Director, NINDB, in 1958. Bailey, 1965; Windle, 1980).

Windle and Bailey returned from Cayo Santiago and met with a number of key individuals about the possibility of negotiating use of the monkey colony for NINDB. They conferred with Dr. E. Harold Hinman, Dean of the Medical School, Dr. Carroll Pfeiffer, Chairman of the Department of Anatomy, Dr. Roberto Buse, the Secretary of the Superior Education Council, Doña Felisa Rincon de Gautier, the Mayor of San Juan, and Luis Muñoz Marin, the venerable Governor of the island of Puerto Rico. (Bailey, 1965).

Definite proposals were made for collaborative work with the University of Puerto Rico based on a congressional mandate (February 9, 1956), from the U.S. House of Representatives Appropriations Committee on Health, Education, and Welfare. Bailey and Windle discussed the proposed work during the days of February 16-19, 1956, but left Puerto Rico without any committment when they learned a decision could not be made about use of the colony until other obligations and plans for it had been clarified. The NINDB proposals were eventually withdrawn (Bailey, 1956b).

Agreement was not reached because some days earlier (February 12) Windle had learned from Dr. Harry Harlow of the University of Wisconsin, that Dean Hinman had also offered the use of the monkey colony to Dr. David Rioch of the Walter Reed Medical Research Institute. Windle had tried to contact Rioch before leaving for Puerto Rico, but was unable to do so. In discussions with Hinman in San Juan, the apparent conflict of interest was raised, but no definitive answers were obtained about any prior committments to Rioch. The meeting with Governor Muñoz Marin confirmed support for the NINDB proposal as did a separate conference with the Chancellor of the University of Puerto Rico, Dr. Jaime Benitez, but Windle told Bailey he did not want to proceed further with developing the project if use of the monkeys had actually been previously promised to another institution (Windle, 1978).

Hinman was dispatched to Washington to meet with Bailey and Rioch and he arrived on February 27. He conferred first with Rioch about utilization of the colony, and Rioch stated that his position was as a motivator of better utilization of the opportunity for the study of monkey behavior. No progress had been made in obtaining funds to support the colony. Rioch told Hinman that if there was an opportunity for cooperation between the National Institutes of Health (NIH) and UPR, then the University should proceed with the collaborative effort (Hinman, 1956a).

Having resolved the conflict of interest, Hinman next met with Bailey at NINDB and reopened negotiations for use of the Cayo Santiago facility. Bailey was enthusiastic and set up an afternoon conference to produce a rough draft of a collaborative agreement. Hinman wanted two items included in the agreement. First, fiscal support was to be guaranteed beyond 1957 and second, the interests of the University of Puerto Rico faculty would have to be protected so that there would be the opportunity for other graduate students and scientists to work on the island, provided such work did not interfere with the basic interests of NIH. Bailey agreed (Hinman, 1956a).

After some preliminary discussion concerning facilities, Hinman said he felt the NINDB program was conceived as an arm of Windle's NIH laboratory and not a cooperative program with the University of Puerto Rico's School of Medicine. Hinman reported that NINDB wished to contract for space and facilities, along with occasional use of local faculty as consultants, but the island was to be under the complete and sole control of Dr. Windle's laboratory. Hinman responded that the University would provide complete autonomy to the laboratory project at the School of Medicine and offered full cooperation, but insisted that the School of Medicine "would not be shoved off the Santiago Island" (Hinman, 1956a). As a result of this turn of events, all negotiations ceased.

Hinman reported to Benitez:

> I believed that the School of Medicine could profit through cooperation with a distinguished group of scientists from the National Institutes of Health. We have received invaluable assistance in the past. From a prestige point of view, it would be helpful. Also, the transfer of full responsibility for operation of Santiago Island would relieve us from certain financial obligations, from many headaches, and probably would result in improvements on the island, and a build-up in the number of primates. On the other hand, our objectives involve the development of Puerto Rican scientists and not the aggrandizement of the National Institutes of Health, and it was inconceivable to me that we should find ourselves in a situation which would place us as outsiders in our own property (Hinman, 1956a).

On March 5, Bailey and Windle met with Rioch at NINDB to again discuss the possibility of work on Cayo Santiago. Rioch informed them that he saw no reason why studies of behavior on the free-ranging macaque population would interfere with the NINDB project and the same day wrote to Benitez in Puerto Rico that he would be happy to assist both NINDB and UPR in finding staff for the study and teaching of problems of group structure and behavior and its influence on individuals in the colony, or alternatively, if no collaboration was agreed upon, he would continue to look for support to establish a biological station at Cayo Santiago for the investigation of primate social behavior and ecology (Rioch, 1956).

Windle (1978) recalled the March 5 meeting with Rioch, noting Rioch told them he could not take advantage of the primate facilities in Puerto Rico because he had been unable to obtain the funds required and graciously relinquished any claim to the facility. He also recommended that Windle hire Stuart Altmann, whom he (Rioch) had hoped to send to Cayo Santiago (Windle, 1978).

Bailey then wrote to Hinman on March 9:

> It is our understanding that the Laboratory of Neuroanatomical Sciences of this Institute would have administrative responsibility in the rehabilitation and use of the monkey colony, but that, at the same time, we will cooperate in encouraging the use of the colony for behavioral and other studies by the faculty of the School of Medicine or other medical faculties, if such projects do not interfere with the prosecution of our objectives (Bailey, 1956c).

Hinman replied on March 12 that the School of Medicine was anxious to collaborate with NINDB providing opportunities for studies of the free-ranging monkeys by UPR faculty and outside investigators could be assured and that the Medical School would have a voice in administration of the colony (Hinman, 1956b).

Windle, Dr. C.J. Bailey—the proposed head of the new NINDB Section of Perinatal Physiology in San Juan— and Echart Wipf, Executive officer for NINDB, went to Puerto Rico on March 19 to make final arrangements with the University of Puerto Rico (Windle, 1978) and by April 16, 1956, a formal agreement between NINDB and the University of Puerto Rico was signed by Pearce Bailey and Jaime Benitez, establishing the Laboratory of Perinatal Physiology and protecting the breeding potential of the Cayo Santiago colony (Bailey and Benitez, 1956). NINDB obtained about 200 macaques along with an indefinite lease of Cayo Santiago, and the new program was formally inaugurated on August 29, 1956 by a conference at the Medical School in San Juan on "Neurological and Psychological Deficits of Asphyxia Neonatorum." As a result of the symposium, additional interest in collaborative work from scientists in Europe and South America followed (Windle, 1978).

The scientific work of Windle and staff subsequently became internationally recognized, dealing with neonatal asphyxia and resuscitation and their neuropathological and behavioral concomitants, induced hyperbilirubinemia, and therapeutic procedures for ameliorating neonatal asphyxia's effects (Bailey, 1965) Note: see Bibliography, Chapter 15, this volume).

As a result of the collaboration between NINDB and the School of Medicine, maintenance of the Cayo Santiago colony was substantially improved. The animals were regularly provisioned with a prepared diet, caging was rebuilt, a field laboratory was established, and four feeding/trapping cages were constructed along with a new cistern for water collection on the Small Cay and a concrete walkway to the pier (Frontera, 1958). The colony became the Behavioral-Ecology Section of the NINDB operation (Carpenter, 1972).

Of great importance was the resumption of the regular census of the population and behavioral observations. Stuart Altmann, then a graduate student of Dr. E.O. Wilson at Harvard, joined Windle's staff on June 11, 1956, and left immediately for Puerto Rico to begin a two-year study of macaque behavior and social organization (Carpenter, 1972; Windle, 1978). He reintroduced the population census and marking program and conducted observations through 1958 which resulted in a classic study of primate biology (Altmann, 1962). Altmann's work set the stage for the longitudinal observations of the colony that continue today.

Dr. Carl Koford, a biologist from the Hopland Field Station of the University of California, succeeded Altmann at Cayo Santiago and began observations of the population at the end of 1958, maintaining the continuity of the marking program, census, and demographic records of the colony (Koford, 1965). It was Koford who learned that Tomilin, living in California, was in need of money and willing to sell all the records he had taken with him when he left Puerto Rico, providing they were placed in safe keeping (Windle, 1978). Eventually the records were purchased by NINDB and Windle kept them housed in Bethesda, but by then they were of little value (Windle, 1978).

Under Koford's stewardship, the Cayo Santiago colony regularly supplied monkeys to the Laboratory of Perinatal Physiology in San Juan and to NIH in Bethesda. Although many animals were removed each year, Koford continued his studies of population dynamics on the island and eventually published a number of significant papers concerning reproductive behavior and demographic trends in the colony (see Chapter 15 for Koford's publications). He left the facility for the Delta Regional Primate Research Center (Tulane University), on a leave of absence in 1965, and did not resume his work as the supervisory biologist and Chief of the Section on Primate Ecology of the Laboratory of Perinatal Physiology in Puerto Rico when the leave expired in August, 1966. It was during Koford's tenure as supervisor that intensive and sustained studies of the reproductive behavior and population dynamics of the colony, of individual growth and development, and behavioral ecology by extramural investigators, such as James Gavan, John Kaufmann, Donald Sade, Clinton Conaway, Margaret Varley, and Elizabeth Missakian began (Carpenter, 1972).

This work was carried out in addition to the principal use of the monkeys as subjects for neurobiological experimentation at the laboratory in San Juan. Despite the removal of animals for laboratory work, the regular provisioning of the monkeys supported a 16 percent net annual increase in the colony (Koford, 1965; Carpenter, 1972), and several groups of monkeys were removed from the island to found other breeding colonies and to provide the animals with a different habitat as a means of testing ecological and behavioral hypotheses generated as a result of work on Cayo Santiago. In 1961, macaques from the Cayo Santiago colony were introduced to the islands of Cueva and Guayacan, near La Parguera on the southwestern coast of Puerto Rico. In 1966, another group was translocated to the island of Desecheo, some 13 miles off the west coast of Puerto Rico (Morrison and Menzel, 1966; 1972). Even with these removals, population pressure due to growth of the Cayo Santiago colony did not abate.

With Koford's departure from Cayo Santiago in 1965, responsibility for management of the colony changed hands many times over the next five years. Newly recruited scientists to the laboratory, such as Halsey Marsden, Steven Vessey, John Morrison, Andrew Wilson, and Margaret Varley all contributed to the continuity of observations on the animals over relatively short periods of time. However, two individuals emerged who would eventually make very important long-term contributions to establishing Cayo Santiago as an invaluable research resource for the study of primate behavior and morphology. These were Angel Figueroa and Donald Sade.

Figueroa, a young man from Punta Santiago, had originally been hired by Koford as a workman and general laborer. After a period of part-time employment, Figueroa was invited to work full time for the facility as an animal caretaker in 1959. He worked closely with Koford and gradually assumed much of the responsibility for the census of the population, as well as the recording of data on rainfall and seasonal changes in vegetation. He also found time to assist with general maintenance of the facility. His growing knowledge of the colony and his ability to identify individual animals on sight made him invaluable.

About the same period, Donald Sade, a student of Dr. Earl Count at Hunter College in New York with an interest in physical anthropology, was asked by James Gavan to participate in a large scale study of growth and development of the Cayo Santiago macaques begun in 1958. Sade arrived on Cayo Santiago in early June 1960, and assisted in trapping animals for measurement and radiography. He became interested in the social interactions of the monkeys in Group F, and devoted as much of his time as possible to observations. He left the island in 1962, when the growth study ended, and returned again during 1963 and 1965 to continue his study of the ontogeny of social relations in Group F, as a graduate student of S.L. Washburn in the Department of Anthropology at the University of California, Berkeley (Sade, 1966). His work established the value of long-term study of primate behavior, over several annual cycles, for improving our understanding of the complexity of social relationships amongst primate troops. It would also eventually lead to a complete reorientation of scientific work at Cayo Santiago.

Figure 1.6 Angel Figueroa, Chief Census Taker.

Figure 1.7 The late C.R. Carpenter, about 1968. Courtesy of Yerkes Regional Primate Research Center.

CAYO SANTIAGO AND THE CARIBBEAN PRIMATE RESEARCH CENTER

As a result of jurisdictional disputes, a change in directorship at NINDB, and a gradual disintegration of the collaborative relationship between the Laboratory of Perinatal Physiology and the UPR School of Medicine, NINDB closed its operations in Puerto Rico and moved its research at San Juan back to Bethesda, Maryland in July 1970 (Carpenter, 1972; Windle, 1980). The UPR School of Medicine resumed control of Cayo Santiago and incorporated it into what is now the Caribbean Primate Research Center (Windle, 1980). Through the efforts of Henry Wagner at NINDB, continued support for the Cayo Santiago colony and its sister colonies near La Parguera was provided by NINDB to the University of Puerto Rico (Carpenter, 1972).

The demise of the Laboratory of Perinatal Physiology in San Juan gravely threatened the existence of the Cayo Santiago colony. Interestingly enough, it was Carpenter who, by then a senior faculty member at the University of Georgia and special advisor to the University president, came to the aid of the facility (Dukelow, 1984).

At the time, Carpenter was involved in an attempt to establish a new macaque colony on a peninsula of land held by the Savannah River Project, a nuclear plant operated by the Atomic Energy Commission in Georgia (Dukelow, 1984). Perhaps acquisition of monkeys from Cayo Santiago for the Savannah project was one of Carpenter's reasons for going to Puerto Rico (a later report—Carpenter, 1969—advocated such a transfer from Cayo Santiago). But his participation, from September 6-11, 1969, in the "Location Study Group," which also included Drs. Pearce Bailey, David Davis, Ronald Myers, and Donald Sade, produced recommendations for the facility which led to continued sponsorship of the colony by NINDB and other granting agencies. Carpenter, some 30 years after founding the Cayo Santiago colony, again played a key role in ensuring its survival.

Once a contractual agreement was reached between NINDB and the University of Puerto Rico for support of the Caribbean Primate Research Center and its constituent colonies of primates, a new period of research emerged at Cayo Santiago. Dr. Clinton Conaway, new Director of the center, named Dr. Sade to the post of Scientist-in-Charge at Cayo Santiago. Sade (1969) had submitted a lengthy prospectus for a research program on the colony which focused on long-term studies of social behavior. The advisory board for the center (Buettner-Janusch, 1971) had recommended the population at Cayo Santiago be substantially reduced, from the more than 685 animals reported alive in 1971 to a more managable level of 200 or 300 monkeys, and endorsed Sade's proposal for future research (Buettner-Janusch, 1971).

As a result, the size of the colony was reduced by a lengthy cull to prepare for longitudinal behavioral work on the island. Prior to formation of the CPRC, behavioral work had not been permitted to impinge upon the supply of monkeys to laboratory units. Selection of individuals was haphazard and many social groups were so severely reduced that, at best, age-sex composition was imbalanced and artificial. Also, the effects of sporadic removals of animals on behavioral patterns was unknown. All but the intact social groups were removed from the island and when the cull was completed in 1972, only four troops remained (Buettner-Janusch et al., 1974). One entire troop (Group K) was removed and sacrificed in 1972 to generate a tissue bank and skeletal collection for morphological studies. Intermittent capture and manipulation of the animals was stopped in an attempt to obtain naturalistic conditions which would support long-term observations of behavior with a minimum of artificial bias (Rawlins, 1979).

Throughout this period of transition, the continuity of the population records was maintained by Angel Figueroa, who remained with the colony after the departure of Koford and the closure of the Laboratory of Perinatal Physiology. Sade and Figueroa, with the support of a large number of "sociometric technicians" hired both by the center and on a National Science Foundation (NSF) grant to Sade, developed a detailed and accurate picture of individual and troop behavior and population dynamics. The goal of maintaining Cayo Santiago as an undisturbed semi-natural colony for research was achieved, and the facility attracted behaviorists from the continental U.S. and Europe who conducted work under the new management protocol.

Although the biomedical community frequently questioned the use of Cayo Santiago macaques in this fashion, the program received sufficient support from funding agencies to proceed unimpeded. Because of continued support from Drs. William Goodwin and Leo Whitehair of the Animal Resources Branch of the Division of Research Resources (NIH), Cayo Santiago was able to emerge as a truly unique site for the study of primate biology under seminatural conditions. The true potential of Carpenter's dream began to be realized.

Between 1970 and 1983, the Caribbean Primate Research Center was directed by Drs. Clinton Conaway, William Kerber, Sven Ebbesson, Gilbert Meier, and Lloyd LeZotte. All supported the use of Cayo Santiago as an unmanipulated population for the study of behavioral ontogeny and microevolutionary genetics and demography.

Sade and his colleagues worked with the population through the Spring of 1976, at which time he returned to academic life at Northwestern University. However, the management policy for Cayo San-

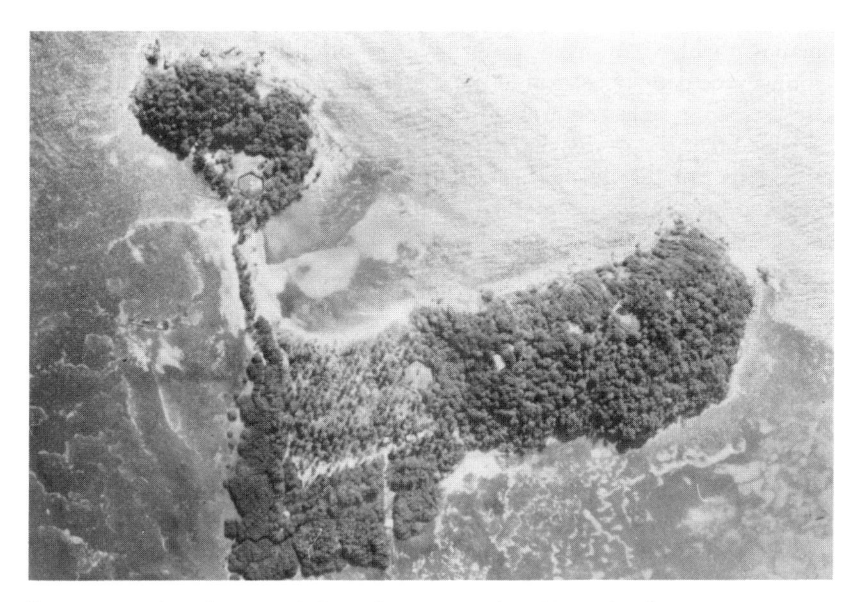

Figure 1.8 Aerial view of Cayo Santiago today. Note the dense vegetation.

tiago, established with formation of the center in 1970, was maintained and the continuity of the demographic data was uninterrupted.

In July 1976, Richard Rawlins took over responsibility for the day-to-day operation of the Cayo Santiago colony and was named Scientist-in-Charge in 1977. With the help of Figueroa, a dedicated crew of animal caretakers, and the many investigators who came to conduct studies on the island, the life history data on the monkeys was routinely collected. In addition to demographic, blood genetic, and behavioral studies, a new emphasis was placed on the investigation of primate functional morphology.

Matt Kessler, a laboratory animal veterinarian with primate experience, was hired as the Director of Veterinary Activities for the CPRC in April, 1977. He and Rawlins were concerned about tetanus in the Cayo Santiago monkeys-Rawlins for the behavioral and demographic and Kessler for the epizootiological aspects of the disease. Their mutual interests in tetanus eventually resulted in a collaborative paper on the subject, and the realization that "behaviorists" and veterinarians could work cooperatively. This understanding fostered a new period at Cayo Santiago, in which the colony was also utilized for noninvasive biomedical research on free-ranging rhesus monkeys during the annual winter roundup of yearlings for marking and genetics research. The biomedical research was done strictly within the confines of the management protocol for the colony and did not disrupt ongoing behavioral studies (Rawlins, 1979).

Rawlins left Cayo Santiago and Puerto Rico in December, 1981, for

Michigan State University and Kessler was made responsible for management of the colony until a new Scientist-in-Charge could be found. The daily operation of the island was assumed by Figueroa. With the aid of a new research assistant, John Berard, and an additional census taker, Edgar Davila, it was possible to maintain the accuracy of the daily census despite the enormous population size which had reached nearly 1200 animals by the end of 1983. Under Kessler's supervision, previous management protocols were maintained, and the behavioral and biomedical research by in-house and guest investigators continued. Of particular interest was the start of surveys of the Cayo Santiago population for health surveillance; viral and bacterial and antibody titers; parasites; dental developmental; anthropometrics; diabetes; and degenerative disorders of the ophthalmologic and musculoskeletal sytems.

Due to the large size of the population, in 1984 three intact social groups were removed from Cayo Santiago to reduce the colony by approximately 50 percent. In January and February, Groups M and J were translocated to Sabana Seca. The former group was released intact into a 2-acre forested hill corral for behavioral and biomedical research (Kessler et al., 1984), and the larger Group J was divided for breeding purposes and placed in small enclosures. The reduction of the colony, the first since 1972, was completed in early April, 1984, with the shipment of Group O to the Deutsches Primatenzentrum (DPZ) in Göttingen, West Germany to provide that primate center with a core of rhesus monkeys for its breeding and research programs. Removal of the 600 monkeys from Cayo Santiago occurred with minimal disruption to the population remaining on the island.

Curt Busse, a sociobiologist from the Yerkes Regional Primate Research Center, arrived in Puerto Rico in March, 1984 to assume the new position of Research Coordinator for Cayo Santiago (formerly Scientist-in-Charge) and begin his research on post-ejaculatory intervals in free-ranging rhesus monkeys.

In 1985, the entire Cayo Santiago population was inoculated against tetanus and Group P, a troop of approximately 65 monkeys which fissioned from Group F, was moved from Cayo Santiago to Sabana Seca in an effort to maintain a stable population size on the island. An annual random cull of the colony is presently under consideration.

Today, as Cayo Santiago completes its 47th year as an international research resource in primate biology, the importance of maintaining existing and establishing new primate colonies has grown. Conditions which led to the development of Cayo Santiago have deteriorated over the last five decades with shrinking numbers of feral primates and continued destruction of their remaining habitats which threaten survival of many species. To halt further reduction of the primate order, alternative animal models for medical testing must be sought and colony-bred monkeys used in research only when appropriate.

Field research stations, such as Cayo Santiago, will continue to enhance our understanding of the ontogeny and organization of primate behavior, biology, and morphology, providing a standard against which manipulative studies in the laboratory and observations on groups of feral monkeys can be assessed. More importantly, these colony reservoirs of primate species may well become a final refuge for many members of this order.

ACKNOWLEDGEMENTS

The authors thank Dr. Norman Maldonado, former Chancellor of the University of Puerto Rico School of Medicine and the late Dr. W.F. Windle for access to documents pertaining to the history of the Cayo Santiago colony. Additional help in locating original articles was provided by Dr. Robert Hinde, Medical Research Council (MRC) Unit, Madingley, Cambridge University, and by Janice Gonzalez, Caribbean Primate Research Center and Ava Gaa Kessler, Department of Zoology, City College of New York. Special thanks are due Dr. W. Richard Dukelow for reviewing earlier drafts. We would also like to acknowledge the assistance of Mr. Leon J. Stout of The University Libraries, The Pennsylvania State University for providing archival photographs from the C.R. Carpenter Papers, Penn State Room used in this chapter. The Caribbean Primate Research Center is supported in part by the Division of Research Resources, Animal Resources Branch, NIH, and by the University of Puerto Rico School of Medicine. We dedicate this history, with many thanks, to Señor Angel Figueroa who has devoted over 25 years of work with the Cayo Santiago rhesus monkey population. Without his efforts over the last quarter century, there would be little history to write.

REFERENCES

Anonymous. Monkeys — a research investment. *The Economic Review.* San Juan, PR, Chamber of Commerce, Summer: 61-77, 1940.

Altmann, S.A. A field study of the sociobiology of rhesus monkeys (*Macaca mulatta*). *Ann. N.Y. Acad. Sci.* 102:338-435, 1962.

Bailey, P. Collaborative program of field investigations related to cerebral palsy and mental retardation. *National Inst. Neurological Disease and Blindness—Fiscal 1956 Report,* 1956a.

Bailey, P. Letter to Jaime Benitez, Chancellor, University of Puerto Rico, February 24, 1956b.

Bailey, P. Letter to E. Harold Hinman, Dean, University of Puerto Rico School of Medicine, March 9, 1956c.

Bailey, P. Early history of the perinatal project of Puerto Rico. Paper presented to the Meeting of Scientific Counsellors, Laboratory of Perinatal Physiology, San Juan, Puerto Rico, October 25, 1965.

Bailey, P. and Benitez, J. Agreement between the National Institute of Neurological Diseases and Blindness and the University of Puerto Rico, April 16, 1956.

Buettner-Janusch, J. Report of the Scientific Advisory Board of the Caribbean Primate Research Center to Dr. Adan Nigaglioni, Chancellor, School of Medicine, University of Puerto Rico and Dr. Clinton Conaway, Director, Caribbean Primate Research Center, March 1, 1971.

Buettner-Janusch, J.; Mason, G.A.; Dame, L.; Buettner-Janusch, V.; Sade, D.S. Genetic studies of serum transferrins of free-ranging rhesus macaques of Cayo Santiago, *Macaca mulatta* Zimmerman, 1780). *Am. J. Phys. Anthropol.* 41:217-232, 1974.

Carpenter, C.R. Rhesus monkeys for American laboratories. *Science* 92:284-286, 1940.

Carpenter, C.R. Sexual behavior of free ranging rhesus monkeys. *J. Compar. Psychol.* 33:113-162, 1942.

Carpenter, C.R. History of the monkey colony of Cayo Santiago. Transcript of lecture given at the University of Puerto Rico Medical School, San Juan, Puerto Rico, August 1959.

Carpenter, C.R. *Naturalistic Behavior of Nonhuman Primates.* University Park, PA, The Pennsylvania State University Press, 1964.

Carpenter, C.R. A report with recommendations on a proposed Caribbean Primate Research Center by the Locations Study Group of the Special Advisory Committee of the Laboratory of Perinatal Physiology, San Juan, Puerto Rico for the National Institute of Neurological Diseases and Blindness Bethesda, Maryland. October 3, 1969.

Carpenter, C.R. Breeding colonies of macques and gibbons on Santiago Island, Puerto Rico. pp. 76-87 in *Breeding Primates.* W. Beveredge ed. Basel: Karger, 1972.

Carpenter, C.R. and Krakower, C.A. Notes on results of a test for tuberculosis in rhesus monkeys (*Macaca mulatta*). *Puerto Rican J. Publ. Hlth. Trop. Med.* 17:3-13, 1941.

Cheslak, S.E. and Taub, D.M. Establishment and maintenance of an island colony of rhesus monkeys (*Macaca mulatta*). Paper presented at 1983 Meetings of the American Assoc. Lab. Animal Sci. San Antonio, TX, 1983.

Dukelow, W.R. Personal communication, 1984.

Frontera, J.G. The Cayo Santiago primate colony. pp. 246-256 in *Neurological and Psychological Deficits of Asphyxia Neonatorum.* W.F. Windle ed., Springfield, Illinois: C.C. Thomas, 1958.

Gericke, H. Personal communication, 1983.

Hinman, E.H. Report to Chancellor Jaime Benitez, February 29, 1956a.

Hinman, E.H. Draft of letter to Pearce Bailey, NINDB, March 12, 1956b.

Kessler, J.J.; Figueroa, A.; Kapsalis, E.; Berard, J.; Davila, E.; Martinez, H.; Gonzalez, J. Trapping, removal and translocation of a group of free-ranging rhesus monkeys (*Macaca mulatta*) from Cayo Santiago to a semi-natural hill enclosure. *Am. J. Primatol.* 6:409, 1984.

Koford, C. Population dynamics of rhesus monkeys on Cayo Santiago, pp. 160-174 in *Primate Behavior: Field Studies of Monkeys and Apes.* I. DeVore ed. New York: Holt, Rinehart and Winston, 1965.

Lebron, A.P. and Otero, P.O. Hemolytic streptococci from the throat of normal monkeys. *Proceedings of the Society for Experimental Biology and Medicine* 45:509-511, 1940.

Locke, C. Peopling an island with gibbon monkeys. *The Illustrated London News*, pp. 290-291, August 13, 1938.

Morrison, J.A. and Menzel, E.W. Adaptation of a rhesus monkey group to artificial fission and transplantation to a new environment (motion picture). *Amer. Zool.*, 6: (Abst. No. 211), 1966.

Morrison, J.A. and Menzel, E.W. Adaptation of a free-ranging rhesus monkey group to division and transplantation. *Wildl. Monogr.* No. 31, 1972.

Morayta, E. Apuntes para la biografia de un edificio (la Escuela de Medicina Tropical de Puerto Rico). Unpublished manuscript on file at the University of Puerto Rico School of Medicine, San Juan, Puerto Rico, 1969.

Nicholas, W.H. and Locke, J. Growing pains beset Puerto Rico. *The National Geographic Magazine* 99:435-436, 1951.

Poindexter, H.S. A study of the intestinal parasites of the monkeys of the Santiago Island primate colony. *Puerto Rican J. Publ. Hlth. Tropl. Med.* 18:175-191, 1942.

Pucak, G.J., Foster, H.L. and Balk, M.W. Key Lois and Raccoon Key: Florida Islands for free-ranging rhesus monkey breeding programs. *J. Med. Primatol.* 11:199-210, 1982.

Rawlins, R.G. Forty years of rhesus research. *New Scientist* 82:108-110, 1979.

Rioch, D.M. Letter to Chancellor Jaime Benitez, University of Puerto Rico School of Medicine, March 5, 1956.

Sade, D.S. Ontogeny of social relations in a group of free ranging rhesus monkeys *(Macaca mulatta)*. Doctoral thesis in anthropology, University of California, Berkeley, 1966.

Sade, D.S. Prospectus for a research program on the Cayo Santiago colony of rhesus monkeys: long term studies of social behavior. Manuscript on file at the Caribbean Primate Research Center, May 8, 1969.

Suarez, R.M.; Diaz Rivera, R.S.; Hernandez Morales, F. Hematological studies in normal rhesus monkeys (*Macaca mulatta*). *Puerto Rican J. Publ. Hlth. Trop. Med.* 18:212-226, 1942.

Windle, W.F. Letter to Dr. Norman Maldonado, Chancellor University of Puerto Rico Medical School, San Juan, Puerto Rico, October 26, 1978.

Windle, W.F. The Cayo Santiago primate colony. *Science* 209:1486-1491, 1980.

Illustration 2.1 Activity around a rainwater pool.

CHAPTER TWO

Demography of the Free-Ranging Cayo Santiago Macaques (1976-1983)

RICHARD G. RAWLINS AND MATT J. KESSLER

INTRODUCTION

The natural regulation of mammalian populations has been intensively studied over the last 50 years, but the population dynamics of nonhuman primates have only recently been examined in any detail. This is due to the difficulty of monitoring feral primate populations for the long periods of time needed to obtain detailed life history data, to the long life span of most primate species, and to the lack of suitable primate populations (or sites) to carry out longitudinal observations. Few field studies of primate species have focused on population biology because it is virtually impossible to establish exact age distributions and to document natality, mortality, emigration, and immigration in primate groups using the short-term observations characteristic of most fieldwork.

The Cayo Santiago colony of free-ranging rhesus monkeys *(Macaca mulatta)* has been continuously monitored since 1956. It provides an excellent resource for the long-term study of primate population dynamics. Previous work by Altmann (1962), Koford (1965), Sade et al. (1977), and Duggleby (Chapter 14, this volume) has established the value of the colony for demographic research and investigation of the genetic microevolution of the population.

This report documents demographic trends and population statistics in the Cayo Santiago colony from July 1976 through June 1983, and examines differential natality, secondary sex ratio adjustment, and mortality as variables contributing to the significant variation in the growth of the component troops of the colony. The utility of population projections based on time-specific life tables is assessed against observed population growth.

MATERIALS AND METHODS

Cayo Santiago has been occupied by an introduced population of free-ranging rhesus monkeys since 1938. Animals have been removed from the island over the years, but new stock has not been added except through births. Between 1972 and 1984, the population was left intact and maintained under seminatural conditions. The monkeys were provisioned daily with 0.23 kg per capita of high (24-26 percent) protein commercial monkey diet (Agway, Inc. Syracuse, NY; Allied Mills Inc., Chicago, IL; Ralston Purina Co., St. Louis, MO). They also foraged on the abundant tropical vegetation and were geophagic (Sultana and Marriott, 1982). Water was available *ad libitum* from automatic waterers distributed about the island. There was no provision for disease prophylaxis, but moribund animals were permanently removed from the colony for treatment or euthanasia (Rawlins and Kessler, 1982).

All monkeys were of known identity, age, sex, and maternal genealogy. Detailed life histories were kept on each animal through a daily census of the population, and data on natality, morbidity, mortality, and group affiliation were recorded by a team of observers responsible for monitoring each social group. In general, new births, missing animals, deaths, and changes in group composition were noted on the day of occurrence or within one to two days thereafter (Rawlins and Kessler, 1983).

During 1976, the population was organized into five naturally formed social groups, Groups I, J, L, F, and M. In 1977, a sixth group, Group O, formed as a result of the fission of Group F along matrilineal lines (Rawlins, 1979). Each group comprised a variable number of adult males associated with one to four matrilines, which in turn consisted of an adult female, her adult female offspring, and the juvenile males and females of each. Females remain with the natal troop for life or until group fission occurs, and maternal genealogies form the nucleus of daily social activity. Adult males were usually non-natal members of the group because most adolescent males disperse from their natal group at puberty (approximately four years of age) and many adult males annually shift group membership during the mating season (see Chapter 6, this volume). The demographic records from 1976 to 1983 comprise the data base for this study.

RESULTS

Population Growth

From July 1, 1976 to June 30, 1983, the Cayo Santiago population increased from 479 to 1161 animals for a net gain of 142.3 percent. Growth of the population is presented in Figure 2.1. Marked fluctuations in the actual number of monkeys alive each month over the years resulted from the significant increases of animals in the population during the annual birth seasons from late December to early July, and the decreases in numbers due to mortality, which peaked during the annual mating seasons from July to December (Rawlins and Kessler, 1982). Growth of the population over seven years was geometric. A linear regression run on the natural log of peak population values for the period showed the rate of increase (the slope) was 0.13, indicating a net increase of 13 percent per annum in

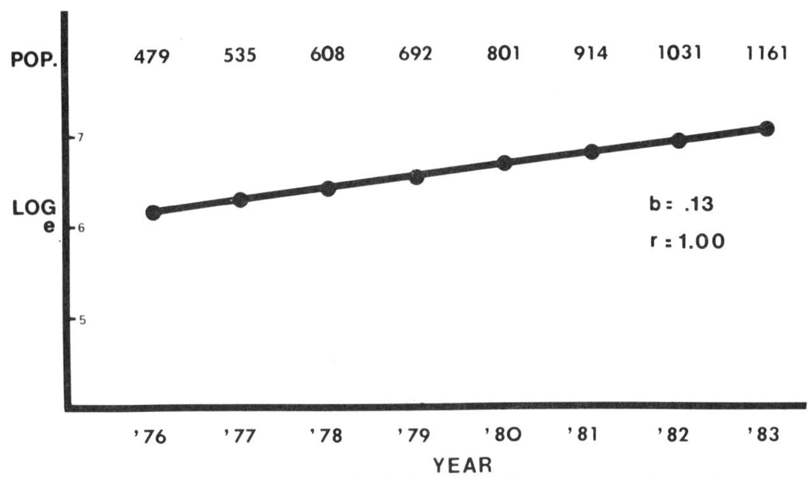

Figure 2.1 Population growth of the Cayo Santiago population (1976-1983).

the colony. The fit of the regression to the logged peak population values was excellent (r = 1.00).

Population growth can also be expressed as the ratio of the absolute size of the population at a specified time *T* divided by its size one time interval earlier, *T-1* (Ricklefs, 1982). This provides a general estimate of the annual increase of the population per individual called the finite rate of increase (Lambda).

$$\lambda = N(T)/N(T-1)$$

Table 2.1 gives the finite rate of increase for the population from 1976 to 1983. The values were essentially constant over the seven-year period and confirm the population increased at the same rate as that obtained from the linear regression on the natural log of peak population values. This is shown by the equation:

$$\log_e \lambda = r \text{ (geometric rate of increase) and } \log_e 1.14 = .13$$

Both measures demonstrate that the Cayo Santiago population was not stationary and continued to grow, but that the colony had reached and maintained a stable age distribution since the major disruption and cull of the population which ended in 1972. For the colony growth rate to remain constant, all age groups must grow or decline at the same rate as the entire population so that the relative proportion of

Table 2.1. Finite Rate of Increase: Cayo Santiago 1976-1983

1976-1977	1977-1978	1978-1979	1979-1980	1980-1981	1981-1982	1982-1983
1.12	1.14	1.14	1.16	1.14	1.13	1.13

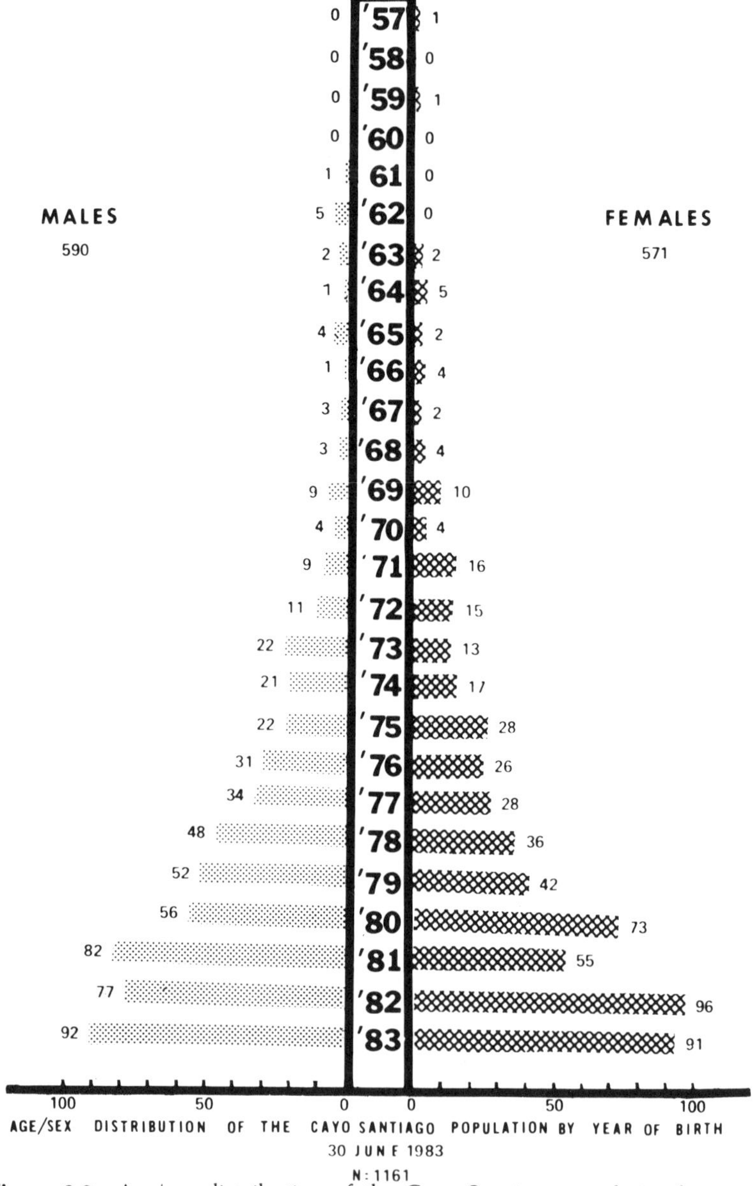

Figure 2.2 Age/sex distribution of the Cayo Santiago population by year of birth.

the population represented by each age group remains the same over time (see Figure 2.2). Having reached a stable age distribution, the growth rate should persist as long as existing management protocols are maintained or until biotic or abiotic events intercede.

The age-sex distribution of the population as of June 30, 1983 is shown in Figure 2.2. The adult (4 years of age and older) secondary sex

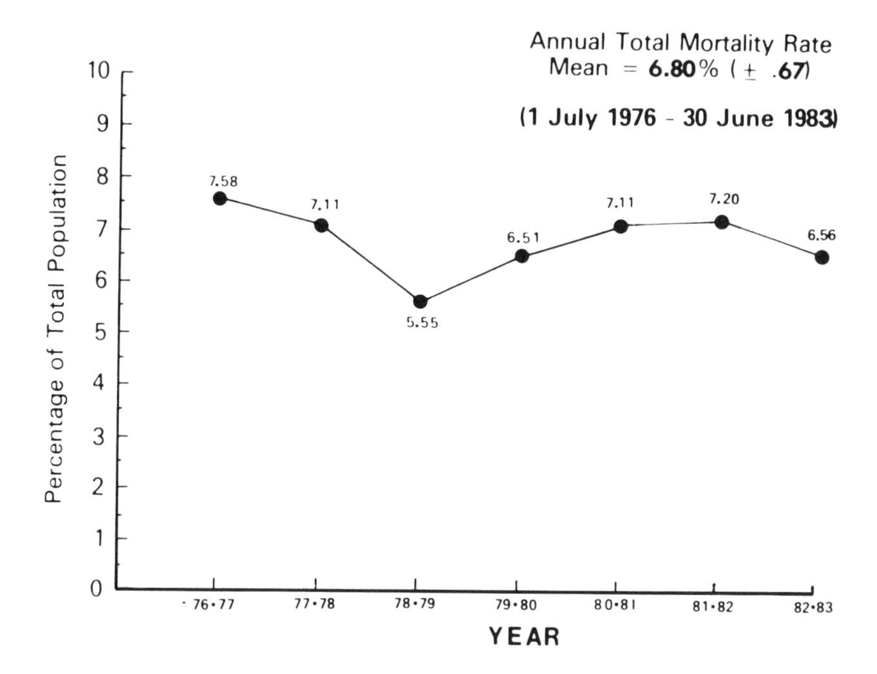

Fig. 2.3 Annual total mortality rate of monkeys at Cayo Santiago, 1976-1983.

ratio at the close of the study was 111 males to 100 females, and the sex ratio of the entire population was 103 males to 100 females. The mean annual mortality rate, expressed as the sum of the monthly mortality rates for the period July 1 to June 30 of a given year, was 6.80 percent (*sd* 0.67 percent). Annual total mortality rates for each year of the study are presented in Figure 2.3. Tetanus was a major cause of death in the colony (Rawlins and Kessler, 1982).

REPRODUCTION ON CAYO SANTIAGO

A total of 1171 births were recorded during the eight birth seasons. The median birth date ranged from February 11 to April 12, but from 1979-1983, the median birth date fell in mid February (18 February ± 5 days). The average interbirth interval for females delivering and weaning offspring in consecutive years was 372 ± days versus 336 ± 29 days for females whose pregnancies terminated in stillbirths or abortions, or whose infants died prior to weaning. The difference was highly significant ($t = 8.06$, $df = 717$, $p. < .0001$). A general summary of the annual reproductive rates for females four years of age and older, and the distribution of normal and abnormal births from 1976 to 1983 is given in Table 2.2.

Table 2.2. Annual Reproductive Rates and the Distribution of Normal and Abnormal Births at Cayo Santiago (1976-1983).

Year	Females[1]	Births[2]	Rate[3] %	Live Births[4]		Aborted	Stillborn		Neonatal Deaths[5]		
				♂	♀	Sex Unknown	♂	♀	Sex Unknown	♂	♀
1976	115	97 (103)	84.3	50	43	4	3	2	1	0	0
1977	121	95 (96)	78.5	45	40	0	3	7	1	0	0
1978	136	112 (116)	82.4	63	46	0	1	6	0	0	1
1979	162	120	74.1	60	53	1	4	2	0	0	0
1980	190	161	84.7	76	85	0	0	0	0	2	1
1981	212	170 (171)	80.2	99	67	0	2	2	1	2	0
1982	239	194 (196)	81.2	86	105	2	2	0	1	0	1
1983	269	208*	77.3	104	99	0	1	4	0	1	0
Total	1444	1157 (1171)	80.3	583	538	7	16	23	4	5	3

[1]. Females ≥ 4 years of age.
[2]. Births to females ≥ 4 years of age. From 1976 to 1983, 14 three-year-old females also gave birth. Birth totals including these animals appear in parentheses.
[3]. Reproductive rate based only on females ≥ 4 years of age.
[4]. Totals for *all* females.
[5]. Neonatal death is defined as mortality within 48 hours post-partum.
* Includes one set of twins (♂, ♀).

Table 2.3. Live Births and Infant Survivorship to Yearling Status

	Males			Females			Total		
	Births	# Alive January 1	%	Births	# Alive January 1	%	Births	# Alive January 1	%
1976	50	47	94.0	43	40	93.0	93	87	93.5
1977	45	42	93.3	40	40	100.0	85	82	96.5
1978	63	59	93.7	46	41	89.9	109	100	91.7
1979	60	60	100.0	53	50	94.3	113	110	97.3
1980	76	63	82.9	85	79	92.9	161	142	88.2
1981	99	92	92.9	67	61	91.0	166	153	92.2
1982	86	83	96.5	105	95	90.5	191	178	93.2
1983	104	90	86.5	99	90	90.9	203	180	88.7
Average Annual Survivorship			92.5			92.8			92.7
		S.D.	5.41		S.D.	3.27		S.D.	3.25

The net reproductive rate for mature females, four years and older, ranged between 74.1 percent and 84.7 percent, with a cumulative mean of 80.3 percent (*sd* 3.60 percent). Live births comprised 95.7 percent of the total and the secondary sex ratio was 107 male to 100 female births. Of all births, 4.3 percent were aborted or stillborn and 0.7 percent of the live births died within 48 hours postpartum. The percentage of stillborn females (of all female births) was 4.1 percent, 1.5 times higher than that of males (2.7 percent). Table 2.3 gives the percentage of infants surviving to yearling status from 1976 to 1983. The mean annual survivorship for male infants was 92.5 percent (*sd* 5.4 percent) and 92.8 percent (*sd* 3.27 percent) for females. The mean annual survivorship for both sexes to yearling status was 92.7 percent (*sd* 3.25 percent).

One set of twins (a male and female) was born in 1983 to an eight-year-old multipara (#633) in Group I. This was the only occurrence of twinning during the study and the first case reported in the colony since 1956 (Koford et al., 1966). Based on 1171 births during the study period, the incidence was 0.09 percent.

Two cases of congenital abnormalities were observed. The first was an anencephalic, acranial female born to a seven-year-old multipara (#411) in Group F in 1978 and the second, a congenitally blind male delivered by an eight-year-old multipara (#585) in Group I in 1982 (Rawlins and Kessler, 1983; and see Chapter 4, this volume). The mothers were matrilineally unrelated and had no prior history of delivering malformed offspring. The incidence of each defect, based on 1171 births, was 0.09 percent, with a cumulative incidence of 0.17 percent for all teratisms seen during the study.

Multiple births of rare, nonpathological and nonlethal hereditary anomalies were also seen. Seven golden macaques were born into three different matrilines in two genetically distinct social groups between 1981 and 1984 (see Chapter 13, this volume). The incidence of this phenotype, based on seven cases, was 0.60 percent, 60 times the expected rate in the wild (Pickering and Van Wagenen, 1969).

Variation in Troop Growth

Though population growth was constant over the study period, the observed rates of increase for the six component groups of the population were not equivalent. Table 2.4 lists the geometric rates of increase for each group, derived from a linear regression run on the natural log of peak troop size for each year of the study. Groups I and J increased at 16 percent and 15 percent per annum respectively, while Groups M and O grew at 13 percent and 14 percent per annum. Group F grew at only 10 percent and Group L showed the lowest rate of increase at only 8 percent, half that of the most rapidly expanding social groups. Table 2.5 presents the relative percentage of the total population represented by each social group from 1976 to 1983.

Table 2.4. Rate of Increase (Geometric*) by Social Groups 1976-1983

Troop[a]	No. Animals July 1, 1976	No. Animals June 30, 1983	$b=$	$r=$
I	88	264	.16	1.00
J	108	306	.15	1.00
O	41[b]	100	.14	.99
M	59	144	.13	.99
F	98[c]	175	.10	.99
L	89	161	.08	.99

* Based on a linear regression of log of the annual troop maximum (June 30). The slope (b) is the geometric rate of increase. The correlation coefficient (r) indicates the goodness of fit of the regression line to the data.

a Extra Group males not included.

b Troop O was part of Troop F until 1977. This value gives maximum size for the troop on June 30, 1977.

c Value for Troop F is for maximum size for the group on June 30, 1977, following the fission of Troop O from Troop F.

The absolute rate of expansion for each group, as for the population as a whole, is limited by the number of females available for breeding in the group for any given year (emigrating males were excluded at the troop level). Recall that females do not disperse from their natal troop at puberty, as do males, so the differential distribution of females by group provides a simple estimate of the potential growth of a group over time. The number of females in Groups J and I increased by 6.0 percent and 3.3 percent, respectively, while Group L declined by 8.1 percent. The relative percentage of all females in the population represented by female members of each social group is given in Table 2.6. The rate of increase for the females in each social group during the study period is given in Table 2.7. Note that the rate at which the female component of Group L expanded was less than one-third that of the fastest growing groups, Groups I and J. Marked differences in the relative growth of the female population per social group may reflect the effects of differential fecundity, differential adjustment of the sex ratio at birth, or differential female survivorship post-partum. Each of these variables was examined as possible factors contributing to the differences in troop growth. Emigration is considered elsewhere (see Chapter 6, this volume).

Table 2.5. Relative Percentage of Total Population Represented by Social Groups 1976-1983

Troop	1976	1977	1978	1979	1980	1981	1982	1983
I	18.3	18.8	20.3	19.4	18.8	20.2	20.7	23.2
L	18.6	18.3	18.2	14.8	14.7	15.0	14.1	14.6
F	27.6	16.6	17.0	17.3	18.1	16.3	15.4	14.7
J	21.1	22.7	21.6	25.3	25.2	25.3	24.1	25.1
M	12.1	12.5	12.2	12.6	12.1	12.9	13.2	12.7
O	*	9.2	8.7	8.6	8.7	8.6	9.3	8.3
Extra Group Males	2.0	2.0	1.5	2.0	2.1	1.8	2.2	1.0

* Group O was part of Group F until 1977.

Table 2.6. Relative Percentage of All Females Represented by Female Members of Each Social Group

Troop	1976	1977	1978	1979	1980	1981	1982	1983
I	19.7	18.6	19.6	21.3	20.4	23.0	21.9	23.0
L	18.7	17.3	17.2	16.5	15.7	12.7	11.4	10.6
F	28.0	17.3	16.8	16.5	17.9	15.3	16.2	15.4
J	22.8	25.0	24.0	24.5	26.6	28.0	28.6	28.8
M	10.9	12.7	13.6	12.8	11.9	13.5	13.8	14.4
O	*	9.1	8.8	8.4	7.5	7.4	8.1	7.9

* Group O was part of Group F until 1977.

Table 2.7. Rate of Increase (Geometric*) of Females by Social Groups 1976-1983

Troop	No. Animals July 1, 1976	No. Animals June 30, 1983	$b=$	$r=$
J	44	142	.17	.99
I	38	113	.16	.99
M	21	71	.16	.99
F	38[a]	76	.12	.99
O	20[a]	39	.12	.94
L	36	52	.05	.93

*Based on a linear regression of the $\log_e$ of the annual maximum number of females (June 30). The slope (b) is the geometric rate of increase. The correlation coefficient (r) indicates the goodness of fit of the regression line to the data.

[a]Troop O was part of Troop F until 1977. Values give maximum number of females for each group after the 1977 fission of Troop O from Troop F.

The fecundity of the females for each troop was determined for each year (1976-83) by dividing the total number of live births in the troop by the number of females alive in that troop at the start of the birth season in January of each year. Table 2.8 gives the yearly percentage of females, four years of age and older, that gave birth to a live infant. Freunds's chi-square test for the difference among proportions was used to test for a significant difference in fecundity between troops which might explain the observed variation in troop growth. A similar test was applied to each troop over the seven year interval. Chi-square values and significance levels are given in Table 2.8. No significant difference in fecundity was seen among troops in any given year, despite the wide variation ($p > .05$). Thus, differential fecundity was ruled out as a factor responsible for variation in troop growth.

Potential differences in the secondary sex ratio at birth among social groups were also examined, since by bearing a higher proportion of female offspring, one troop might out-reproduce another over time. Table 2.9 gives the proportion of each birth crop represented by female births for each social group for each year of the study. The chi-square test for the difference among proportions was used for statistical testing of the results. Values are given in Table 2.9. No significant difference was found in the secondary sex ratios among the groups in any given year, and no significant difference was found within a group over the duration of the study. Thus, although there were wide fluctuations in the sex ratio, both within a group and among groups for each year of the study, no statistically significant evidence was found for sex ratio adjustment as a factor influencing the differential growth of one troop over another. Of particular interest is the fact that a comparison of the sex ratio of all infants born showed no significant variation over the years, indicating that although there was variation within and among groups, the net ratio was smoothed to a nearly constant value for the population as a whole.

Table 2.8. Fecundity of Females $\geq$ 4 Years of Age by Social Group and Year*

Troop	1976	1977	1978	1979	1980	1981	1982	1983	x^2	p
I	72.7	66.7	67.9	62.5	84.2	80.0	77.4	79.3	7.48	>.05
L	79.2	61.9	72.7	75.0	79.3	80.6	75.0	78.8	3.25	>.05
F	79.3	72.2	90.0	75.0	71.9	87.9	79.5	66.7	7.83	>.05
J	91.7	71.0	84.2	74.4	84.3	73.2	88.9	80.3	10.36	>.05
M	75.0	82.4	82.4	57.9	95.8	69.0	77.4	77.1	10.41	>.05
O	**	70.0	81.8	75.0	87.5	77.8	66.7	81.8	3.03	>.05
X^2	3.04	2.08	4.92	3.41	6.26	4.21	6.16	3.71		
p	>.05	>.05	>.05	>.05	>.05	>.05	>.05	>.05		

* Values give percent of births per troop which were female.

** Group O was part of Group F until 1977.

Table 2.9. Secondary Sex Ratio* by Social Groups 1976-1983

Troop	1976	1977	1978	1979	1980	1981	1982	1983	Pooled	X^2	p
I	37.5	50.0	63.2	45.0	65.6	36.1	63.4	50.0	52.2	10.95	>.05
L	36.8	69.2	56.3	38.9	34.8	28.0	45.8	53.8	43.9	9.37	>.05
F	39.1	46.2	44.4	52.4	47.8	48.3	48.4	46.7	46.8	.92	>.05
J	50.0	36.4	28.1	56.3	58.1	48.8	51.8	42.6	47.2	9.92	>.05
M	66.7	42.9	35.7	36.4	65.2	45.0	58.3	48.1	51.0	6.55	>.05
O	**	42.9	22.2	33.3	28.6	35.7	64.3	72.2	45.5	12.04	>.05
Pooled	44.6	47.1	41.7	46.5	53.2	41.2	54.7	49.5	48.0	10.68	>.05
X^2	3.69	3.78	9.09	3.24	10.55	4.05	3.36	5.19	3.65		
p	>.05	>.05	>.05	>.05	>.05	>.05	>.05	>.05	>.05		

* Values give percent of births per troop which were female
** Group O was part of Group F in 1976.

The data suggest that the difference in absolute growth among the troops was due primarily to variation in survivorship for both males and females, rather than differential rates of recruitment through fecundity or secondary sex ratio adjustment. This finding is consistent with the results of a five-year study of mortality in the colony from 1976 to 1981 (Rawlins and Kessler, 1982).

Total mortality was differentially distributed among the social groups. Table 2.10 gives the combined distribution of mortality for both sexes by social group and for Extra Group (solitary) males for the study period. Mortality in Group L (10.16 percent) was 2.0 times that observed in the largest group on the island, Group J, (5.04 percent) and 2.5 times that of the group with the lowest mortality rate, Group M (4.13 percent). Mortality in Group L was high throughout all years of the study, but was significantly greater than that of the other groups during only three years (see Table 2.10). Note also that the very low mortality rates in Group F for 1976-77 and in Group M for 1979-80 contributed to the statistical significance of the differences observed, but when factored out of the analysis, it was the mortality rate of Group L which was the chief variable accounting for the significance. The higher mortality rate in Group L was the cause of the reduced growth rate for this group relative to others on the island. The difference relates to the fact that the mortality rate due to tetanus in Group L was 2.6 times that of the largest troop (Group J) and 9.6 times that of the smallest (Group O), and to the fact that tetanus accounted for 33.3 percent of all deaths in Group L. A discussion of the differential susceptibility of Group L members to tetanus infections and the epizootiology of the disease on Cayo Santiago has been presented elsewhere (Rawlins and Kessler, 1982).

The high attack rate in Group L was thought to be linked to the area of the island ranged by the group, the Small Cay. During the late 1960s, a horse was maintained and corralled on the Small Cay for use as a pack animal to distribute chow. This proved unsuccessful and the horse was removed. Because horses excrete high concentrations of the vegetative form of *Clostridium tetani* (Gillespie and Timoney, 1981), the nearly doubled incidence of tetanus fatalities on the Small Cay in Group L was considered to have resulted in part from prolonged exposure of the group to the pathogen resulting from the animals limiting their range to the highly contaminated corral area. A natural experiment to test this hypothesis occurred in 1977-78, when Group F moved across the isthmus of the island to occupy the Small Cay with Group L. From 1972 through 1983, Group L had been the sole occupant of the Small Cay and was rarely joined by the other groups, even though access was unrestricted. Group F remained on the Small Cay throughout the mating season and mortality in the group tripled (see Table 2.10). Much of this increase was due to intergroup fighting and the annual mortality rate due to tetanus nearly doubled

Table 2.10. Annual Mortality Rate (%) by Social Groups 1976-1983

Troop	1976-1977	1977-1978	1978-1979	1979-1980	1980-1981	1981-1982	1982-1983	$\overline{X}$	S.D.
I	5.55	6.98	3.57	2.41	4.57	5.31	7.37	5.11	1.77
L	14.02	8.70	8.40	11.19	12.34	7.19	9.25	10.16	2.44
F	3.00	10.08	4.65	7.69	5.59	7.06	6.42	6.36	2.27
J	6.15	4.79	3.91	5.50	5.06	5.18	4.67	5.04	.70
M	4.22	3.75	4.54	0.00	4.88	5.67	5.88	4.13	1.97
O	4.65	3.64	7.69	5.63	4.94	10.20	2.91	5.67	2.51
Extra Group Males	10.00	15.38	0.00	10.00	0.00	20.83	15.38	10.23	7.90
χ^2	12.77*	5.73	4.86	18.95*	12.11*	3.96	6.14		
p	$<.05$	$>.05$	$>.05$	$<.01$	$<.05$	$>.05$	$>.05$		

* = Statistically significant.

(Rawlins and Kessler, 1982). The increased mortality in Group F lent support to the idea that the Small Cay presented a greater risk for death due to tetanus. The following year, Group F returned to the Big Cay and mortality in the group dropped by half. It is interesting to note in Table 2.10 that the only other animals in the population with mortality rates comparable to those of Group L were the Extra Group (solitary) males who also spent considerable amounts of time on the Small Cay.

DISCUSSION

Koford (1965; 1966) reported that between 1959 and 1964, the Cayo Santiago population increased at an annual rate of about 16 percent and that the annual mortality rate was 6.5 percent (excluding infants) with no indication of an increased mortality as a function of population growth. The mean reproductive rate for mature females was 78.5 percent and the secondary sex ratio was 96 male to 100 female births. The median birth date was March 21 (*sd* 14 days). Of all live births, the mean annual infant mortality rate was 8.5 percent. During this period, animals were routinely removed from the colony for biomedical experimentation and the population was subjected to intermittent trapping. Koford (1965) felt the removals had little effect on the colony.

From 1972-83, the Cayo Santiago population was left intact and undisturbed, except for an annual trapping of the population during January of each year to capture and mark the previous year's birth crop and to obtain blood samples from them for ongoing studies of genetic microevolution. Over the study period, from 1976 to 1983, provisioning and management practices were kept constant to minimize the effects of abiotic intervention on population dynamics.

From 1976 to 1983, the growth of the Cayo Santiago colony was 13 percent per annum, down from the rate Koford observed. The lower rate compared to the 1959-1964 period probably reflected the fact that during the years Koford observed the colony, the adult population contained twice as many females as males, compared to 1.11 adult males for each female in 1983. The colony was also undergoing rapid expansion due to the resumption of regular provisioning of the animals two years earlier, after a period of many years of marginal or no provisioning at all (see Chapter 1, this volume). Animals were routinely being taken off the island which would also contribute to rapid growth of a population. There was little difference in the mean annual mortality rates for the population between the two studies (6.5 percent versus 6.8 percent) and the disparity would likely disappear if mortality data for infants had been included in Koford's (1965) calculation of the total mortality. The fact that mortality rates did not increase as a function of population size during either study contradicts the concept that increasing population densities alone result in higher mortality rates in mammalian populations (see Geller and Christian, 1982). The reproductive rates for mature females were similar (78 percent versus 80 percent), as was infant survivorship to yearling status (91.5 percent versus 92.7 percent). There was no difference in the secondary sex ratio, but adult male survivorship has increased over the years. The incidence of congenital anomalies was equivalent in both studies (Rawlins and Kessler, 1983). Thus, after a period spanning 25 years, the vital statistics for the colony have remained remarkably consistent under relatively similar management protocols. The achievement of a stable age distribution by this expanding colony is a measure of the consistency with which the management program was applied after completion of the population reduction in 1972.

In recent years, a number of excellent studies of the dynamics of macaque populations have been published, which, in conjunction with data from Cayo Santiago, have begun to document the individual life history patterns and longitudinal changes in nonhuman primate groups necessary for understanding demographic variability amongst populations of the same or related species.

An iterative census of 17 groups of rhesus monkeys in the Aligarh region of western Uttar Paradesh, India was begun in 1959 by Southwick and Siddiqi (1977) and continues at present (Southwick et ad., 1980; Southwick, Siddiqi and Oppenheimer, 1983). Significant variation in troop growth has been seen between semiprotected and unprotected troops as a result of differences in natality and mortality. The population has generally declined as habitat destruction has continued and social groups in contact with humans have been exposed to disease and capture. A recent report (Malik, Seth and Southwick, 1984) on one population in India showing remarkable increase was encouraging. Regardless, the data provide a detailed description of population change, based on estimated rates of natality and mortality derived from comparative analysis of existing age-sex distributions of monkeys relative to those previously observed in the same social groups over the years. While very valuable, census work alone grants little insight into the proximate mechanisms underlying the dynamics of change observed in these groups.

Long-term observations of temple dwelling groups of rhesus monkeys in Nepal have added a behavioral component to the understanding of population regulation (Teas et al., 1982). In the absence of human interference, the study troops did not increase in size over seven years. Equilibrium was maintained through decreased fecundity and increased mortality resulting from nutritional deprivation, disease, and high levels of intratroop aggression. Regulation of population growth in these troops was attributed to natural biosocial control, but there is no direct evidence for it, only correlational support.

Dittus (1975; 1977; 1981) reported similar regulation of social groups of toque macaques (*Macaca sinica*) in Sri Lanka. Equilibrium was the result of balanced natality and mortality in the troops, but in Sri Lanka, infant mortality was the primary regulator of population growth, while reduced fecundity and high adult mortality hold Nepalese rhesus monkeys in check. In India, Southwick's (1980) data showed that fecundity was extremely high, and that post-partum and adult mortality offset recruitment.

The studies of feral macaque populations point to density-dependent regulation of growth, but the factors responsible differ. The extrinsic agent thought to induce an increase in mortality or decrease fecundity is limitation of nutritional resources, yet in one location fecundity is depressed and in another it remains high, while juvenile and adult mortality act to control animal numbers. Resource limitation is said to induce regulation by behavioral mechanisms such as increased rates of aggression, increased stress and stress-related depression of fecundity, and enhanced susceptibility to disease (Geller and Christian, 1982). Perhaps such constraints are operating in the feral macaque populations that have been studied.

In a provisioned, free-ranging troop of translocated Japanese macaques (*Macaca fuscata*), Fedigan et al. (1983) documented an expanding, rather than a stationary, population. As a result of the transfer of the group from Japan to Texas, the Arashiyama West troop experienced a temporary decline in population resulting from lower natality and high infant mortality. The change of environment had it's greatest impact on the young, but the impact of translocation on other age-sex classes was minimal. Within two years, the population returned to reproductive and survivorship rates essentially equivalent to those observed prior to the removal of the troop from its native habitat. The data strongly support the idea that any troop or population of primates experiences phases of expansion, decline, or no net growth and that longitudinal observations, rather than cross-sectional estimates based on a standing age-sex distribution are required for the development of a characteristic demographic profile of a population or species (Fedigan et al., 1983).

At Cayo Santiago, the regular and constant per capital rate of provisioning has minimized the impact of differential distribution of resources on population growth. Both females reproductive rates and infant survivorship are uniformly high for the population and for its component groups. Despite a constant mortality rate for the colony as a whole, it is the differential expression of adult mortality among the social groups which accounts for the significant variation in troop growth observed on the island. There is no evidence of sex-ratio adjustment as a means of internal regulation of population density.

An important reason for the collection of demographic data is its utility in the development and assessment of management protocols for primate colonies. Immediately following the completion of the cull of the Cayo Santiago colony in 1972, one of the few life tables for a primate population was prepared using data collected on the colony during 1973 and 1974 (Sade et al., 1977). The resulting time-specific life table combined natality and mortality data over two years and estimates of age-specific mortality and natality rates were calculated from the standing age-sex distribution of the female population at that time. Form the $l_x m_x$ of the life table, the demographic parameter, the Intrinsic Rate of Natural Increase *(r)*, was calculated to express the instantaneous rate of increase per individual (Sade et al., 1977). This value functions as a retrospective measure of the success achieved by a population in negotiating its environment, an indicator of environmental quality, and a predictor of future growth under stable environmental conditions (Sade et al., 1977).

The results of the analysis showed the Cayo Santiago population was in a stable age distribution and a partition of the colony by social groups showed the rates of increase among the troops were unequal and very different. The *r* value for Group J (.170) indicated its potential for growth was greatest, while Groups I and F were equivalent (.051, .050) but much lower than Group J. Group L (r = .031) was projected for an even slower rate of increase, and Group M (r = -.236) was to decline to less than 10 percent of its size in the next ten years (Sade et al., 1977).

The data from the current study do not support these earlier findings and call into question the utility of using the Euler-Lotka *r* (Intrinsic Rate of Natural Increase) as an empirical tool in colony management. Based on the 1973-74 time-specific life table, the finite rate of increase (λ) for females in the troops were, in descending rank order: Group J = 1.185, I = 1.052, F = 1.051, L = 1.032, M = .790 (Sade et al., 1977). The *actual* finite rate of increase for the females of each social group from 1976 to 1983, without any change in management protocol for the colony, were: Group J = 1.184, I = 1.173, F = 1.127, L = 1.056, M = 1.194, and for Groups F and O combined = 1.111. A Spearman's rank correlation run on the two sets of data showed no correlation (r_s = 0.00) between the projection from the life table and the actual growth of the social groups as measured by the finite rate of increase for females, even when the data for Group F and Group O were pooled to approximate Group F prior to its fission in 1977. The predicted growth for Group J was good, but the results for Group I and Group M were radically different. As a consequence, the projected growth for the female population λ = 1.073, Sade et al., 1977) significantly underestimated the actual growth for females (λ = 1.139) and for the entire colony (λ = 1.140) (see also Rawlins, Kessler, and Turnquist, 1984).

Some possible explanations for the failure of a widely used population parameter (Euler-Lotka's *r*) to correctly project population growth are that the life table from which the parameter was calculated is in error or that the data upon which it was based are idiosyncratic and do not accurately reflect long-term trends in the population. This is a major risk in the use of time-specific life tables, which rely on a cross-sectional view of the life histories of the individual animals in the population to estimate survivorship. For most calculations of 1_x, the survivorship of each age class is estimated by dividing the cohort by the total number of animals in the youngest age class. This method assumes that the birth cohorts of all of the older age classes were proportional equivalents of the most recently born group of animals in the population. Significant errors in estimates of survivorship can frequently result (see Caughley, 1971; 1977).

Use of a standing age distribution averaged over 1973 and 1974, immediately after the 1972 disruption of the colony, to estimate survivorship was the probable source of error in projections of future population growth at Cayo Santiago. Figure 2.4 plots the 1_x curve for the population based on the time-specific life table (Sade et al., 1977) against the actual (age-specific) survivorship of females in each cohort, based on the number of females alive from each birth crop from 1972 to 1982. The proportion surviving in each age class for each birth crop was averaged over 10 years to obtain the mean survivorship (1_x) for each age group. Only the first 10 years of life were examined due to limitations in our data set. This method represents the first attempt to develop an age-specific (cohort based) life table for a nonhuman primate population. The results show the time-specific estimates of survivorship underestimated actual female survivorship in the population by 46.2 percent (7 percent projected versus 13 percent actual).Table 2.11 compares 1_x values obtained from the time and age-specific estimates. These values are plotted in Figure 2.4. The population grew at a much greater rate than was predicted, as did the component groups. Since the actual 1_x values were not given for the partition of the 1973-74 population by social group, no direct comparison can be made between estimated and actual female survivorship within each group. There is no apparent explanation for the projected decline in the growth of Group M. Discrepancies in natality (m_x) are currently being analyzed as an additional source of error in time-specific estimates.

The development of an empirically derived schedule of female survivorship (l_x), coupled with accurate information on age-specific fecundity (m_x), for the Cayo Santiago population from 1973 to 1983 permits assessment of the accuracy and utility of using estimates of population growth based on time-specific life tables for colony management and comparison of colony growth with other primate

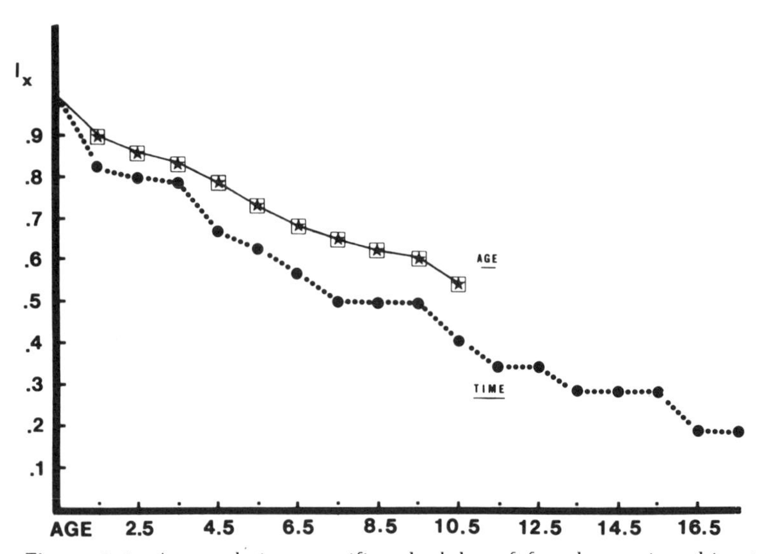

Figure 2.4 Age and time specific schedules of female survivorship at Cayo Santiago.

Table 2.11. Time and Age Specific Schedules of Female Survivorship (l_x)

Cohort Age	Time Specific[a] l_x	Age. Specific[b] l_x
.5	1.000	1.000
1.5	.824	.904
2.5	.804	.866
3.5	.788	.834
4.5	.671	.784
5.5	.628	.729
6.5	.573	.680
7.5	.499	.649
8.5	.499	.626
9.5	.499	.605
10.5	.405	.542
11.5	.343	
12.5	.343	
13.5	.286	
14.5	.286	
15.5	.286	
16.5	.190	
17.5	.190	
18.5	.143	
19.5	.071	
20.5	.071	
21.5	.071	
22.5	.000	

a From Sade et al., 1977.
b Mean survivorship for each age group, averaged over the maximum number of years available for each cohort.

populations. Results of the Cayo Santiago study show traditional time-specific schedules of survivorship underestimated actual survivorship and therefore predicted lower net population growth and rates of increase than those actually observed. Although the age-specific schedule of female survivorship developed in this study encompasses only the first trimester of the rhesus life cycle, it can still provide a direct test of the accuracy of logistically determined projections. The finding that time-specific schedules of survivorship underestimated actual survivorship in the colony calls into question recent publications (Smith, 1982) which have used deterministic equations (Euler-Lotka *r*) to obtain values for the intrinsic rate of natural increase in order to compare the productivity of caged colonies against that of the free-ranging population at Cayo Santiago. Smith (1982) has suggested that production is greater in caged colonies in comparison to Cayo Santiago for the same species, but the work could not take into account 1) the difference between estimated and actual survivorship and its effects on projected growth of the population, 2) the fact that high numbers of males are maintained in the Cayo Santiago colony relative to the minimum number of males required to support production in a breeding colony, 3) that less productive females are not culled from Cayo Santiago, and 4) the fact that there was no provision for disease prophylaxis, especially tetanus, in the Cayo Santiago colony because the management protocol specified that the animals be maintained with an absolute minimum of intervention to avoid additional bias in the study of the natural regulation of population growth. Our findings on the Cayo Santiago macaque population from 1976 to 1983 provide strong evidence that long-term investments of both time and resources are required to obtain the data necessary for an adequate understanding of the population dynamics of long lived species, such as *Macaca mulatta*, even under seminatural conditions. The results further show the difficulty which is encountered in using data for demographic studies derived from short-term observations of feral primate populations.

Finally, a major concern of continued population growth in captive, provisioned colonies is the increased density of animals and concomitant behavioral and economic effects. The density of the Cayo Santiago colony (76 animals per hectare) in 1983 was about six times that observed in natural macaque populations. The difference is offset by the realization that resources affect animal density greatly and that low numbers of monkeys are found in sparse resource zones, but interindividual differences in spacing within the troop are not greatly different. The troop remains a cohesive unit, regardless of the size of its range. The key factor is the discrepancy in the number of monkeys that can survive per hectare without supplementary feeding. Provisioning of the Cayo Santiago macaques permits the island to support greater numbers of monkeys than would be able to survive on the natural semi-tropical vegetation alone. It offsets the density problem relative to resource availability, but the exact effects with respect to behavioral biases introduced by artificial supply of food have not been shown. Only the gross effects on net increase in population size have been documented in this and other colonies. Perhaps most significant is the economic cost of maintaining an expanding population, such as that on Cayo Santiago, without systematic culls to hold numbers in check. Again, the behavioral impact of culling has not been documented, but the costs of supporting a large and growing population are staggering. The projected yearly chow consumption required to maintain provisioning at 0.23 kg per animals per day, when plotted against minimum population values was 81 tons for 1983 (January to January). This estimate translates to a budgetary allotment of $55,000 for food alone. Clearly, available wildlife management tools must be introduced to primate colony management in future years so that accurate population projections can be obtained (see Caughley, 1977), and reductions can be carried out with a minimum of disturbance to the behavioral dynamics of the population.

ACKNOWLEDGEMENTS

The authors would like to thank A. Figueroa, C.J. DeRousseau, D. Decarie, J. Hickman, M. McGill, E. Garcia, J. Berard and numerous guest investigators for contributions to the census of the Cayo Santiago population over the study period. We also thank E. Tolentino, H. Vasquez, E. Davila and I. Pena for their dedicated work in maintenance of the colony. Special thanks are also due to attorney Carlos M. Torres and Irma Martinez for their administrative skills which were so essential to keeping the operation viable. The efforts of the former directors of the Caribbean Primate Research Center, Drs. C. Conaway, W. Kerber, S. Ebbesson, G. Meier, and L. LeZotte Jr. at securing support for the colony are gratefully acknowledged. The Cayo Santiago colony and the Caribbean Primate Research Center were supported under contract NO1-RR-7-2115 and by grant RR-01293 from the Animal Resources Branch, Divison of Research Resources (DRR), NIH, and by the University of Puerto Rico School of Medicine. Drs. W.T. Goodwin and L. Whitehair provided crucial support from the Animal Resources Branch, DRR, NIH. Finally, we thank Ms. L.M. Cleeves and Ms. S. Deacon of the Endocrine Research Center, Michigan State University, for their assistance in preparation of the manuscript. Dr. Donald Hall of the Department of Zoology and Dr. W.R. Dukelow, Director of the Endocrine Research Center, Michigan State University, provided invaluable help with development of the draft.

REFERENCES

Altmann, S.A. A field study of the sociobiology of rhesus monkeys (*Macaca mulatta*). *Ann. N.Y. Acad. Sci.* 102:338-435, 1962.

Caughley, G., Birch, L.C. Rate of increase. *J. Wildlife Management* 35:658-663, 1971.

Caughley, G. *Analysis of Vertebrate Populations* John Wiley & Sons, New York. 1977.

Dittus, W.P.J. Population dynamics of the toque monkey, *Macaca sinica*. In R.H. Tuttle ed. *Socioecology and Psychology of Primates.* The Hague, Netherlands Mouton Publishers, P. 125-152, 1975.

Dittus, W.P.J. The evolution of behaviors regulating density and age-specific sex ratios in a primate population. *Behaviour* 69: 265-302, 1979.

Dittus, W.P.J. Primate population analysis. In *Techniques for the Study of Primate Population Ecology,* Subcommittee on Conservation of Natural Populations, National Academy Press, Washington, D.C. pp. 135-175, 1981.

Fedigan, L.M., Gouzoules, H., Gouzoules, S. Population dynamics of Arashiyama West Japanese macaques. *Int. J. Primatol.* 4:307-321, 1983.

Geller, M.D., Christian, J.J. Population dynamics, adrenocortical function and pathology in *Microtus pennsylvanicus. J. Mammal.* 63:85-95, 1982.

Gillespie, J.H., Timoney, J.F. Clostridium tetani. In *Hagan and Bruner's Infectious Diseases of Domestic Animals.* 7th Edition. Cornell University Press, Ithaca, NY. pp. 200-201, 1981.

Koford, C.B. Population dynamics of rhesus monkeys on Cayo Santiago. In *Primate Behavior: Field Studies of Monkeys and Apes.* I. DeVore ed. Holt. Rinehart and Winston, New York. pp. 160-164, 1965.

Koford, C.B. Population changes in rhesus monkeys: Cayo Santiago, 1960-1964. *Tulane Studies in Zoology* 13:1-7, 1966.

Koford, C.B., Farber, P.A., Windle, W.F., Twins and teratisms in rhesus monkeys. *Folia Primatol.* 4:221-226, 1966.

Malik, I., Seth, P.K., Southwick, C.H. Population growth of free-ranging rhesus monkeys at Tughlaqabad. *Am. J. Primatol* 7:311, 1984.

Pickering, D.E., Van Wagenen, G. The 'golden' mulatta macaque (*Macaca mulatta*): developmental and reproduction characteristics in a controlled laboratory environment. *Folia Primatol.* 11:161-166, 1969.

Rawlins, R.G. Forty years of rhesus research. *New Sci.* 82:108-110, 1979.

Rawlins, R.G. and Kessler, M.J. A five year study of tetanus in the Cayo Santiago rhesus monkey colony: Behavioral description and epizootiology. *Am. J. Primatol.* 3:23-39, 1982.

Rawlins, R.G. and Kessler, M.J. Congenital and hereditary anomalies in the rhesus monkeys (*Macaca mulatta*) of Cayo Santiago. *Teratology* 28:169-174, 1983.

Rawlins, R.G., Kessler, M.J., Turnquist, J.E. Reproductive performance, population dynamics and anthropometrics of the free-ranging Cayo Santiago rhesus macaques. *J. Med. Primatol.* 13:247-259, 1984.

Ricklefs, R.E. *Ecology* Chiron Press, Concord. 1979.

Sade, D.S., Cushing, P., Dunaif, J., Figueroa, A., Kaplan, J., Lauer, C., Rhodes, D. and Schneider, J. Population dynamics in relation to social structure on Cayo Santiago. *Ybk. Phys. Anthropol.* 20:253-262, 1977.

Smith, D.G. A comparison of the demographic structure and growth of free-ranging and captive groups of rhesus monkeys (*Macaca mulatta*). *Primates* 23:24-30, 1982.

Southwick, C.H., Siddiqi, M.F. Population dynamics of rhesus monkeys in northern India. In *Primate Conservation,* H.S.H. Prince Rainier and G. Bourne eds. Academic Press, New York. pp. 339-362, 1977.

Southwick, C.H., Richie, T., Taylor, H., Teas, J., and Siddiqi, M.F. Rhesus monkey populations in India and Nepal: patterns of growth, decline and natural regulation. In *Biosocial Mechanisms of Population Regulation,* M.N. Cohen ed. Yale University Press, New Haven. pp. 151-170, 1980.

Southwick, C.H., Siddiqi, M.F., Oppenheimer, J.R. Twenty-year changes in rhesus monkey populations in agricultural areas of northern India. *Ecology* 64:434-439, 1983.

Sultana, C.J., Marriott, B.M. Geophagia and related behavior of rhesus monkeys (*Macaca mulatta*) on Cayo Santiago, Puerto Rico. *Int. J. Primatol* 3:338, 1982.

Teas, J., Southwick, C.H. and Siddiqi, M.F. Natural population regulation in rhesus monkeys of Nepal. Depart. Anthropol., Univ. Delhi, India. *SAP* 4:1-7, 1982.

CHAPTER THREE

Maternal Lineages as Tools for Understanding Infant Social Development and Social Structure

CAROL M. BERMAN

INTRODUCTION

This chapter reviews a few examples from my research on Cayo Santiago to illustrate the ways in which maternal kinship lineages have been valuable in shedding light on the interface of developing relationships with social structure. How are relationships between young individuals influenced by the social structure in which they live? How do individuals become integrated into that structure?

When I first began studying infant rhesus monkeys on Cayo Santiago in late 1973, a great deal was already known about certain aspects of their social development from a number of pioneering studies, particularly those by Robert Hinde and by Harry Harlow and their coworkers. The general course of development of mother-infant relationships had been described in detail for a few nonhuman primate species (Hinde, Rowell and Spencer-Booth, 1964; Hinde and Spencer-Booth, 1967a; Jensen et al., 1967; Struhsaker, 1971; Chalmers, 1972). The importance of close mother-infant relationships for the development of normal social, sexual, and maternal development had also been demonstrated through studies of infants which had been raised in varying degrees of social deprivation (Harlow and Harlow, 1965) and of infants which had experienced brief separations from their mothers (Hinde and McGinnis, 1977). In addition, there were indications at that time that infant development was strongly influenced by members of the infant's social group other than the mother (Rowell et al., 1964; Hinde and Spencer-Booth, 1967b; Spencer-Booth, 1968; Wolfheim et al., 1970; Kaplan, 1972; Rosenblum, 1974a; 1974b) and by the social structure of the group (Rosenblum, 1971a; 1971b). For example, the extent to which mothers restricted their infants depended in part on the availability of other group members, on their propensities to interact positively or negatively toward the infant, and on the mother's ability to inhibit their behavior.

Illustration 3.1 An infant at rest between its mother and aunt.

There was, however, little detailed information on the early relationships of infants with either their mothers or other group companions in naturally structured groups. Most studies had been carried out in captivity, where social groups were typically smaller than in the wild, where their compositions were generally artificial and/or unstable, and where important features of naturally formed groups were usually missing (e.g., kinship lineages). For these reasons, opportunities to understand the influence of normal social structure on maternal behavior and infant development were limited.

Cayo Santiago was chosen as a study site because it appeared to be well suited for approaching these problems. The free-ranging population lived in large, naturally formed, species-typical social groups which were relatively free from manipulation. The monkeys were well habituated and comparatively easy to observe, making it possible to collect large amounts of data in relatively short periods of time. Most importantly, the individuals had known histories and known maternal kinship relationships.

I was concerned at first about the fact that the animals were provisioned, predator-free, and more densely populated than were wild rhesus monkeys. Thus, natural ecological influences on social development could not be studied directly. However, the same conditions resulted in the development of particularly large and extended kinship lineages which proved to be extremely useful tools for probing kin-related behavioral propensities in monkeys and their roles in social development. The presence of these lineages has not only allowed researchers to identify kinship as an important organizing principle of behavior with more certainty than in the wild or in captivity, but also to probe more deeply into its nature.

METHODS

The data were collected between late December 1973 and early November 1975. The study group (Group I) was formed naturally by fission in 1961 and was organized in the manner typical of multimale macaque groups (Sade, 1972). Like the other five groups on the island, it was composed of a permanent core of adult females and their immature offspring, along with a number of more transient adult males. Male offspring typically leave their natal groups sometime after puberty and transfer to another group. Female offspring remain in their natal groups for life and continue to associate closely with their maternal kin and offspring. In this way, large maternal kinship lineages develop that also represent subunits of social organization. There were three such lineages in Group I, with members spanning as many as four generations. The total membership of the group varied from 53 to 77 individuals during the course of the study.

Stable linear dominance hierarchies could be constructed separately for the adult males and females from the directions of submissive interactions within pairs of individuals (Sade, 1967). The female hierarchy was closely related to the lineage structure of the group. At adolescence, daughters usually acquired ranks immediately below their mothers and immediately above their next elder sister. Thus adult sisters came to rank in reverse order of age. Lineages as a whole came to rank in the same order as their matriarchs.

The Infant Sample

Twenty infants who were born into Group I during the 1974 and 1975 birth seasons were observed from birth to at least 30 weeks of age. They came from all three lineages. Ten were male and ten were female. Six had primiparous mothers and 14 had multiparous mothers.

Data Collection

Observation conditions were excellent; it was possible to observe intimate details of interaction almost continuously without the use of binoculars by following infants at a distance of a few meters. Focal-animal sampling methods (Altmann 1974) were used to record the social interaction of infants with their mothers and with other members of the group. Each infant was observed individually for three hours during each fortnight age period. All interactions involving the infant were listed chronologically on checksheets along with the identities of the initiator and the recipient of the interaction and the time it occurred. At two-minute intervals, point (instantaeous) time samples were taken by tape recorder that identified the individuals in contact with the infant, within 60 cm, and between 60 cm and 5 m. All observations were made between 7:00 a.m. and 12 noon Atlantic Standard Time (AST), hours which included all major group activities: feeding at the food bins, foraging, traveling, socializing, and resting. However, time spent near the feeder bins was not included in the observations. The sequence of infants observed was determined by a randomized order arranged at the beginning of each week.

RESULTS AND DISCUSSION

Mother-infant Relationships and Social Organization

In my initial study of infant social development (Berman, 1980a), I compared mother-infant interaction on Cayo Santiago with that observed in the Madingley Colony in Cambridge, England. The

Madingley Colony was made up of six pens, each composed of an outdoor area (5.30 x 2.40 x 2.40 m) communicating with an inside room (1.80 x 1.35 x 2.25 m). Each pen was the permanent home for about eight rhesus monkeys—one adult male, and two to four adult females and their offspring. The original monkeys forming the colony in the early 1960s were obtained from a number of sources. Since that time most had been born and raised to adulthood within the colony. By 1974, this resulted in the formation of small clusters of kin in some of the pens.

The general course of development of mother-infant relationships was remarkably similar across the two colonies both qualitatively and quantitatively. This was shown by comparing directly a number of measures of interaction, including the percentage of time mothers and infants spent out of contact (time off), the proportion of attempts to get on the nipple that were prevented by the mother (relative rejections) and Hinde's well known proximity index (% Api - %li). The proximity index is a measure of the infant's relative role in maintaining proximity with the mother over a 60 cm limit. It is the percentage of approaches made by the infant to the mother over that limit minus the percentage of departures by the infant from the mother over that limit (Hinde and Atkinson, 1970).

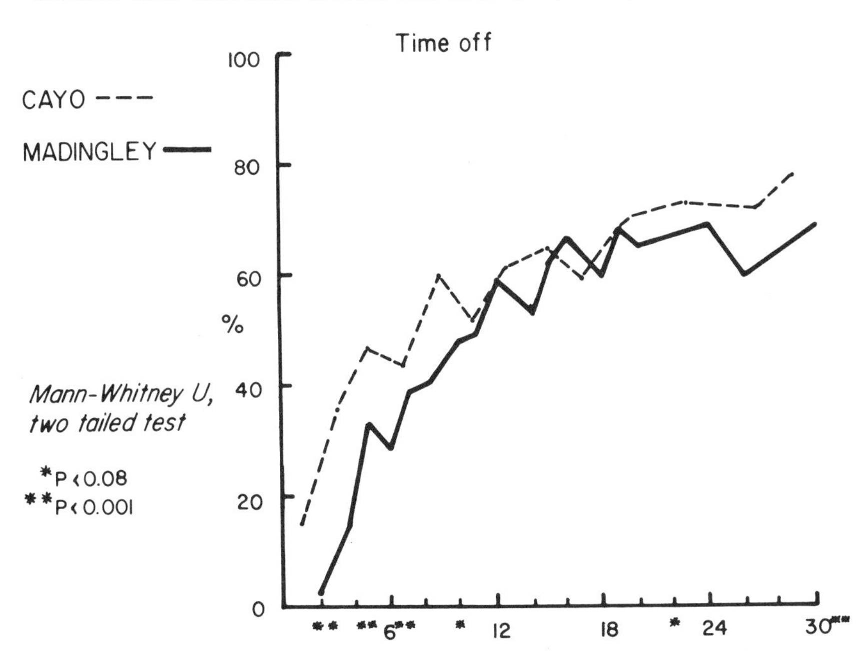

Figure 3.1 Median percentages of time spent off the mother for Cayo Santiago pairs and Madingley pairs. Data are derived from 90 two-minute point time samples per fortnight age period for each infant from birth to 30 weeks of age. Asterisks refer to the significance of differences between colonies (from Berman, 1980a).

In both colonies mothers and infants initially spent nearly all their time in ventro-ventral contact. The mother rarely rejected the infant and was primarily responsible for maintaining contact and proximity with it (shown by a negative proximity index). Over time, however, the mother and infant spent less time in contact, the mother rejected the infant more, and the infant became primarily responsible for maintaining contact and proximity (shown by a positive index). In both groups, changes in mother-infant relationships over time were due primarily to decreases in mothers' rather than infants' propensities to seek contact and proximity.

In spite of these similiarities, there were small, but consistent differences between free-ranging and captive pairs that appeared to be related to the kinship structures of the groups. As Figure 3.1 illustrates, free-ranging pairs spent slightly more time off than captive pairs in 13 of 15 fortnight age periods. Figure 3.2 shows that free-ranging mothers also rejected their infants slightly more in all 15 fortnight age periods. Finally, free-ranging infants took slightly larger roles in maintaining proximity with their mothers in all but one age period (Figure 3.2, top row). Using Hinde's (1969; 1974) methods for assessing interindividual differences in relationships, it was apparent that free-ranging mothers were seeking contact and proximity with their infants less than captive mothers, and in this sense, could be described as more rejecting and more encouranging of independence.

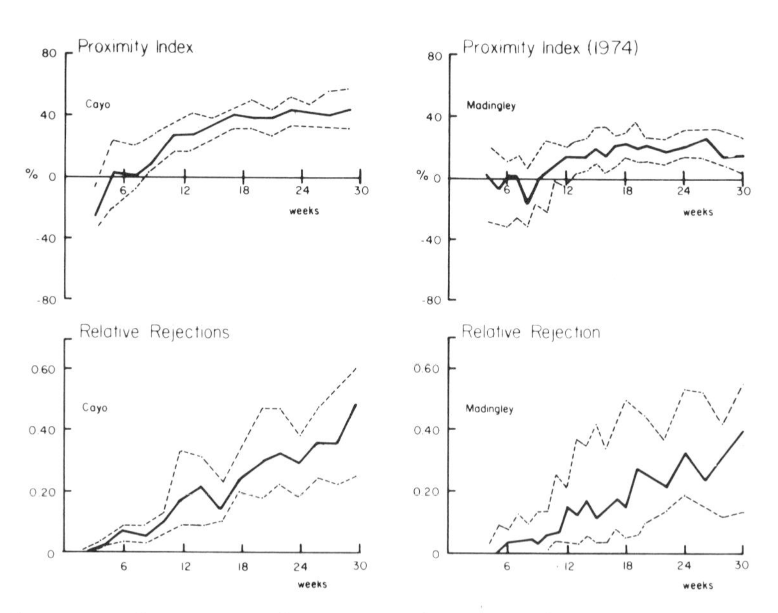

Figure 3.2 Comparison of measures of mother-infant interaction at Cayo Santiago (left) and Madingley (right). Medians and interquartile ranges for the proximity index (top), and the relative frequency of rejections by the mother (bottom) (from Berman, 1980a).

Obviously, many aspects of the two environments differed and it was not immediately clear which differences were responsible for the observed differences between the mothers. A clue was provided by the observation that the Madingley mothers seemed to be becoming more like free-ranging mothers over the years. For example, Figure 3.3 shows the proximity invoices for Madingley pairs in 1968, for Madingley pairs in 1974, and for the Cayo Santiago pairs. In 1968, the index was negative for a longer period of time than in 1974 and it leveled out at a lower value. In other words, Madingley infants before 1968 began to play the primary role in maintaining proximity to the mother at a later age than subsequent Madingley infants, and even after they took on the primary role, they took less responsibility for maintaining proximity than did subsequent infants. In both cases, Madingley infants played smaller roles in maintaining proximity to their mothers than did Cayo Santiago infants. This suggested that important differences between the two colonies were likely to be related to factors which were changing at Madingley over time.

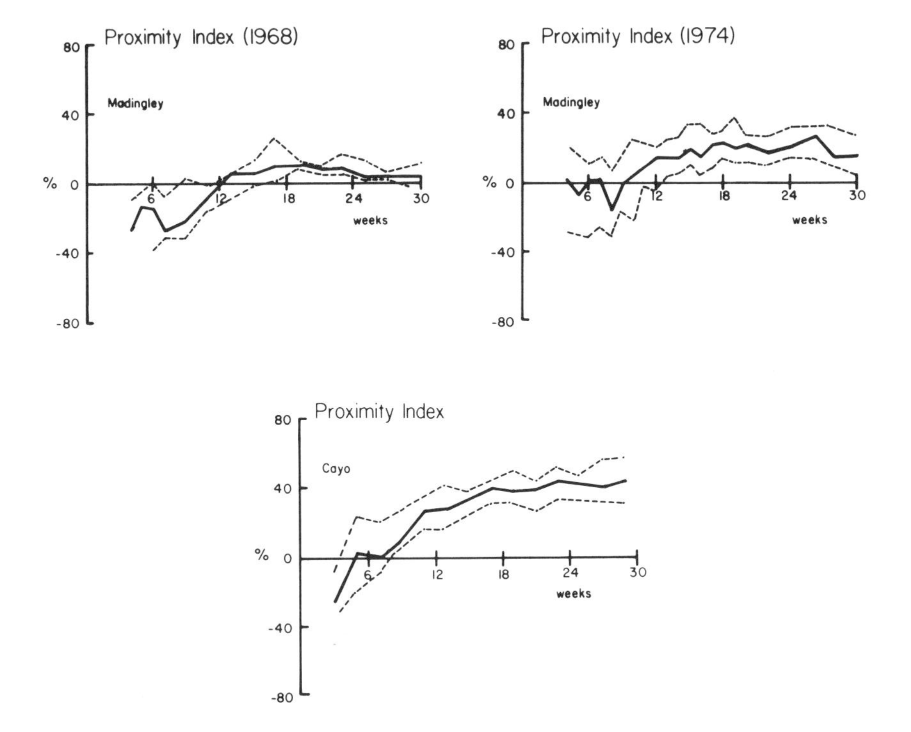

Figure 3.3 Medians and interquartile ranges for the proximity index for: a) Madingley pairs (compiled in 1968); b) Madingley pairs (compiled in 1974); and c) Cayo Santiago pairs (from Berman, 1980a).

After eliminating a number of possible explanations (e.g., differences in environmental complexity, differences in the ages of the colonies, differences in the parities of the mothers, differences in the sexes of the infants, differences in the origins of the mothers, and differences in the amounts of human disturbance), I was left with the hypothesis that free-ranging mothers may have been more relaxed with their infants because they were able to raise them among their kin. Changes in the captive mothers over time appear to have been related to the development of small kinship clusters in their pens.

The Infant's Developing Social Network

Extended lineages also proved useful for describing the infant's developing social network with group companions (Berman, 1982). Within the first few days of life, rhesus monkey infants interact with a wide range of group members in addition to the mother (Rowell et al., 1964; Kaufmann, 1966; Hinde and Spencer-Booth, 1967a; Lindburg, 1971; Berman, 1978). The mother-infant pair is attractive to others, especially females (Spencer-Booth, 1968; Hinde and Proctor, 1977), presumably because the infant displays a number of distinctive physical characteristics (Alley, 1980). New mothers also appear to be attractive, particularly those with high dominance status (Fady, 1969; Gouzoules, 1975; Seyfarth, 1976; Cheney, 1978). At first, the mother closely regulates the infant's interactions with others by taking primary responsibility for maintaining close proximity and contact and by actively and selectively thwarting attempts by others to interact with the infant or vice-versa (Rowell et al. 1964; Kurland, 1977; Hrdy, 1976). As a result, when the infant does interact with others, its social network—i.e., its individual pattern of distributing social interaction among various group members—mirrors that of its mother (Berman, 1978; see also Simonds, 1974); mothers interact more with close kin than with distant kin of unrelated individuals, more with female companions than male companions, and more with younger immatures than with older immatures. Also females in high-ranking lineages interact more intensively with their own kin than do females in other lineages (Seyfarth, 1976; McMillan, 1982). However, what occurs as the mother and infant become more mutually independent? Does the infant's network change, or does the mother's early influence set the infant's patterns for the future?

In order to investigate these questions, I focused on several measures of affiliative interaction between infants and finely-divided categories of group companions. Each measure was examined as a function of degree of relatedness with the companion, the sex of the companion, the age of the companion and the lineage rank of the infant. Figure 3.4 shows the distribution of the percentages of time the infant spent within 5 m of group companions as a function of

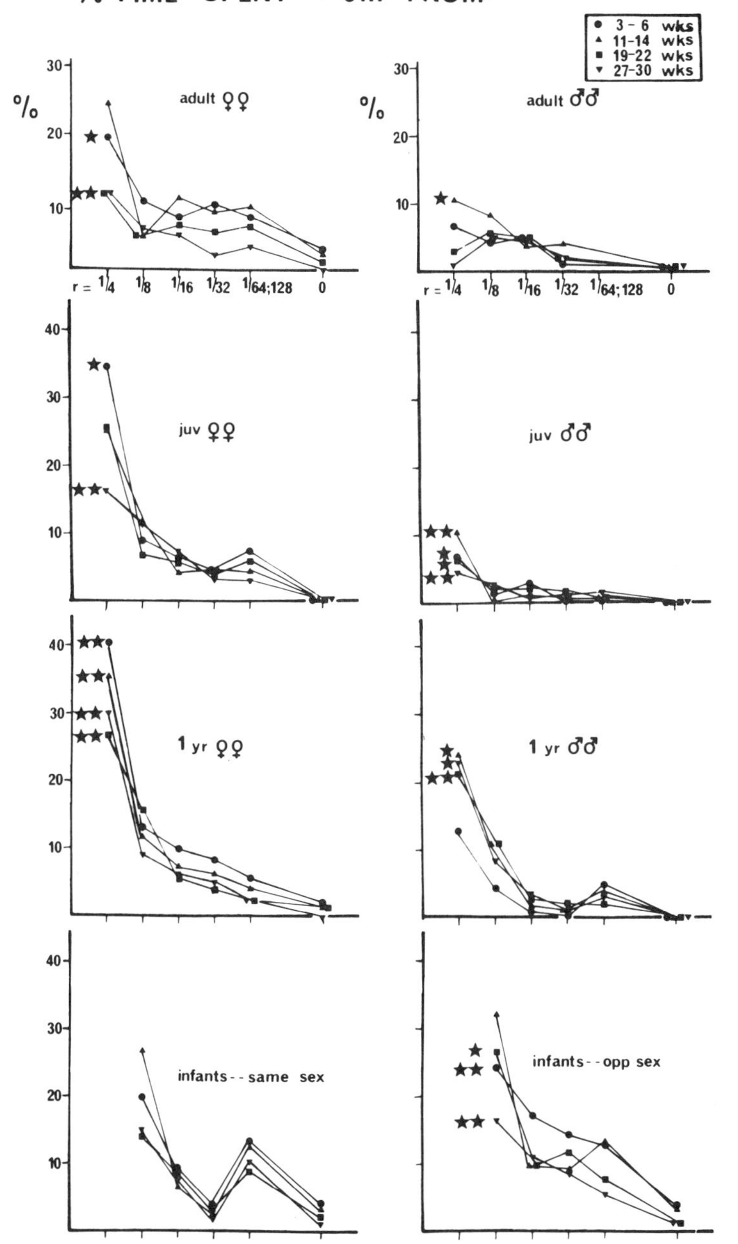

Figure 3.4 Time infants spent within 5 m of group members as a function of degree of relatedness (*r*) through maternal lines. Median scores are shown for time infants spent near companions in each age-sex-kinship category. Sample sizes within age-sex-kinship categories range from 3 to 20 infants with a mean of 12. Separate curves are shown for eight age-sex classes of companion and for four 4-week age periods. One star and two stars indicate $p < 0.05$ and $p < 0.01$, respectively, Separman rank order coefficients for time spent within 5 m with degree of relatedness within each age-sex class (from Berman 1982).

degree of relatedness through maternal lines. The degrees of relatedness are underestimates and must be considered only as relative measures of relatedness through maternal lines, because group members may be inbred to a certain extent, and because some individuals may share paternal genes with infants (Berman, 1978; Altmann, 1979). Each graph in the figure represents median scores for a single age-sex class. The four curves in each graph represent four, four-week age periods: 3-6 weeks, 11-14 weeks, 19-22 weeks and 27-30 weeks.

As with adult females, maternal kinship emerges as a strong factor associated with the structure of the infant's network; median scores for all companions with infants of all ages increased with degree of relatedness. Since the group was large and the lineages were extended, this could be shown separately for each age-sex class of companion, thus eliminating possibly confounding factors of companion's age and sex found in other studies. In addition, the relationship of the companion's age and sex could be described separately from each other and from kinship. Like their mothers, infants of all ages associated more with females than with male companions of the same age and degree of relatedness. In each case, infants associated more with younger immatures than with older immatures of the same sex and degree of relatedness. Moreover, these patterns persisted when the analyses were repeated with a number of more intimate measures of affiliative interaction, including approaches and departures, friendly contact, grooming and embracing (Berman, 1978; 1982). In Figure 3.5 the data for adult female companions are broken down further to show infants from each lineage separately. Again, infants of all ages appeared to follow their mother's patterns; infants in the top-ranking lineage associated more intensively with their own kin than did infants in the middle and bottom-ranking lineages.

In summary, with Cayo Santiago data one could 'see' several major aspects of the developing structure of the infant's social network at a glance. The infants' network appeared to mirror those of their mothers both in the earliest weeks of life, when they were almost completely under their mothers' control, and after 30 weeks, when they were considerably more independent. This implies a number of things. First, infants appear to function as members of their own lineages virtually from the beginning. Second, the mother's early influence on the infant's associations does indeed appear to have long-term effects on the structure of the infant's social network. Finally, the component relationships making up the infant's network develop through a process resembling differentiation out of the mother's relationships. At first they may be described as mere associations reflecting the mother's relationships; the infants relationships not only share many qualities with those of the mother, they are

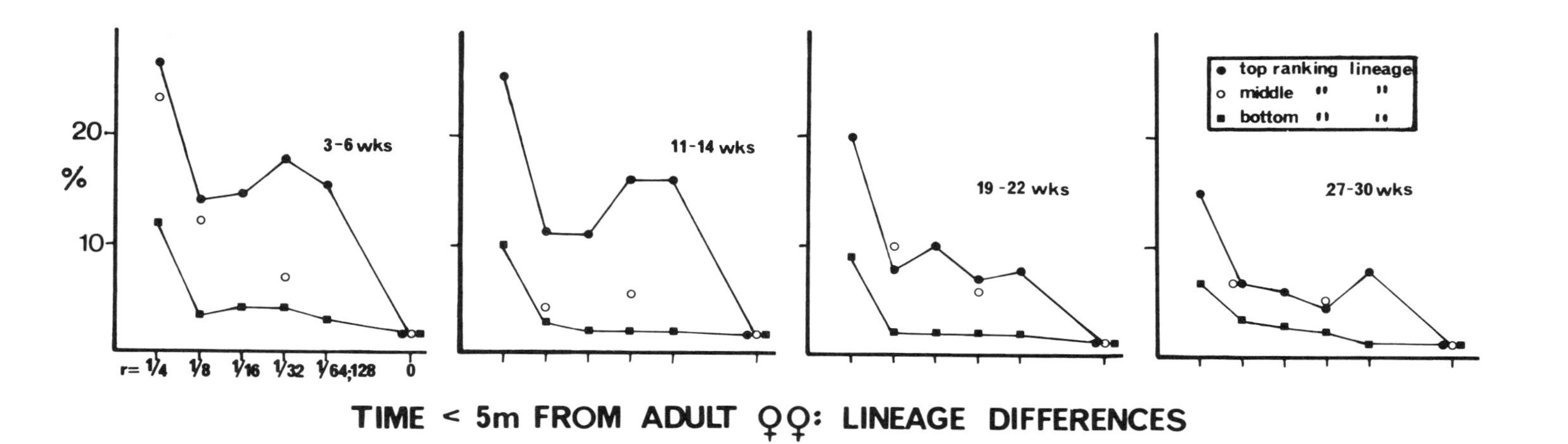

Figure 3.5 Time infants spent within 5 m of adult females for infants belonging to each lineage. Median of scores for infants are shown as a function of degree of relatedness (r) through maternal lines. Sample sizes are 5 to 11 infants (top-ranking lineage), 1 to 2 infants (middle-ranking lineage), and 4 to 9 infants (bottom-ranking lineage) (from Berman, 1982).

sustained primarily through interaction between the mother and the companions. However, by 30 weeks, the infant's relationships with the same companions show signs of independence. They are sustained primarily through interaction between the infant and the companions, and they may have different qualities from those of the mother (Berman, 1978).

Peer Rank Aquisition

Finally, extended lineages on Cayo Santiago were useful in describing early agonistic experiences of infants, and some probable mechanisms by which infants acquire ranks vis-a-vis one another (Berman, 1980b). Infants appear to acquire peer-peer ranks during their second six months of life and these ranks typically correlate with those of their mothers (see also Sade, 1967). However, before these ranks are fully established, infants of high and low-ranking mothers differ markedly in agonistic experience in ways which tend to lead to their acquiring ranks like their mothers. From the earliest weeks, infants of high-ranking mothers tend to be threatened less frequently by other group members than infants of lower-ranking mothers. They are also less likely to be threatened by unrelated, and hence unfamiliar group members. By 27 to 30 weeks of age these infants are more likely to receive protection when threatened. Protectors intervene by approaching the threatened infant, making contact or carrying it away, or by screaming at, threatening, or chasing the threatener. Interventions are also counted when the threatened infant approaches and contacts a protector who neither avoids or breaks contact with the infant. When infants are protected, the protectors of infants of high-ranking mothers are less likely to emit fearful gestures to the infant's threatener than are the protectors of other infants. In this sense infants of high-ranking mothers are protected more successfully and are perhaps presented with better models to copy. The protectors of infants are primarily their mothers and other close female relatives (CFR's) such as sisters and grandmothers. Interestingly, however, high and low-ranking mothers protect their infants at comparable rates. Differences in the protection of their infants are due primarily to differences in protection received from other protectors.

More detailed questions could be answered about these differences because the lineages were known and extended. Did infants of high-ranking mothers have different agonistic experiences because they had more close female relatives, because they spent more time near them where they could be more easily protected, and/or because their CFR's were high-ranking and thus better able to inhibit aggression and win fights?

KENDALL RANK ORDER CORRELATION COEFFICIENTS FOR INFANTS 27 to 30 WEEKS

	Mother's Rank	Rel. Freq. of Threats	Prob. of Intervention by others	Success Rate of Intervention
No. of CFR's (Sisters & Grandmothers)	.44	-.63*	.59*	.55
Time < 5m from CFR's	.56*	-.64*	.58*	.59
N(infants)=	11	9	10	8

*$p < .05$

Table 3.1

Table 3.1 shows Kendall rank order correlation coefficients for numbers of CFR's and for time spent near CFR's with mother's (and CFR's) rank, frequencies of threat, probabilities of intervention by individuals other than the mother, and success rates of intervention for 1975 infants, aged 27 to 30 weeks. At this age, threats were the most common and infants were beginning to interact with one another according to maternal rank. The data for 1974 infants were in complete harmony with those for 1975.

Infants of high-ranking mothers did indeed spend more time near CFR's, partly because they tended to have more CFR's (top row of Table 3.1) and partly because they spent more time near each of them (bottom row of Table 3.1). These results probably reflect both the greater social cohesiveness of high-ranking lineages (see Cheney, 1978) and their faster rates of reproduction (Sade et al., 1977).

Frequencies of threat were negatively correlated with both the number of CFR's and with the amount of time spent near them. Kendall partial correlation tests suggested that these correlations were virtually independent of mother's rank, suggesting in turn that the presence of CFR's may inhibit others from threatening infants regardless of their ranks. Even so, infants of high-ranking mothers were probably at an advantage because they tended to associate with CFR's more than other infants.

Infants who were more likely to be protected by others were those who had more CFR's and who spent more time with them. These correlations were partly dependent on variations in mother's (CFR's) ranks, suggesting that differences in probabilities of protection were

KENDALL RANK ORDER CORRELATION COEFFICIENTS FOR INFANTS 27 to 30 WEEKS

	Mother's Rank	Rel. Freq. of Threats	Prob. of Intervention by others	Success Rate of Intervention
Time < 5m from Mother	.02	.03	.19	.04
No. of kin (not CFR's)	.34	-.38	-.17	.53
Time < 5m from kin (not CFR's)	.20	-.50	.36	.25
Time < 5m from top ranking ♂♂	.26	-.54*	.39	.51
N(infants)=	11	9	10	8

*$p < .05$

Table 3.2

due both to relative amounts of association with CFR's and to differences in their ranks and/or the mother's ranks.

Finally, infants whose protectors were less likely to emit fearful gestures during intervention were also those who had more CFR's and who spent more time near them. However, these correlations were highly dependent on variations in mother's (CFR's) ranks, suggesting that differences in success rates were due largely to differences in mother's and/or CFR's ranks.

These analyses were repeated for time spent near mothers, for numbers of other close kin (brothers, aunts, uncles, nephews, and nieces) and for time spent near other close kin (Table 3.2, rows 1-3). In contrast, measures of agonistic interaction were not strongly related to these factors, strengthening the suggestion that CFR's play a special role in protecting infants and in determining differences in early agonistic experience.

The bottom row of Table 3.2 shows the same analysis for time spent near the four highest-ranking adult males of the group who were not necessarily related to the infants. Although there were no tendencies for infants with high-ranking mothers to associate more with these males, there were moderate tendencies for infants who did spend a lot of time near them to be threatened less frequently and to have higher success rates. This suggests that association with high-rank itself may benefit infants regardless of their mother's ranks. Nevertheless, infants of high-ranking mothers were generally in

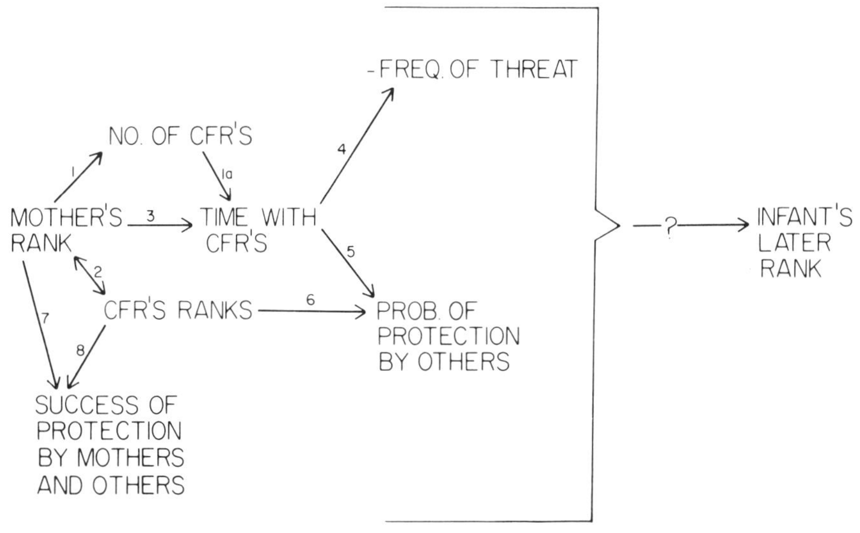

Figure 3.6 Hypothetical causal relationships linking variation in mothers' ranks with variation in infant's early agonistic experiences (from Berman, 1980b).

better positions to enjoy this advantage because they spent more time near members of the top-ranking lineage and/or because their principal protectors were also high-ranking.

Figure 3.6 summarizes some hypothetical causal relationships derived from these data which may be involved in the process of peer-peer rank acquisition. These data suggest first that more than one mechanism may contribute to peer-peer rank acquisition. For example, infants may learn their ranks through differences in amounts and sources of agonistic interaction directed toward them. They may also learn their ranks through intervention in agonistic interaction by other group members. Infants may also learn by observing the responses of their protectors toward their threateners. Finally, certain demographic factors (e.g., that high-ranking infants have more CFR's) may reinforce the process of rank acquisition. Second, the data suggest that mothers' ranks have important influences on infants, but that this influence may be indirectly expressed through the social structure and demographic processes taking place in the group. Mother's rank directly influences the number of protectors the infant has, the amount of time spent with them, and their ranks. Differences in frequencies of threat appear to be related directly to the amount of time infants spend near the CFR's and only indirectly to the mother's and CFR's ranks. The probability of protection by others may be influenced both by the time spent near CFR's and by their ranks. Finally, rates of successful intervention may be influenced by both the mother's rank and the close female relative's ranks. Perhaps because the mother's influence can be indirect, rank acquisition can take place occasionally in her absence.

GENERAL DISCUSSION

Two frequent criticisms of work on Cayo Santiago will end this discussion. The most obvious is that the work is necessarily correlational in nature because the monkeys cannot be manipulated. This is a general criticism of observational fieldwork which I believe is compensated by the advantages of being able to describe behavior in all its complexity and within a naturalistic context, and by the increasing ability to bring natural phenomena into laboratory settings subsequently for experimental verification (Plimpton et al. 1981). The fertile exchange of hypotheses, insights, and methodology between the field and the laboratory is well recognized today. Indeed, now that detailed studies of infant development are also being done in the wild (e.g., Nash, 1978; Altmann, 1980; Nicholson, 1982), studies on Cayo Santiago may serve with increasing effectiveness to bridge gaps between feral and captive studies. Opportunities to gain new insights and testable hypotheses should be becoming increasingly available through comparison across studies in different settings (Berman, 1984).

The second criticism focuses on the idea that kin-related patterns of behavior seen on Cayo Santiago are not likely to be typical of those taking place in the wild because Cayo Santiago lineages are more elaborate than those generally found in the wild (e.g., Altmann and Altmann, 1979). While this may be true, it is not necessarily a disadvantage providing that the problem is recognized and that comparable data are also available from a variety of settings, including the wild. So far, such data suggest that rhesus monkeys and several other species of nonhuman primates in a wide range of settings also have strong tendencies to respond positively toward kin. However, in many settings these propensities are less obvious to the observer or more difficult to describe in detail. For example, Walters (1980) reports that adolescent female baboons (*Papio cynocephalus*) in the Amboseli Game Reserve tend to solicit intervention primarily from high-ranking females and not necessarily from kin, particularly if the adolescents are low-ranking. However, this is probably because close kin are considerably less abundant there than on Cayo Santiago. Although kin perform only nine percent of the interventions and coalitions on behalf of adolescent females, this proportion is greater than their representation in the group. Thus one of Cayo Santiago's best uses may be to delineate the underlying propensities of primates to respond with respect to kinship (or other aspects of group structure) that cannot be described to the same degree or with the same precision in other groups. Subsequently researchers can gain further insight into the constraints on the expression of these propensities by comparing Cayo data with those from other settings.

Related to this is the idea that conditions on Cayo Santiago allow one to compare social structure in an expanding population with that found in more commonly stable or declining populations, and thus to elucidate the effects of demographic processes on behavior. Studies on Cayo Santiago may be thought of as modeling macaque behavior during periods when populations were expanding, diversifying, spreading over wide areas, and probably laying down many of the behavioral patterns seen today (Duggleby, 1980).

In conclusion, Cayo Santiago is a uniquely valuable resource not only because it resembles the wild in some ways, but also because it differs from it. In the examples given here, known and extended maternal lineages have made possible studies which have shed light on: 1) the role of social structure in mother-infant relationships; 2) the process(es) by which infants develop independent relationships with a variety of group companions; and 3) the possible mechanisms by which infants both receive protection and acquire rank vis-a-vis one another. By comparing behavior in groups with extended lineages with that in groups with more typically abbreviated lineages, one can delineate both the propensities of nonhuman primates to respond toward kin, and the constraints imposed on such propensities in wild and captive settings.

ACKNOWLEDGEMENTS

I am grateful to Donald Sade and the Directors of the Caribbean Primate Research Center for permission to collect data on Cayo Santiago and for advice and encouragement. I am greatly indebted to Robert Hinde for guidance during this phase of the study and to the following institutions for financial support: National Institute of Mental Health (NIMH), National Science Foundation (NSF), the Sigma Xi Society, the Wenner-Gren Foundation, and the Explorer's Club. Cayo Santiago was supported by NIH contracts RR 71-1003 and 7-2115 and grant RR-01293. Finally I thank the organizers of this symposium, Richard Rawlins, and Matt Kessler.

REFERENCES

Alley, T.R. Infantile colouration as an elicitor of caretaking behavior in old world primates. *Primates* 21:416-429, 1980.

Altmann, J. Observational study of behavior: sampling methods. *Behaviour* 49:227-267, 1974.

Altmann, J. Age cohorts as paternal sibships. *Behavioral Ecology and Sociobiology* 6:161-164, 1979.

Altmann, J. *Baboon Mothers and Infants.* Cambridge, Harvard University Press, 1980.

Altmann, S.A.; Altmann, J. Demographic constraints on behavior and social organization, pp. 47-63 in *Primate Ecology and Human Origins.* I.S. Bernstein and E.O. Smith, eds. New York, Garland STPM Press, 1979.

Berman, C.M. Social relationships among free-ranging infant rhesus monkeys. Ph.D. thesis. UK, University of Cambridge, 1978.

Berman, C.M. Mother-infant relationships among free-ranging rhesus monkeys on Cayo Santiago: a comparison with captive pairs. *Animal Behaviour* 28:860-873, 1980a.

Berman, C.M. Early agonistic experience and rank acquisition among free-ranging infant rhesus monkeys. *International Journal of Primatology.* 1:152-170, 1980b.

Berman, C.M. The ontogeny of social relationships with groups companions among free-ranging infant rhesus monkeys I. Social networks and differentiation. *Animal Behaviour.* 30:149-162, 1982.

Berman, C.M. Variation in mother-infant relationships: Traditional and nontraditional factors. In *Female Primates—Studies by Women Primatologists.* M. Small, ed., New York, A.R. Liss, Inc., pp. 17-36, 1984.

Chalmers, N.R. Comparative aspects of early infant development in some captive Cercopithecines. pp. 63-82, in *Primate Socialization,* F. Poirier, ed. New York, Random House, 1972.

Cheney, D.L. Interactions of immature male and female baboons with adult females. *Animal Behaviour* 26:389-408, 1978.

Duggleby, C.R. Genetic variability and adaptive strategies: Cayo Santiago. Paper delivered at American Society of Primatologists meeting, 1980.

Fady, J.C. Les jeux sociaux: Le comportement de jeux chez les jeunes. Observations chez macaca irus. *Folia Primatologica* 11:134-143, 1969.

Gouzoules, H. Maternal rank and early social interactions of infant stumptail macaques, *Macaca arctoides. Primates* 16:405-418, 1975.

Harlow, H.F.; Harlow, M.K. The affectional systems. pp. 287-334 in *Behavior of Nonhuman Primates,* Vol. 2, A.M. Schrier; H.F. Harlow; F. Stollnitz, eds. New York, Academic Press, 1965.

Hinde, R.A. Analysing the roles of the partners in a behavioral interaction—mother-infant relations in rhesus macaques. *Annals of the N.Y. Academy of Sciences* 159:651-667, 1969.

Hinde, R.A. *Biological Bases of Human Social Behavior.* New York, McGraw Hill, 1974.

Hinde, R.A.; Atkinson, S. Assessing the roles of social partners in maintaining mutual proximity, as exemplified by mother-infant relations in monkeys. *Animal Behaviour* 18:169-176, 1970.

Hinde, R.A.; Rowell, T.E.; Spencer-Booth, Y. Behaviour of socially-living rhesus monkeys in their first six months. *Proceedings of the Royal Society of London,* 143:609-649, 1964.

Hinde, R.A.; Spencer-Booth, Y. Behaviour of socially living rhesus infants in their first two and a half years. *Animal Behaviour* 15:169-196, 1967a.

Hinde, R.A.; Spencer-Booth, Y. The effects of social companions on mother-infant relations in rhesus monkeys. pp. 267-286 in *Primate Ethology,* D. Morris, ed. London, Weidenfeld and Nicholson, 1967b.

Hinde, R.A.; McGinnis, L. Some factors influencing the effect of temporary mother-infant separation: some experiments with rhesus monkeys. *Psychological Medicine* 7:197-212, 1977.

Hinde, R.A.; Proctor, L.P. Changes in the relationships of captive rhesus monkeys on giving birth. *Behaviour* 62:304-321, 1977.

Hrdy, S.B. The care and exploitation of non-human primate infants by conspecifics other than the mother. pp. 101-158 in *Advances in the Study of Behavior,* Vol. 6. J.R. Rosenblatt; R.A. Hinde; E. Shaw; C. Beer, eds. New York, Academic Press, 1976.

Jensen, G.D.; Bobbitt, R.A.; Gordon, B.N. The development of mutual independence in mother-infant pigtailed monkeys; *Macaca nemestrina.* pp. 49-53 in *Social Communication Among Primates,* S.A. Altmann, ed. Chicago, University of Chicago Press, 1967.

Kaplan, J.N. Differences in the mother-infant relations of squirrel monkeys housed in social and restricted environments. *Developmental Psychobiology* 5:43-52, 1972.

Kaufmann, J.H. Behavior of infant rhesus monkeys and their mothers in a free-ranging band. *Zoologica* 51:17-27, 1966.

Kurland, J.A. Kin selection in the Japanese monkey. *Contributions to Primatology 12,* London, S. Karger, 1977.

Lindburg, D.G. The rhesus monkey in north India; An ecological and behavioral study. In L.A. Rosenblum (ed.), *Primate Behavior: Developments in Field and Laboratory Research.* New York, Academic Press, pp. 1-106, 1971.

McMillan, C. Factors affecting mating success among rhesus macaque males on Cayo Santiago. Ph.D. thesis, SUNY/Buffalo, 1982.

Nash, L.T. The development of the mother-infant relationship in wild baboons *(Papio anubis). Animal Behaviour* 26:746-759, 1978.

Nicholson, N.A. Weaning and the development of independence in olive baboons. Ph.D. thesis. Cambridge, Mass., Harvard University, 1982.

Plimpton, E.H.; Swartz, K.B.; Rosenblum, L.A. The effects of foraging demand on social interactions in a laboratory group of bonnet macaques. *International Journal of Primatology* 2:175-185, 1981.

Rosenblum, L.A. Kinship interaction patterns in pigtail and bonnet macaques. *Proceedings of the Third Congress of Primatology, Zurich* 3:79-84, 1971a.

Rosenblum, L.A. The ontogeny of mother-infant relations in macaques. pp. 315-367 in *Ontogeny of Vertebrate Behavior,* H. Moltz, ed. New York, Academic Press, 1971b.

Rosenblum, L.A. Sex differences, environmental complexity and mother-infant relation. *Archives of Sexual Behavior* 3:117-128, 1974a.

Rosenblum, L.A. Sex differences in mother-infant attachment in monkeys. pp. 123-145 in *Sex Differences in Behavior,* P.C. Freedman; R.M. Riehart; R.L. Van de Weile, eds. New York, John Wiley, 1974b.

Rowell, T.E.; Hinde, R.A.; Spencer-Booth, Y. 'Aunt'-infant interaction in captive rhesus monkeys. *Animal Behaviour* 12:219-226,1964.

Sade, D.S. Determinants of dominance in a group of free-ranging rhesus monkeys pp. 99-114 in *Social Communication Among Primates,* S.A. Altmann, ed. Chicago, University of Chicago Press, 1967.

Sade, D.S. A longitudinal study of social behavior of rhesus monkeys. pp. 378-398 in *The Functional and Evolutionary Biology of Primates,* R. Tuttle, ed., Chicago, Aldine-Atherton, 1972.

Sade, D.S.; Cushing, K.; Cushing, P.; Dunaif, J.; Figueroa, A.; Kaplan, J.R.; Lauer, C.; Rhodes, D.; Schneider, J. Population dynamics in relation to social structure on Cayo Santiago. *Yearbook of Physical Anthropology* 20:235-262, 1977.

Seyfarth, R.M. Social relationships among adult female baboons. *Animal Behaviour* 24:917-938, 1976.

Simonds, P.E. Sex differences in bonnet macaque networks and social structure. *Archives of Sexual Behavior* 3:151-166, 1974.

Spencer-Booth, Y. The behavior of group companions toward rhesus monkey infants. *Animal Behaviour* 16:541-557, 1968.

Struhsaker, T. Social behavior of mother and infant vervet monkeys *(Cercopithecus aethiops). Animal Behaviour* 16:541-557, 1971.

Walters, J.R. Interventions and the development of dominance in female baboons. *Folia Primatologica* 34:61-89, 1980.

Wolfheim, J.H.; Jensen, G.D.; Bobbitt, R.A. Effects of group environment on the mother-infant relationship in pig-tailed monkeys *(Macaca nemestrina). Primates* 11:119-124, 1970.

CHAPTER FOUR

Social Development in a Congenitally Blind Infant Rhesus Macaque

CATHERINE E. SCANLON

INTRODUCTION

The question of how a physically handicapped animal relates to its environment is one that has been seldom considered. The relative infrequency of naturally occurring detectable handicaps (e.g., Collins et al., 1984; Fedigan and Fedigan, 1977; Furuya, 1966; Nakamichi et. al., 1982; Rasa, 1976; 1983; Rawlins and Kessler, 1983; Kessler and Rawlins, 1985), and the ethical considerations involved in imposing a physical handicap on an animal (e.g., Berkson, 1970; 1973; 1975; Talmage-Riggs et al., 1972; Taub and Berman, 1968) are two reasons which account for the lack of studies which attempt to explore this issue. It is important to monitor the development of a handicapped animal to ascertain the extent to which it can overcome a physical handicap and how far it may be treated differently by other animals in its group, whether favorably, indifferently or unfavorably.

An opportunity to monitor the development of an infant rhesus disabled by a sight defect occurred when, in February 1982, a blind infant male rhesus macaque (Figure 4.1) was born into one of the free-ranging groups on Cayo Santiago, Puerto Rico. His condition was diagnosed as congenital bilateral keratitis with possible congenital cataracts probably due to an infectious agent (Kaufman and Kessler, personal communication, 1983; Rawlins and Kessler, 1983). The implications of this condition were that the monkey could see nothing when in bright light due to constriction of the irises, but that in dim light he could probably see shadows. This provided a unique opportunity to observe a natural experiment of sight impairment, comparing the development of this visually handicapped infant with that of normal infants.

Rhesus macaques rely on sight for orientation to their physical and social environment. It must play an important role in the location and identification of food and conspecifics as well as aiding in the observational learning of appropriate behaviors and responses during

Illustrations 4.1 A young mother cradles her infant.

Figure 4.1 The blind infant with his mother, tattoo number 585.

the course of interaction within the social group. Since sight would seem to be of great importance to rhesus macaques, it was hypothesized that a congenital visual defect would constitute a severe handicap.

Two questions were investigated in this study: first, whether the relationship of the blind infant with his environment differed from that of normal infants, and second, whether any differences detected between the blind infant and the peer group might be due to factors other than his handicap.

METHODS

The study troop, Group I, was a large, high-ranking group. At the beginning of this study, in April 1981, it had 184 members. The group increased in size and, in September 1982, when the study was terminated, it had 230 members. The adult females and their offspring formed the core of the group and relationships could be traced through three matrilines. Most adult males were non-natal. A linear dominance hierarchy was established for both males and females using *ad libitum* observations of interactions involving a gesture of submission. During the course of this study a small group of female kin, their offspring and some adult males formed a subgroup which was less closely associated with the main body of the group. The monkeys were provisioned with monkey chow. This was generally supplied every morning, and was emptied into food hoppers in one of three food corrals.

Ad libitum and focal-animal sampling techniques (Altmann, 1974) were used. Focal animals were the blind infant and 20 normal infants. Observations were made on these animals from 17 to 32 weeks of age. The normal infants were born and observed in 1981, whereas the blind infant was born and observed the following year. The control group infants came from all three matrilines in the group and their mothers ranged from high through to low rank. The mother of the blind infant ranked sixteenth in the dominance hierarchy of 37 adult females. She belonged to the largest, highest-ranking genealogy and was a member of the subgroup (Figure 4.2). She had tattoo number 585. Her blind infant was untattooed during the course of this study and is referred to as I585B (Infant 585 Blind) in this paper. Her infant of 1981, tattoo D03, was one of the infants observed for the control group.

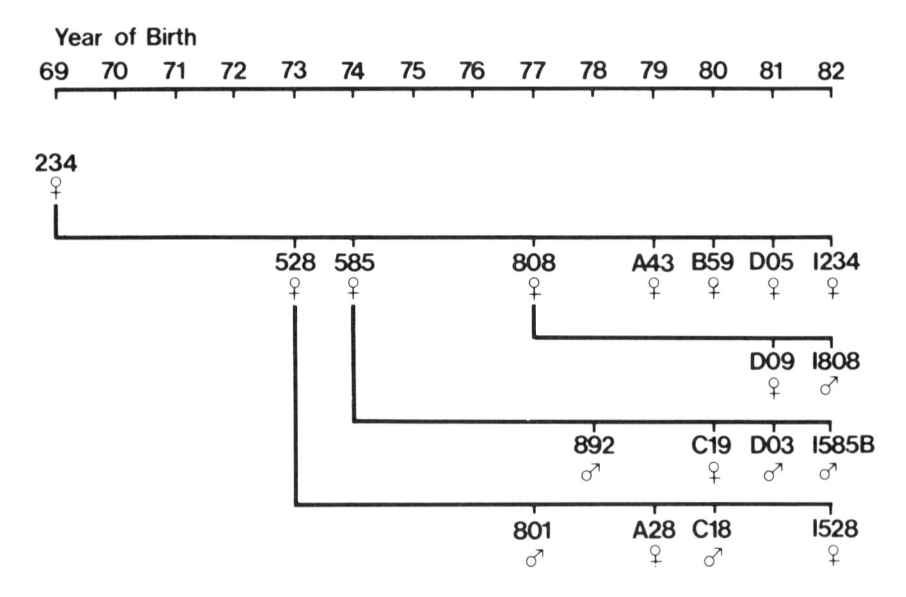

Figure 4.2 Subgenealogy of the blind infant. Mother, 585; blind infant, I585B; yearling brother, D03.

Each focal sample lasted 15 minutes. Animals were sampled in random order, ensuring that each animal was sampled approximately three times a week and that the distribution of samples over the day was the same over a four-week time period. During sampling, interactions with the mother, playful interactions, and competitive interactions over food were recorded as they occurred. At intervals of one minute, a scan was taken of the activity of the infant, the distance from his mother, and his position relative to the ground.

The question of whether and how the blind infant differed from the normal infants was explored by comparison of the frequencies of various behaviors for the blind infant with the median and range of the frequencies exhibited by the peer group. When the frequency of the blind infant's behavior was outside the range for the group, his behavior was considered to be abnormal. The question of whether any differences detected were due to family effects was investigated by determining whether the frequency of the behaviors for D03, the blind infant's sighted brother, followed the same trend as did those for the blind infant.

RESULTS

Relationship with the mother

Infants spend virtually all their time after birth with their mothers, without whose physical and social support they are unlikely to survive. The quantity and quality of care shown by the mother to her handicapped infant was hypothesized to be of paramount importance to the probability of his survival.

The proportion of observation time that the blind infant spent in contact with his mother was normal (Table 4.1), although overall responsibility for maintenance of body contact, as shown by the make/break index devised by Hinde and White (1974) was not (Figure 4.3) In the case of the blind infant the index was negative, indicating that his mother was responsible for contact. By contrast, the index for the relationship between his mother and his older brother, D03, when at the same age, was positive, indicating that in this case D03 was responsible for contact. The difference between the two indexes indicates that the mother played a greater role in maintaining contact with her blind infant than she did with her normal infant. The contact behavior of the mother to her blind infant was unusual for an animal of this age and was more typical of maternal behavior toward younger offspring (Berman, 1980). It may be that the blind infant's mother behaved toward him as if he were at an earlier stage of development, in response to his own behavior.

The reason for the difference between the make/break indexes for the blind infant and for the peer group is shown by examining the

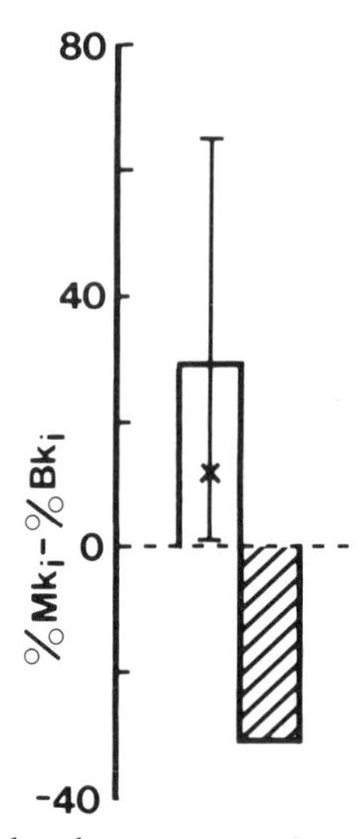

Figure 4.3 Make/break index between mothers and infants. The median index shown for the normal infants (open column). The blind infant's brother is represented by a cross on the line joining the lowest and the highest index for the control group. The index is shown for the blind infant (hatched column).

Table 4.1. Percentage of time sampled that blind and control infants spent on the mother

blind infant I585B	brother D03	Peer Group median	range low	range high
45	39	35	25	47

frequencies of the behaviors contained in this index: the number of contacts made by the infant and by the mother, and the number of contacts broken by the infant and by the mother (Figure 4.4). The index for the blind infant is lower than for control infants because the mother made contact with the blind infant more frequently, and because the blind infant broke contact from the mother more often than did controls. Because the measures of making and breaking contact for D03 did not differ from controls, the difference would appear to be a result of the blind infant's handicap, and not a result of other effects. Similar trends to those shown for the contact data were shown for the time the infants spent less than two meters from their mothers and for the responsibility for the maintenance of a proximity of one meter.

A reasonable explanation for the difference between the behavioral mechanism by which contact and proximity were maintained between the blind infant and his mother as compared with the peer group and their mothers would seem to be that the blind infant would break contact from or leave his mother more frequently, thus inadvertently

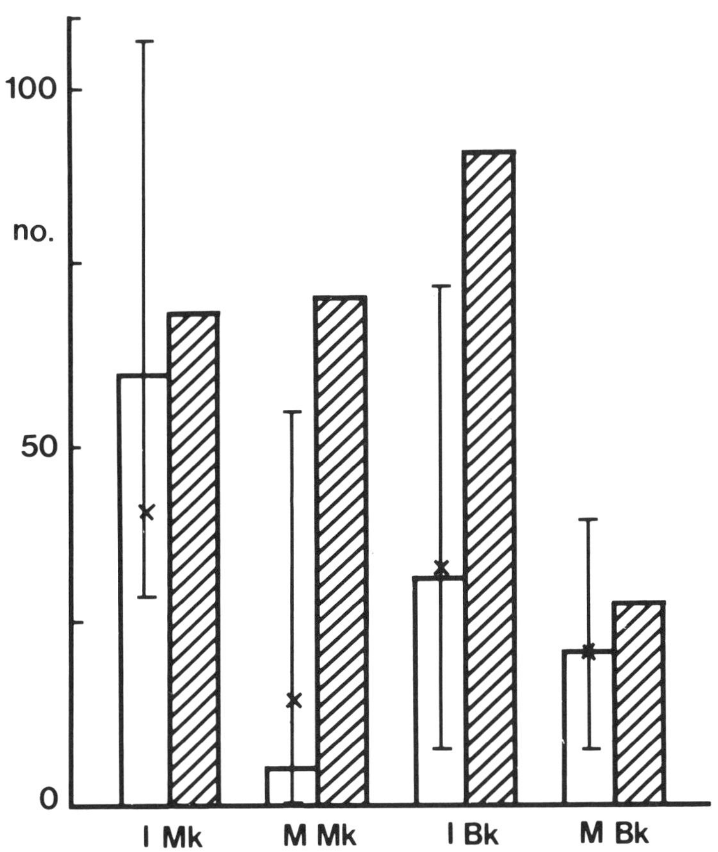

Figure 4.4 Mother-infant relationship. The number of contacts made and broken (IM, the number of contacts made by the infant; MM, the number of contact made by the mother; IB, the number of contacts broken by the infant; MB, the number of contacts broken by the mother). Conventions are as for Figure 4.3.

putting himself into situations which he could not recognize as potentially dangerous due to his impaired sight. His mother could see potential danger, and hence she may have been stimulated to behave in a restrictive way toward her infant. An alternative explanation is that the infant broke contact so frequently as a result of the restrictive behavior of his mother. A restrictive mother would be expected to both make contact very frequently and to break contact very infrequently. Thus this latter explanation seems less likely since the mother did not break contact unusually infrequently.

Play

Over the first six months of life the amount of time spent on the mother decreases and infants spend more time exploring their

Table 4.2. Play

Description of behavior	blind infant I585B	brother D03	Peer Group median	range low	range high
(a) Percentage of time sampled spent in play	2.5	0	5	0	9
(b) Number of initiations of play:					
(i) successful bv focal infant	9	7	27.5	7	46
(ii) unsuccessful by focal infant	6	7	13	3	24
(iii) successful by partner	13	1	24.5	1	38
(iv) unsuccessful by partner	2	4	10	2	14
(c) Percentage of play components in which:					
(i) focal infant directed play at partner	25	31	14	7	31
(ii) direction of play was reciprocal	58	54	70	54	86
(iii) partner directed play at focal infant	16	15	15	7	21

physical environment where they may come into proximity with other infants and juveniles with whom they will frequently play. Play may be important in the learning of physical and social skills (Chalmers and Locke-Haydon, 1984) and hence the presence or absence of it in the blind infant may have had an important role in determining his subsequent ability to forage and to interact successfully within the social group.

The blind infant was observed to participate in play to a normal extent (Table 4.2(a)). No difference was detected between the blind infant and the peer group in the relative success at initiation of play by the focal animal or by his partner (Table 4.2(b)). It was hypothesized that the quality of play would differ in two respects: that the blind infant would have play directed at him relatively more than he directed play at others because his visual disability would not impair others from play-mauling or play-chasing him, whereas it could prevent him from behaving similarly. However, the blind infant directed more play at partners than some of the peer group, and had less play directed at him than did some of the peer group (Table 4.2(c)). The second hypothesis was that the blind infant would engage in contact play more than approach-leave or chasing play, since contact play (Figure 4.5) involves tactual, visual, and possibly olfactory cues, whereas approach-leave play (Figure 4.6) involves predominantly visual cues, and may frequently involve a more intricate and vigorous movement around the physical environment. The blind infant showed a normal amount of contact play, but no approach-leave play, whereas all the peer group engaged in at least some such playful interactions (Figure 4.7).

Access to the Food Corral

The food corrals constituted a highly concentrated, high-quality food resource and there was often much intragroup and intergroup aggression while the animals competed for access to it.

The blind infant spent no less time in the food corral than did his sibling or the peer group (Table 4.3(a)). Thus is appeared that access to the corral for the blind infant was not impaired by his visual defect. However, the relationship with the mother when in the food corral did appear to be abnormal (Figure 4.8); the blind infant spending 78 percent of his time in the food corral in contact with his mother, whereas the median value for the peer group with their mothers was only 5.5 percent. D03, the blind infant's sibling, spent 58 percent of his time in the food corral in contact with his mother, thus spending at least 36 percent more of his time in contact than any other of the peer group, and 20 percent less than his blind sibling at the same age. Contact with the mother in the food corral may have had two effects. The infant may have had more chance to forage on chow because his

Figure 4.5 Contact play between two infants.

Figure 4.6 Approach-leave play between two infants.

Table 4.3. Food corral and non-aggressive food resource competition

Description of behaviour	blind infant I585B	brother D03	Peer Group median	range low	range high
(a) Percentage observation time spent in food corral	9	3	4	0	9
(b) Percentage of successful attempts by focal infant to obtain or retain a food resource from another monkey	32.5	44	33.5	0	62.5
(c) Number of interactions involving nonaggressive food competition	36	18	13	8	24

mother would carry him up to the food bins from where he may have been able to grasp food from the bin or where he may have had more chance to take chow from his mother. The second effect may have been that he was kept safely from any conflicts occurring in the corral. An explanation in harmony with the data is that the blind infant spent extra time on his mother in the food corral partly because of his handicap, and partly because of family effects. It may be that the subgroup mothers and infants were more frequently in contact in the food corral than mothers and infants from the main body of the group.

Nonaggressive Resource Competition

Infants may also obtain food such as leaves or monkey chow indirectly from other animals in nonaggressive situations. For example, an infant on his mother may reach up and grasp a piece of

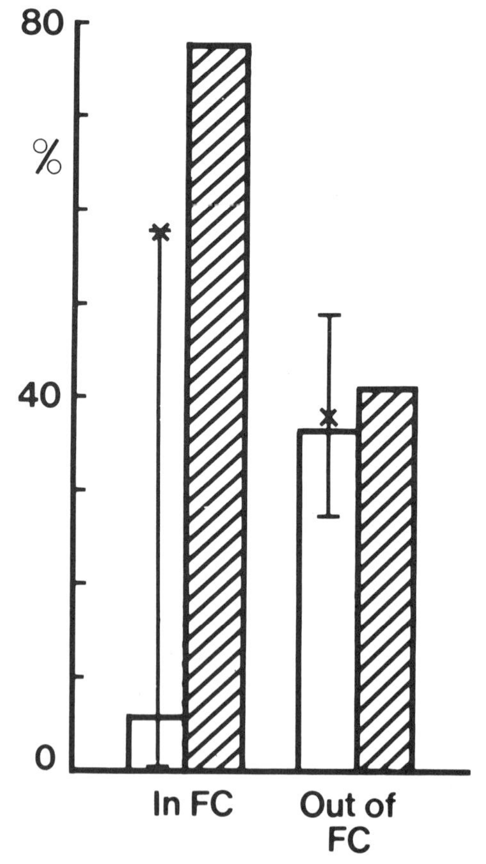

Figure 4.8 Percentage of time in the food corral (FC) which was spent in contact with the mother. Conventions are as for Figure 4.3.

food she is eating which she may, or she may not, tolerate him taking. The percentage of attempts made by the blind infant to obtain or retain a food resource from another monkey that were successful was within the range of variation exhibited by the peer group (Table 4.3 (b)). However, the blind infant was involved in such situations more frequently than the peer group (Table 4.3(c)). An explanation in harmony with these data is that animals may use different strategies to meet the same needs, in this case food, depending on their individual situation. In the case of the blind infant, it seems likely that he would have found it very difficult to obtain food directly by foraging for it, and so it may have been more efficient for him to attempt to obtain it from another animal, usually his mother, despite the fact the he was not always successful. By contrast, the peer group could forage directly more efficiently, in which case to take food from another monkey may have been less attractive and less efficient because a proportion of all such attempts were liable to be unsuccessful.

Canopy Layer

The canopy provides a place in which young macaques may escape from aggression, and feed and play undisturbed. The blind infant appeared clumsy in his explorations of the canopy, and several times was observed to bump into branches, an event which never occurred in unhandicapped infants. However, it was very striking that he was observed exploring as much as he did, spending similar proportions of his time on the ground, in the lower canopy, and in the upper canopy as did the peer group (Table 4.4(a)). The time the blind infant spent in contact with his mother while in the different canopy levels (Table 4.4(b)) did not fall outside the range for the peer group, indicating that the blind infant could move between the different canopy layers normally, although the extent to which he explored within a layer may have differed.

DISCUSSION

To summarize these results, the blind infant was shown not to differ from his sibling and the peer group in many important respects: overall time spent in contact with the mother, time spent in play number and success of initiations of play, direction of play, access to the food corral, relative success at gaining or retaining food, and movement around the canopy. However, the mechanisms by which some of these frequencies were achieved did appear to differ: the blind infant contributed less to the overall maintenance of contact with the mother, since he broke contact and his mother made contact more frequently than did the other infants and their mothers. The blind infant was never involved in approach-leave play. He spent more time in contact with the mother while in the food corral. He was involved in

Table 4.4. Canopy level

Description of behavior	blind infant I585B	brother D03	Peer Group median	range low	range high
(a) Percentage of observation time spent:					
(i) on the ground	58	55	61.5	44	73
(ii) in lower canopy	32	32	30	19	43
(iii) in upper canopy	14	17	14.5	4	19
(b) Percentage of time on the mother in relation to canopy level:					
(i) on the ground	56	54	48	29	66
(ii) in lower canopy	20	20	20	6	33
(iii) in upper canopy	44	44	15.5	0	46

nonaggressive food competition more frequently.

The apparent normality of so much of the blind infant's behavior may have been brought about in large part by the extra care received from the mother. Casual observation suggested that the mother gave D03, her *yearling* infant, less care than was normal, at least during the first six months of her blind infant's life. During this time, D03 appeared inactive and spent much time in contact with his three-year-old brother, 891, who was observed to carry him ventrally. In addition, the behavior of his subgenealogy toward the blind infant appeared to involve extra vigilance and protective behavior toward him. On several occasions the blind infant was seen in the lower branches of a tree between two and five meters from close kin: the approach of an unrelated animal was observed to result in an unusually high number of threats from one or more of these kin, in particular his mother, 585, his aunts, 528 and 808, and his five-year-old male cousin, 801 (Figure 4.2).

These data are generally in harmony with other data obtained for handicapped infants. Fedigan and Fedigan (1977) reported the development of a Japanese macaque infant born with a condition resembling that of cerebral palsy. He was reported as being virtually unable to care for himself although he did vocalize frequently and followed his mother persistently when he could locate her. He received extra care from his mother, and to a lesser extent from some other group members. However, his disability was probably a greater one than for the blind rhesus infant observed in this study since he was involved in all forms of social activity, except those with the mother, much less than the healthy infant with whom he was compared, whereas the blind infant rhesus in this study attained normal levels of several activities. Both these handicapped infants died at about one year of age, possibly reflecting the inability or reluctance of their mothers to contiue giving their infants extra care, added to the fact that an older infant will have greater nutritional requirements. Neither mother had a new infant when her handicapped infant died, although 585 was beginning to wean her blind infant (Kessler personal communication, 1983).

A series of studies were carried out by Gershon Berkson who deliberately blinded infant macaques in feral (Berkson, 1970), free-ranging (Berkson, 1973), and captive (Berkson, 1975) situations, and subsequently monitored their survival and social development. Many of his results are similar to those obtained in this study although there was a trend toward his animals being less proficient than the Cayo Santiago blind infant. For example, he reported a lower incidence of play initiation (Berkson, 1973) and a lower proportion of time spent in play in his handicapped infants as compared with his controls (Berkson, 1970; 1975). However, some of his focal handicapped animals also survived until at least three years of age in the free-ranging situation on La Cueva, Puerto Rico (Berkson, 1973), which was rather longer than did the Cayo Santiago blind infant. In that study, the mothers of two blind infants had new infants when their handicapped infants were one year old. Berkson (1973) reported that this did not affect the care given to their handicapped infants, and that the mothers would carry both infants when necessary.

CONCLUSIONS

1. A congenitally blind male infant rhesus macaque was observed to maintain frequencies of certain behaviors within the range exhibited by a control group of 20 infants.
2. There were differences in the way these frequencies were maintained, and in some cases they appeared to involve additional care from the mother of the handicapped infant.
3. Differences appeared to be due to the infant's handicap and not because of his subgenealogy membership or mother's rank.

ACKNOWLEDGEMENTS

This study was carried out while the author was funded by the Science and Engineering Research Council. I am grateful to Dr. Richard Rawlins and the directors of the Caribbean Primate Research Center for permission to gather data on Cayo Santiago. Dr. Paul L. Kaufman, an opthalmologist from the University of Wisconsin Medical School, and Dr. Matt J. Kessler, veterinarian for the Caribbean Primate Research Center, carried out the examination and diagnosis of the blind infant's condition in January 1983. My thanks are due to Neil Chalmers, David Hill, Jan Locke-Haydon, Julie Roberts, Jonathan Silvertown, and Claudine Teyssedre for commenting on an earlier draft of this chapter.

REFERENCES

Altmann, J. Observational study of behavior: sampling methods. *Behaviour* 49:227-267, 1974.

Berkson, G. Defective infants in a feral monkey group. *Folia Primatologica* 12:284-189, 1970.

Berkson, G. Social responses to abnormal infant monkeys. *American Journal of Physical Anthropology* 38:583-586, 1973.

Berkson, G. Social response to blind infant monkeys, pp. 49-57 in *Aberrant Development in Infancy.* N.R. Ellis, ed. Hillsdale, New Jersey, Lawrence Erlbaum Associates, 1975.

Berman, C.M. Mother-infant relationships among free-ranging rhesus monkeys on Cayo Santiago: a comparison with captive pairs. *Animal Behaviour* 28:860-873, 1980.

Chalmers, N.R.; Locke-Haydon, J. Correlations among measures of playfulness and skillfulness in captive common marmosets (*Callithrix jacchus jacchus*). *Developmental Psychobiology* 17:191-208, 1984.

Collins, D.A.; Busse, C.D.; Goodall, J. Infanticide in two populations of savanna baboons, pp. 193-215 in *Infanticide: Comparative and Evolutionary Perspectives.* G. Hausfater; S. Blaffer Hrdy, eds. New York, Aldine Publishing Company, 1984.

Fedigan, L.M.; Fedigan, L. The social development of a handicapped infant in a free-living troop of Japanese monkeys, pp. 205-221 in *Primate Bio-Social Development.* S. Chevalier-Skolnikoff; F.E. Poirier, eds. Garland Publishing, 1977.

Furuya, Y. On the malformation occurred in the Gagyusan troop of wild Japanese monkeys. *Primates* 7:488-492, 1966.

Hinde, R.A.; White, L.E. Dynamics of a relationship: rhesus mother-infant ventro-ventral contact. *Journal of Comparative and Physiological Psychology* 86:8-23, 1974.

Kessler, M.J., Rawlins, R.G. Congenital cataracts in a free-ranging rhesus monkey. *Journal of Medical Primatol.* 14: 225-228, 1985.

Nakamichi, M.; Fujii, H.; Koyama, T. Behavioural development of a malformed infant in a free-ranging group of Japanese monkeys. Poster presented at the IXth Congress of the International Primatological Society, Atlanta, 1982.

Rasa, O.A.E. Invalid care in the dwarf mongoose (*Helogale undulata rufula*). *Zeitschrift Fur Tierpsychologie* 42:337-342, 1976.

Rasa, O.A.E. A case of invalid care in wild dwarf mongooses. *Zeitschrift Fur Tierpsychologie* 62:235-240, 1983.

Rawlins, R.G.; Kessler, M.J. Congenital and hereditary anomalies in the rhesus monkeys (*Macaca mulatta*) of Cayo Santiago. *Teratology* 28:169-174, 1983.

Talmage-Riggs, G. et al. Effect of deafening on the vocal behaviour of the squirrel monkey (*Saimiri sciureus*). *Folia Primatologica* 17:404-420, 1972.

Taub, E.; Berman, A.J. Movement and learning in the absence of sensory feedback, pp. 173-192 in *The Neuropsychology of Spatially Oriented Behaviour.* S.J. Freedman, ed. Dorsey Press, Homewood, 1968.

Illustration 5.1 An adult male vocalizes a threat.

CHAPTER FIVE

Vocal Communication: A Vehicle for the Study of Social Relationships

HAROLD GOUZOULES, SARAH GOUZOULES,
PETER MARLER

INTRODUCTION

Until very recently little attention had been devoted to the vocal communication of rhesus monkeys living under free-ranging conditions. After more than 20 years, Rowell and Hinde's (1962) classic paper has continued to be the standard reference and source for researchers, and it remains the only extensive catalog of vocalizations analyzed spectrographically for this species. Rowell and Hinde's recordings came from the then recently established captive colonies at Madingley, Cambridge and, as they pointed out, did not represent the full range of contexts in which rhesus monkeys use calls under natural conditions. On the basis of judgements by ear, Lindburg (1971) subsequently found equivalence between many calls he heard during a field study of rhesus monkeys in northern India and those which had been described by Rowell and Hinde.

The apparent neglect of rhesus vocal behavior is attributable in part to a widely held view that monkey vocalizations convey information only about an animal's emotional state or motivation (e.g., Lancaster, 1975; Smith, 1977). This interpretation has recently been shown to be an overly simplistic explanation of the vocal behavior of some primate species. For instance, there is evidence that a number of calls of the vervet monkey function *representationally*; that is, they convey information primarily about specific external objects and even abstract social relationships (Seyfarth et al., 1980a; 1980b; Cheney and Seyfarth, 1982). Intriguingly, the early work of Rowell (1962) and Lindburg (1971) contained hints that certain vocalizations of rhesus monkeys, notably those used in agonistic situations, such as screams and screeches, might also function in a more complex manner. For instance, Rowell (1962, p. 95) observed that rhesus agonistic vocalizations made "a complete scheme of communication, so that the progress of an encounter (could) be followed *without being able to see the animals involved*," (emphasis ours). Also, Lindburg (1971, p. 50) noted that a juvenile's "screeches" often elicited its mother's intervention in a fight. In order to examine the possibility that rhesus monkeys might use their agonistic vocalizations in a more complex, perhaps representational manner, we conducted a study of the vocal behavior of the free-ranging population on Cayo Santiago, Puerto Rico.

Rhesus Monkey Scream Vocalizations

Rhesus monkeys on Cayo Santiago give five acoustically different types of scream vocalizations when they are involved in agonistic encounters (Gouzoules et al., 1984) (Figure 5.1). The particular type of scream given in a fight depends on the identity of a monkey's opponent as well as the severity of the aggression. For example, acoustically different screams are given when an opponent is a matrilineal relative of the caller as opposed to when it is an unrelated monkey. Different screams are also given when an unrelated opponent is higher-ranking than the caller and when it is lower-ranking. A number of studies have documented attempts by monkeys involved in agonistic encounters to recruit support from other group members (Cheney, 1977; Kaplan, 1977; 1978; Seyfarth, 1976; Walters, 1980; Watanabe, 1979), and suggest that scream vocalizations are effective in the recruitment of allies (de Waal, 1978; de Waal, et al., 1976; 1981; Cheney, 1977; Cheney and Seyfarth, 1980). We conducted a series of experiments to test whether the different types of rhesus scream vocalizations might function in the recruitment of allies by conveying to them information about a signaler's opponent in an agonistic event.

During field experiments mothers were played tape-recorded examples of the different types of scream vocalizations recorded previously from their immature offspring. The mothers were not provided with any other information available during the original agonistic encounter. The responses of the females were filmed and later subjected to frame-by-frame analysis. If the acoustically distinct scream vocalizations conveyed different information, then mothers should respond differently, depending on the class of opponent being signaled. For example, mothers should respond more strongly when their offspring's opponent was a higher-ranking unrelated individual than when it was a matrilineal relative. This is, in fact, one pattern of agonistic aiding that Kaplan (1977) had reported for rhesus monkeys on Cayo Santiago. The results of the experiments did indeed indicate that mothers responded differently to playback of different types of screams from their immature offspring. The strength of their responses to the different screams was predictable from knowledge of the pattern of alliance formation in rhesus monkeys (Gouzoules et al., 1985). We concluded that rhesus monkey scream vocalizations, like

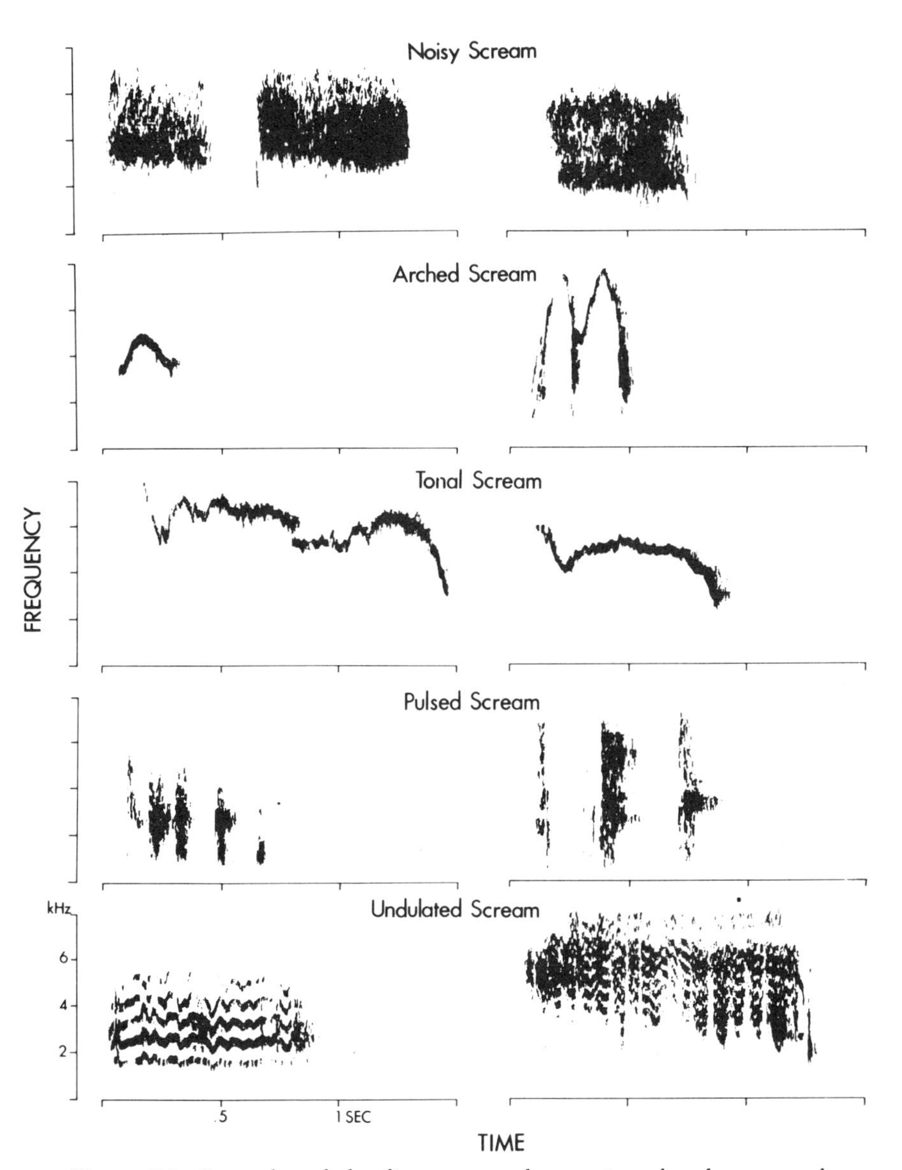

Figure 5.1 Examples of the five scream classes given by rhesus monkeys during agonistic encounters: noisy, arched, tonal, pulsed, and undulated. x-axis shows time; y-axis shows frequency in units of 2 kHz.

vervet monkey alarm calls and grunts, functioned representationally. They can convey information about external objects and social relationships in the absence of any other signal modalities.

While it is known that aid to immature offspring is the dominant pattern of the alliance system in female rhesus and Japanese monkeys (*M. fuscata*) (Kaplan, 1977, 1978; Watanabe, 1979), aid to other, more distant, matrilineal kin is also common. Several studies of agonistic alliance formation in macaques have suggested that female aid to matrilineal relatives is distributed in correlation with the degree of relatedness between ally and aid recipient (Massey, 1977; Kurland, 1977; Kaplan, 1978; Datta, 1983a). As these studies indicate, primate alliance behavior can be particularly useful for testing theories of kin selection and altruism. Other studies have emphasized the importance of alliance behavior in the ontogeny and maintenance of female dominance relationships and imply that degree of relatedness may be less crucial to the formation of rank-related alliances than has been supposed (Cheney, 1977; Kaplan, 1978; Seyfarth, 1976; 1977; Chapais and Schulman, 1980; Walters, 1980; Datta, 1983a; 1983b). Results from these latter studies suggest that there may be both short and long-term costs and benefits to alliance formation and that these may vary with degree of relatedness between ally and aid recipient. However, in none of the studies was it possible to submit predictions from either theory of alliance formation to an experimental test.

Here, we suggest that knowledge of the role vocalizations play in the recruitment of allies in rhesus monkeys provides a unique means for testing these theories experimentally and for analyzing the dynamic processes underlying alliance formation and assessing the consequences for individuals participating in different alliance networks. This approach to the study of alliance formation is based on field experiments in which different types of calls from known individuals are played to specific group companions. The strength of a subject's response to playback of these calls provides an empirical assessment of the nature and strength of the agonistic alliance between two individuals. In the remainder of this paper we report on a study of rhesus monkey scream vocalizations in which both the vocal component and the function of rhesus matrilineal alliance formation were examined. Field playback experiments were used to compare: 1) the extent to which females respond to the tape-recorded screams of immatures of different degrees of relatedness; and 2) the pattern of responses of adult females to immatures' acoustically distinct vocalizations that signal different classes of opponents.

METHODS

The study was conducted on one troop of rhesus monkeys (Group L) on Cayo Santiago, Puerto Rico. A detailed description of the island and a history of the colony can be found in Chapters 1 and 2 of this volume. Group L ranged predominantly on the Small Cay and had virtually exclusive access to one of three feeding corrals provisioned daily with commercial monkey chow. Consequently, the rate of intergroup encounters between Group L and the other five troops on Cayo Santiago was low. Group L contained 118 members at the beginning of the study and was composed of four matrilineages with female kinship documentation extending back to 1956.

The study was conducted during two periods: November 1980-May 1981, and October-December 1981. Vocalizations collected during the first phase of the project were analysed using a Real-time Spectrum Analyzer (Princeton Applied Research, model 4512, analysis range: 10 kHz) at the Rockefeller University Field Research Center. Playback experiments were then conducted during the second phase of fieldwork. Vocalizations were tape recorded by two observers, using either a Nagra III or a Nagra SN tape recorder and a Sennheiser directional microphone (model 804 or 805). Identities of vocalizers, agonistic opponents, and other information about the focal event were noted at the time of the recording (see Gouzoules et al., 1984 for more details of observation methods). The methods were designed to maximize the overall number of recorded screams rather than to equalize recording opportunities among group members.

Selection of subjects for playback experiments was based upon female dominance rank, matrilineage, and, especially, the availability of recorded scream vocalizations from both immature offspring and other more distantly related immature kin. Ten females received playbacks of screams recorded from their offspring and other immature relatives. Table 5.1 provides demographic profiles of subjects and vocalizers.

Playback Experiments

As reported elsewhere (Gouzoules et al., 1984), two scream classes were directed significantly more often to matrilineally unrelated opponents than to matrilineal kin: *noisy* screams were directed primarily to higher ranking opponents and were frequently given in response to physical contact; *arched* screams were directed to lower ranking opponents in the absence of contact (Figure 5.1). Most subjects were the focus of four trials (see Table 5.1), receiving playbacks from each of the two scream classes recorded from both an immature offspring and a more distantly related immature. One female received only a noisy exemplar from her immature relative while another female received noisy and arched exemplars from two immature relatives. The order in which the playback stimuli were presented to each subject was varied. Choice of a particular subject for a playback trial was determined by rotation through a fixed schedule on which subjects were listed randomly. We attempted to complete one rotation before returning to a subject.

Target animals were monitored for variable lengths of time (up to four hours) prior to each trial to insure that neither the vocalizer nor, when the vocalizer was a more distant relative, its mother, was nearby. A large part of the island was heavily wooded and most

Table 5.1 Description of Subjects and Callers Used in Playback Experiments.

Female Subject	Rank	Matriline	Offspring (age)	Other Relative (age)	Degree of Relatedness between Subject & Relative
258	17	3	A55 (2)	A58 (2)	1/16
278	11	1	A99 (1)	A66 (2)	1/4
404	15	2	A75 (2)	A63 (2)	1/8
432	9	1	A65 (2)	A98 (1)	1/4
483	19	3	965 (3)	A55 (2)	1/16
572	23	4	A97 (1)	A94 (1)	1/16
606	13	1	A66 (2)	A99 (1)	1/4
660	18	3	A58 (2)	965 (3)	1/16
670	10	1	A98 (1)	A65 (2)	1/4
				A92 (1)	1/4
671	26	4	A94 (1)	A97 (1)	1/16

Dominance rank is based on the direction of agonistic encounters. Degrees of relatedness between females and non-offspring (other) relatives are calculated minimally—i.e., through maternal lines only.

playbacks were conducted in these areas. Screams were played to subjects from a Sony cassette recorder (TCM 5000) through a Nagra DSN speaker, both placed inside a wooden speaker box. The speaker box was hidden in foliage either to the right or left of a subject at distances of 3.7 to 13.7 meters. Amplitudes for each playback trial were set according to the subject's distance from the speaker. Previously, all playback tapes had been assigned volumes for each call at distances of 3, 6, 9, and 12 meters on another part of Cayo Santiago (away from Group L). These settings were used as standards for the amplitude setting during a trial. In general, we attempted to keep volumes low, so that calls might appear to originate from a distance.

Subjects were filmed continuously for the 10 seconds before and 15 seconds following a playback on a sound movie camera (Elmo model 350 SL), placed at a 90° angle in relation to the speaker and the subject. The camera was positioned at distances ranging from 2.5 to 7.5 meters from subjects. Equipment placement was such that, at the beginning of a trial, subjects were either facing toward the camera or in the direction opposite the speaker. If the subject turned its body to face the speaker before the trial began the experiment was terminated. Overall, 31 percent of the trials were aborted for this or other reasons, such as the appearance of the animal whose call was to be played. Subjects were frequently filmed at times other than during playback experiments and the monkeys paid little attention to observers carrying equipment or filming. For each playback we recorded, and later transcribed, a verbal description of subjects' responses. A map of the equipment placement, landscape features, and monkeys in each trial was also drawn. To avoid habituation to playback stimuli we played an average of only two playback calls a day (range 0-4 playback calls per day). We allowed at least an hour between playbacks and always moved away from the vicinity of the previous trial. Another trial was performed only after verifying that no subjects or vocalizers from a preceding test were nearby. Stimulus recordings from the same vocalizer were never played more than once during a day.

Responses were scored by counting the number of film frames (18 fps) a subject looked toward the speaker following playback of the call (*duration*) and the number of frames after the playback call before she looked in the direction of the speaker (*latency*). Recording of the playback call on the film sound track permitted precise detection of its time of occurrence. If the subject looked in the direction of the speaker before the playback a correction factor, equivalent to the proportion of frames looked to the speaker before playback, was subtracted from the number of frames she looked after playback.

RESULTS

Response of females to playbacks of screams of non-offspring relatives. Females responded to the playbacks of both noisy and arched screams from non-offspring immature relatives despite the absence of those immatures' mothers. In other words, females responded to immature relatives' screams in the absence of any cues received from the juveniles' mothers. Subjects looked in the direction of the speaker significantly longer following a playback than before it (noisy: $T = 0$, $p = 0.008$, $N = 7$; arched: $T = 3$, $p = 0.0391$, $N = 7$; one-tailed Sign tests). There was no significant difference in subjects' probability of response to playbacks of the two scream classes (Fisher exact probability test, $p > 0.05$). Nor was there a significant difference in the strength of female response to screams of the two classes as measured by: 1) duration of response ($T = 16$, $p > 0.10$, $N = 8$); or 2) latency to respond ($T = 11$, $P > 0.10$, $N = 7$; Wilcoxon signed ranks matched pairs tests, two-tailed).

Responses of females to playback of screams of offspring and more distant relatives. Females showed a significantly greater probability of response to playback of noisy screams from their immature offspring than from their other immature relatives ($p = 0.03$, Fisher exact probability test). However, there was no significant difference in response probability to arched screams given by offspring and those given by other relatives ($p > 0.05$). The duration of a female's response to noisy screams of her offspring was also significantly longer than to those of other immature relatives ($T = 6$, $p < 0.05$, $N = 10$, Wilcoxon signed ranks matched pairs test, two-tailed, Figure 5.2a), but there was no significant difference in the duration of response to the arched screams of offspring compared to those from other relatives ($T = 19$, $p > 0.05$, $N = 9$, Figure 5.2a). These results suggest that female response to noisy and arched screams varies differentially with the relatedness of the caller.

The latency of females to respond when screams of offspring and other relatives were played was also examined. There were no significant differences in the latency of response to either the noisy or arched screams of offspring and other relatives (noisy: $T = 11$, $p > 0.05$, $N = 10$; arched: $T = 16.5$, $P > 0.05$, $N = 9$; Wilcoxon signed ranks matched pairs test, two-tailed, Figure 5.2b).

Degree of relatedness and strength of response. We next tested whether females discriminated more finely among non-offspring kin depending upon their degree of relatedness to the caller. For this analysis female/other relative dyads were divided into two kinship categories based upon degrees of relatedness (calculated through maternal lines only): 1) closely related ($r = 1/4$); and 2) distantly related ($r = 1/16$). One dyad was related by an intermediate value, $r = 1/8$ (see

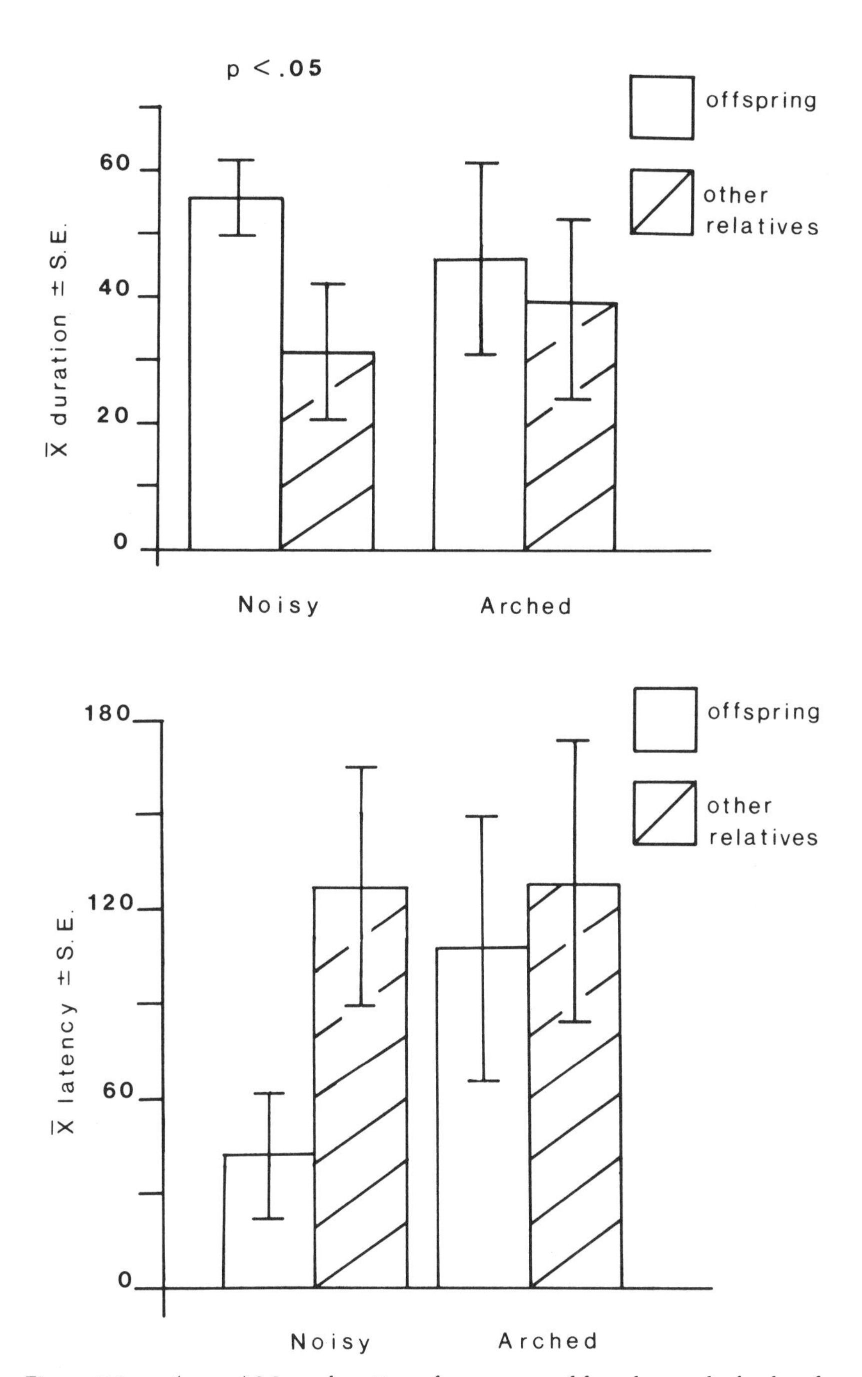

Figure 5.2 a. (upper) Mean duration of responses of females to playbacks of noisy and arched screams from offspring and other immature relatives. b. (lower) Mean latency to response of females to playbacks of noisy and arched screams from offspring and other immature relatives.

Table 5.1). Since the mother of the immature in this dyad had recently died and behavioral observations suggested that the juvenile kept close contact with the subject, this dyad was classified into the closely related category.

The duration of female response to non-offspring relatives' noisy screams was stronger if these relatives were closely related to her ($U = 5.5$, $p < 0.05$, Mann-Whitney U-test, one-tailed, Figure 5.3a). There was no significant difference in female latency to respond to the noisy screams of closely and distantly related immatures ($U = 8.5$, $p > 0.10$, Figure 5.3b). Nor was there a significant difference in either the duration of, or latency to, female response to close and distant relatives' arched screams (duration: $U = 14$, $p > 0.10$, Figure 5.3a; latency: $U = 8$, $p > 0.10$, Figure 5.3b). These results suggest that females tend to respond more strongly to noisy screams when given by a closely related immature than when given by a more distantly related juvenile. In contrast, female response to arched screams of immature relatives does not appear to vary with degree of relatedness.

Female dominance rank and strength of response to playback of immature relatives' screams. The effect of dominance rank on the strength of a female's response to non-offspring relatives' screams was analyzed in two ways. First the correlation between the dominance rank of a subject and the duration and latency of her response to a non-offspring relative's screams was calculated. We found no significant correlation between dominance rank and either response duration or latency to respond to noisy and arched screams (noisy: duration, $r_s = 0.508$, $p > 0.05$; latency, $r_s = 0.410$, $p > 0.10$, $N = 11$; arched: duration, $r_s = 0.032$, $p > 0.10$; latency, $r_s = 0.188$, p 0.10, $N = 10$). Also analyzed was whether or not females responded more strongly to a more distant relative's scream if its mother ranked above the subject in the dominance hierarchy. There were no significant differences in female response to screams of either type when immatures' mothers ranked above and below the subject (noisy: duration, $U = 14$, $p > 0.20$; latency, $U = 15$, $p > 0.20$; arched: duration, $U = 13$, $p > 0.20$; latency, $U = 14.5$, $p > 0.20$; Mann-Whitney U-tests, two-tailed, Figure 5.4).

Strength of female response to male and female juvenile relatives' screams. The strength of female response to the screams of immature non-offspring relatives of different sex was also analyzed. The duration of females' responses to arched screams was significantly longer if the caller was a female ($U = 4$, $p = 0.048$, Mann-Whitney U-test, one-tailed, Figure 5.5a). There was no significant difference in the duration of subject response to the noisy screams of immature male and female relatives ($U = 7$, $p > 0.10$, Figure 5.5a). While there was a significant difference in the latency of females to respond to the arched screams from male and female relatives there was no significant difference in

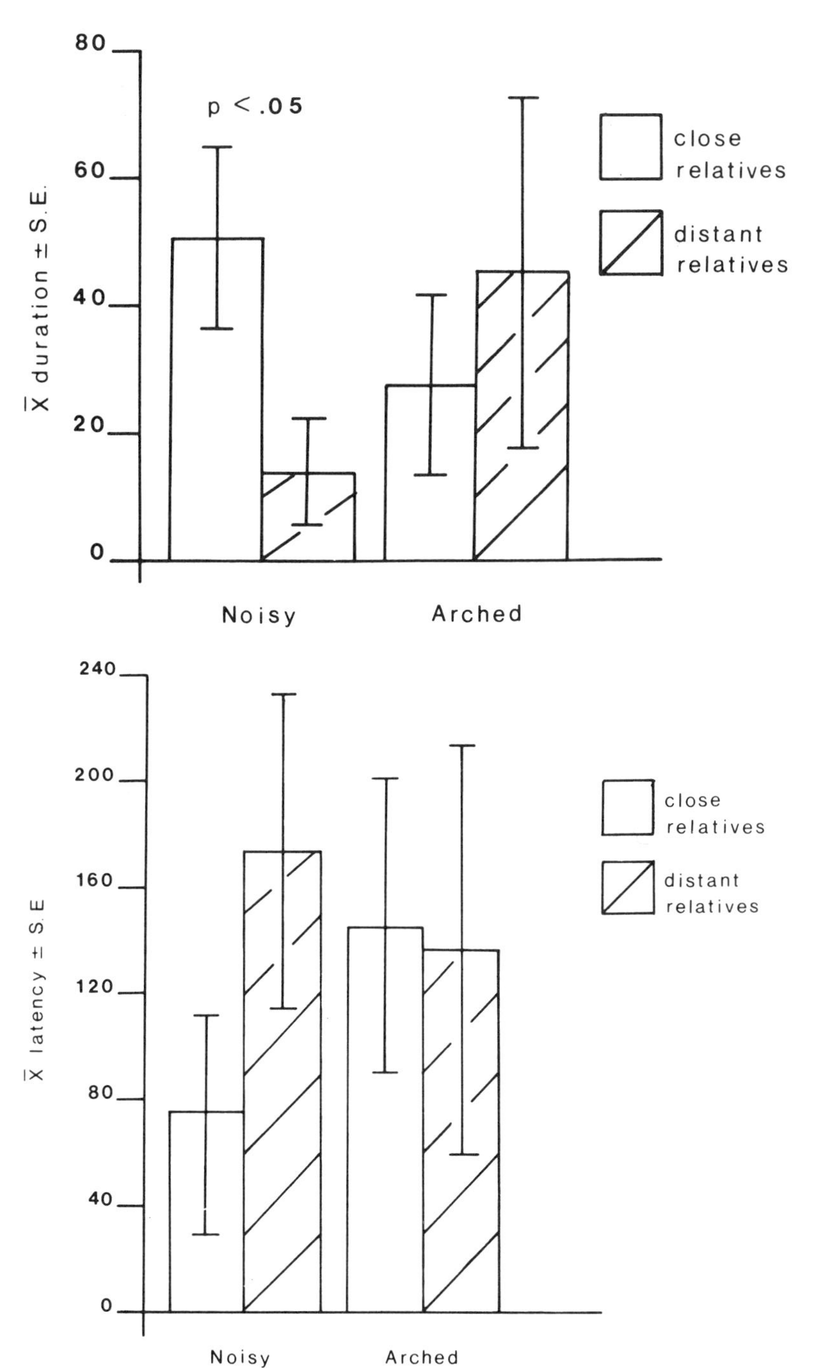

Figure 5.3. a. (upper) Mean duration of responses of females to playbacks of noisy and arched screams from close (non-offspring) and distant relatives. b. (lower) Mean latency to response of females to playbacks of noisy and arched screams from close (non-offspring) and distant relatives.

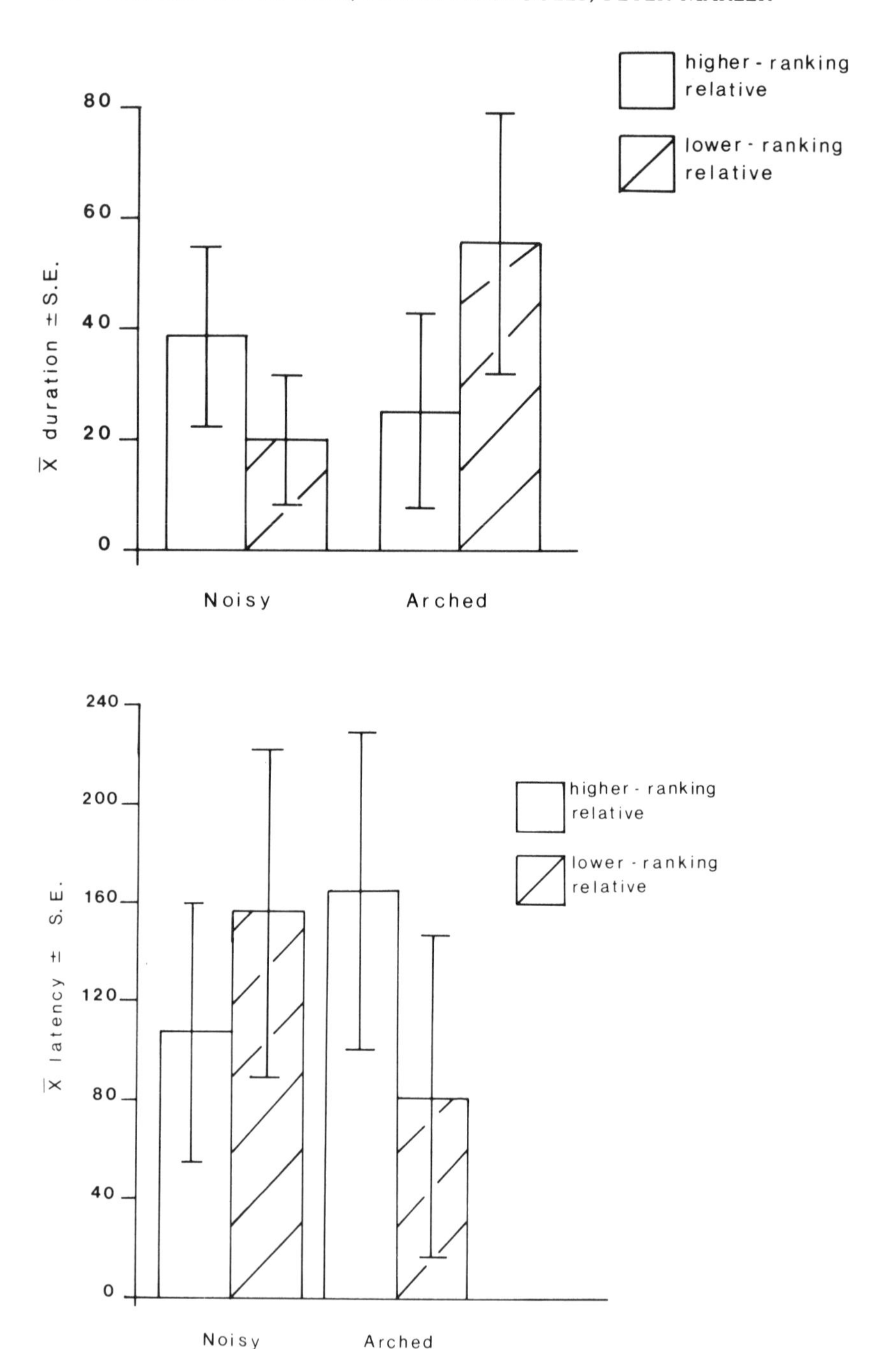

Figure 5.4 a. (upper) Mean duration of responses of females to playbacks of noisy and arched screams from the offspring of higher and lower-ranking relatives. b. (lower) Mean latency to response of females to playbacks of noisy an arched screams from the offspring of higher and lower-ranking relatives.

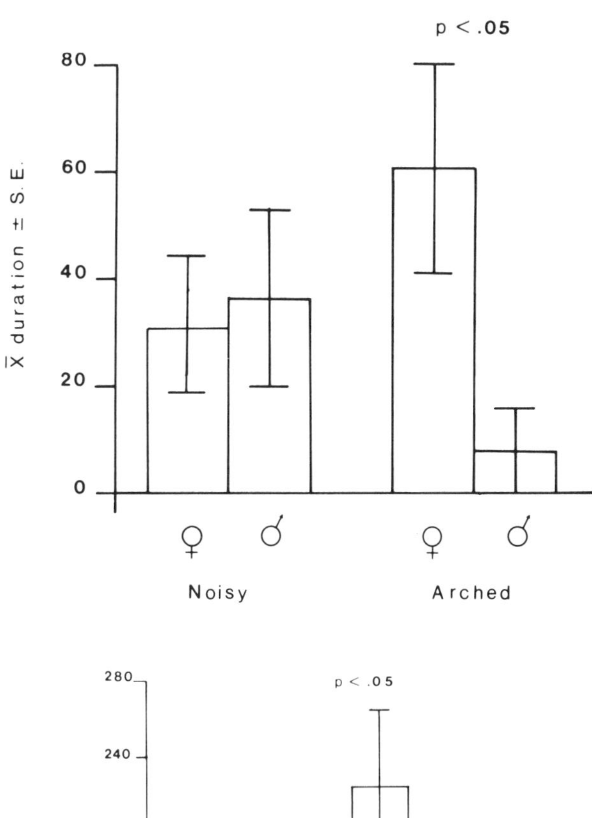

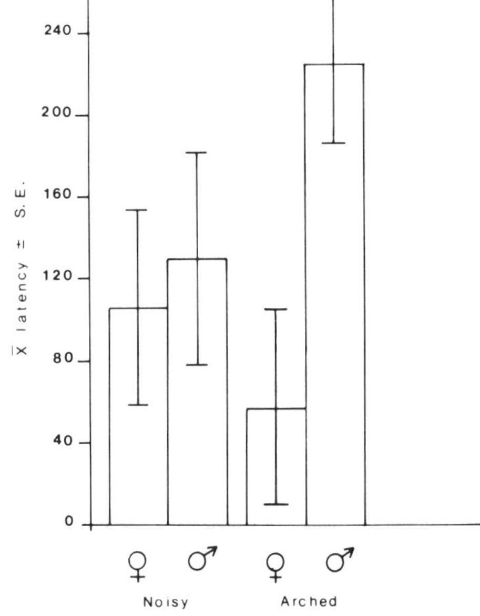

Figure 5.5 a. (upper) Mean duration of responses of females to playbacks of noisy and arched screams from male and female immature relatives. b. (lower) Mean latency to response of females to playbacks of noisy and arched screams from male and female immature relatives.

latency to respond to male and female noisy screams (noisy: $U = 16$, $p > 0.10$; arched: $U = 1$, $p < 0.05$, Figure 5.5b). The latency of females to respond to immature relatives' arched screams was significantly shorter if the caller was female. Thus, female response to the noisy and arched screams of immature relatives varied differentially with the sex of the caller.

Discussion

Playback of scream vocalizations recorded from immature rhesus monkeys consistently elicited a response, not only from their mothers, but also from their other adult female matrilineal relatives. Responses from these latter relatives occurred in the absence of any cues from the immatures' mothers as well as any other contextual features originally available during the agonistic encounter. The pattern of female response to playback of immatures' screams closely resembled that of female aid to juvenile relatives described by Kaplan (1978) and further argues that scream vocalizations alone can function as a principal mechanism in recruitment. Moreover, the patterning of female response to screams of different classes lends additional support to the contention that screams primarily convey information about external objects or events rather than caller internal state (Gouzoules et al., 1984). Screams convey information about: 1) the opponent and the nature of the agonistic event; and 2) the caller and, by extension, its relationship to the recipient.

Cheney and Seyfarth (1980; 1982) suggested, on the basis of playbacks of juvenile vervet monkey screams to adult females, that vervets might classify group members into a hierarchical taxonomy based on matrilineal kinship. Our results strongly support their suggestion that individual vocal recognition extends well beyond the mother-offspring dyad to more complex categorizations of group members. Of particular interest is the fact that such kinship classification can apparently occur solely on the basis of vocal cues, in the absence of any other contextual information. The potential, in primate groups, for precise communication about external events and social interactions through the vocal modality alone has only recently been realized (Seyfarth et al., 1980; Cheney and Seyfarth, 1982; Seyfarth and Cheney, 1982; Gouzoules et al., 1984), although previous studies have also stressed that macaque vocal systems demonstrate the capacity to signal both specificity and detail of context (Green, 1975, p. 98).

Our results thus tend to contradict Altmann's (1981) suggestion that, when nonhuman primates respond in a particular way in an agonistic encounter, they do so on the basis of immediate and observable "physical" (behavioral) properties, and are incapable of responding on the basis of any inferred abstract relational properties among the participants. In contrast, the results of this study, as well as those of other recent studies of vocal communication under natural conditions (see above), suggest that nonhuman primates may in fact routinely process information and communicate about complex relationships. For example, the receiver of a rhesus monkey scream vocalization, which designates an opponent of a particular relationship to the caller, can decode the scream only through knowledge of the *caller's* set of relationships (Gouzoules et al., 1985). This process is analogous to that used by humans to decode terms describing kinship relations, and it implies complex cognitive abilities.

The results of this study also lend support to the suggestion that the different networks of alliance formation observed in primate groups (e.g., aid to juvenile offspring, aid to closely ranked individuals aid to unrelated individuals) serve different functions and are probably the consequence of multiple selective pressures (Walters, 1980; Datta, 1983a). For instance, Kaplan (1978) found that one important pattern of agonistic aiding in rhesus monkeys could be predicted by kin selection hypotheses, whereas others could not—e.g., alliances among unrelated adult females (see also Seyfarth, 1976; 1977; 1980). Similarly, Cheney (1977) suggested that, in baboons, aids to immatures from females who ranked adjacently to the immatures' mothers (and were possibly their kin) may have functioned primarily to perpetuate the existing rank relations among matrilines. Walters (1980) also noted that the intervention of unrelated individuals in agonistic disputes was a function of rank relations and not kin, or even other social relationships.

In the present study, maternal response was significantly greater than was that of other relatives to those screams (noisy) associated with more severe agonistic situations (higher-ranking opponents, risk of physical contact). In response to noisy screams, females also apparently discriminated, not only between offspring and other matrilineal kin, but also within this latter category of kin by the degree of relatedness. In contrast to this pattern, females' responses to immatures' screams associated with an altercation involving a member of a low-ranking matriline (arched screams, with little risk of physical contact) were equally high, regardless of how closely related the caller was. In fact, the response of nonmaternal relatives to immatures' arched screams did not differ significantly from that of mothers, suggesting that rank-related alliance formation is a particularly important pattern of rhesus agonistic aiding among more distantly related individuals. Though the social factors that account for these rank-governed patterns of alliance formation must ultimately relate to individual reproductive success, precisely how they do so remains unclear and is the subject of considerable debate (Kaplan, 1978).

The differential response of females to the noisy and arched screams of offspring and more distant relatives found in our study also corresponds to the results of a recent study of rhesus alliance formation by Datta (1983a). She observed that close relatives, particularly mothers, were more likely than distant relatives to interfere when they were subordinate to the opponent; this is the characteristic context of a noisy scream (Gouzoules et al., 1984). On the other hand, less closely related individuals were likely to come to the aid of monkeys when they were dominant to their opponent, a context we found associated with arched screams. Datta concluded that agonistic aiding is "not a unitary phenomenon: a variety of patterns and functions may coexist," (1983a, p. 290). Additional support for this interpretation of alliance formation is provided by the fact that, in our study, nonmaternal kin responded significantly longer to the arched screams of their immature female, rather than male, relatives. Rank-related agonistic support of female relatives is especially important in rhesus macaques. Pubescent males generally emigrate from their natal group, whereas young females remain and enter the adult female dominance hierarchy. The process of rank acquisition by young females often involves challenges from females of lower-ranking matrilines, and the support of mothers and other female kin is probably crucial (Chapais and Schulman, 1980; Walters, 1980; Datta, 1983b). The support of young females may also confer a long-term benefit on the aider, by helping to maintain her own dominance rank with respect to lower-ranking females in the hierarchy. This benefit may explain the finding from some studies that females may sometimes aid unrelated immatures in rank disputes (e.g., Walters, 1980).

It has been suggested that high-ranking females are more likely than low-ranking ones to support their relatives in agonistic encounters and that high-ranking matrilines should be more cohesive than low-ranking ones (Cheney, 1977; Silk and Boyd, 1983). In this study there was no significant tendency for high-ranking females to respond more strongly than low-ranking ones to screams of either type from their immature relatives. Nor do high-ranking mothers respond more strongly than low-ranking mothers to the screams of their offspring (Gouzoules et al., 1984). We suggest that the maintenance of dominance rank vis-a-vis lower-ranking individuals has considerable importance for females regardless of their relative dominance rank.

CONCLUSION

Griffin (1978) has suggested that studies of animal communication can make valuable contributions to some of the challenging questions about animal cognition that, only a few years ago, seemed hopelessly unanswerable. He has encouraged, for example, the devising of experiments in which communication could be used to explore questions about the thought processes of animals. Field experiments on the vocal communication of vervet and rhesus monkeys have begun to show how these animals categorize and classify the objects and events that constitute their physical and social environment (Cheney and Seyfarth, 1982; Gouzoules et al., 1984). The results of the present study show that it is feasible to study animal communication experimentally, under field conditions, to analyze the nature of the complex social relationships that are the essence of primate societies.

ACKNOWLEDGEMENTS

We thank the Caribbean Primate Research Center for permission to conduct research on Cayo Santiago, and we especially thank R. Rawlins for logistical assistance during the study. Research was supported by NIMH postdoctoral fellowship F32 MH08533 to Harold Gouzoules, NIMH postdoctoral fellowship F32 MH08473 to Sarah Gouzoules, NSF grant BNS 8023423T to Peter Marler and Biomedical Research Grant PHS RR07065-15 to The Rockefeller University.

REFERENCES

Altmann, S. Dominance relationships: the Cheshire cat's grin? *The Behavioral and Brain Sciences* 4:430-431, 1981.

Chapais, B.; Schulman, S.R. An evolutionary model of female dominance relations in primates. *Journal of Theoretical Biology* 82:47-89, 1980.

Cheney, D.L. The acquisition of rank and the development of reciprocal alliances among free-ranging immature baboons. *Behavioral Ecology and Sociobiology* 2:303-318, 1977.

Cheney, D.L.; Seyfarth, R.M. Vocal recognition in free-ranging vervet monkeys. *Animal Behaviour* 28:363-367, 1980.

Cheney, D.L.; Seyfarth, R.M. Recognition of individuals within and between groups of free-ranging vervet monkeys. *American Zoologist* 22:519-529, 1982.

Datta, S.B. Relative power and the acquisition of rank. pp. 93-103 in *Primate Social Relationships: An Integrated Approach.* R.A. Hinde, ed. Boston, Sinauer, 1983a.

Datta, S.B. Relative power and the maintenance of dominance. pp. 103-112 in *Primate Social Relationships: An Integrated Approach.* R.A. Hinde, ed. Boston, Sinauer, 1983b.

Gouzoules, H.; Gouzoules, S.; Marler, P. External reference and affective signaling in mammalian vocal communication. pp. 77-101 in *The Development of Expressive Behavior: Biology-Environment Interactions,* G. Zivin, ed. New York, Academic Press, 1985.

Gouzoules, S.; Gouzoules, H.; Marler, P. Rhesus monkey (*Macaca mulatta*) screams: representational signaling in the recruitment of agonistic aid. *Animal Behaviour* 32:182-193, 1984.

Green, S. Variation of vocal pattern with social situation in the Japanese monkey (*Macaca fuscata*): a field study. pp. 1-102 in *Primate Behavior*, Vol. 4, L.A. Rosenblum, ed. New York, Academic Press, 1975.

Griffin, D.R. Prospects for a cognitive ethology. *The Behavioral and Brain Sciences* 4:527-538, 1978.

Kaplan, J. Patterns of fight interference in free-ranging rhesus monkeys. *American Journal of Physical Anthropology* 47: 279-288, 1977.

Kaplan, J. Fight interference and altruism in rhesus monkeys. *American Journal of Physical Anthropology* 49:241-250, 1978.

Kurland, J. Kin selection in the Japanese monkey. *Contributions to Primatology* No. 12. Basel, Switzerland, S. Karger, 1977.

Lancaster, J.B. *Primate Behavior and the Emergence of Human Culture*. New York, Holt, Rinehart and Winston, 1975.

Lindburg, D.G. The rhesus monkey in northern India: an ecological and behavioral study. pp. 1-106 in *Primate Behavior Vol. 2*, L.A. Rosenblum, ed. New York, Academic Press, 1971.

Massey, A. Agonistic aids and kinship in a group of pigtail macaques. *Behavioral Ecology and Sociobiology* 2:31-40, 1977.

Rowell, T.E. Agonistic noises of the rhesus monkey (*Macaca mulatta*). *Symposia of the Zoological Society, London* 8:91-96, 1962.

Rowell, T.E.; Hinde, R.A. Vocal communication by the rhesus monkey (*Macaca mulatta*). *Proceedings of the Zoological Society, London* 138:279-294, 1962.

Seyfarth, R.M. Social relationships among adult female baboons. *Animal Behaviour* 24:917-938, 1976.

Seyfarth, R.M. A model of social grooming among adult female monkeys. *Journal of Theoretical Biology* 65:671-698, 1977.

Seyfarth, R.M. The distribution of grooming and related behaviors among adult female vervet monkeys. *Animal Behaviour* 28: 798-813, 1980.

Seyfarth, R.M.; Cheney, D.L.; Marler, P. Vervet monkey alarm calls: semantic communication in a free-ranging primate. *Animal Behaviour* 28:1070-1094, 1980a.

Seyfarth, R.M.; Cheney, D.L.; Marler, P. Monkeys' responses to three different alarm calls: evidence of predator classification. *Science* 210:801-803, 1980b.

Seyfarth, R.M.; Cheney, D.L. How monkeys see the world: a review of recent research on East African vervet monkeys. pp. 239-252 in *Primate Communication*. C.T. Snowdon, C.H. Brown, M.R. Petersen, eds. Cambridge, Cambridge University Press, 1982.

Silk, J.B.; Boyd, R. Cooperation, competition, and mate choice in matrilineal macaque groups. pp. 316-347 in *Social Behavior of Female Vertebrates*. S.K. Wasser, ed. New York, Academic Press, 1983.

Smith, W.J. *The Behavior of Communicating*. Cambridge, Harvard University Press, 1977.

de Waal, F.B.M. Exploitative and familiarity-dependent support strategies in a colony of semi-free-living chimpanzees. *Behaviour* 66:268-312, 1978.

de Waal, F.B.M.; van Hooff, J.A.R.A.M.; Netto, W.J. An ethological analysis of types of agonistic interaction in a captive group of Java monkeys (*Macaca fascicularis*). *Primates* 17:257-290, 1976.

de Waal, F.B.M.; van Hooff, J.A.R.A.M. Side-directed communication and agonistic interactions in chimpanzees. *Behaviour* 77:164-198, 1981.
Walters, J. Interventions and the development of dominance relationships in female baboons. *Folia Primatologica* 34:61-89, 1980.
Watanabe, K. Alliance formation in a free-ranging troop of Japanese macaques. *Primates* 20:459-474, 1979.

Illustration 6.1 A young male at the age of emigration.

CHAPTER SIX

Proximate Causes of Male Emigration at Puberty in Rhesus Monkeys

JOHN D. COLVIN

INTRODUCTION

In many animal species there is dispersal of adults away from the natal site (Baker, 1978). In a number of species of birds and higher mammals this dispersal entails emigration from the natal social group to another group, in which breeding may occur (Greenwood, 1980; 1983; Greenwood and Harvey, 1982). This initial emigration has been termed "natal emigration" as further emigrations from non-natal groups may later occur (cf. "natal dispersal" versus "breeding dispersal" Greenwood et al., 1979). In primates, as in other mammals, it is males that emigrate, while females remain with their natal group throughout their lives. Thus, in 25 of 29 species studied, almost all males born within the group transfer to neighboring groups before full adulthood (Wrangham, 1980). The rhesus macaque is no exception to this rule, and on Cayo Santiago male emigration patterns within the population have been documented in detail over the past ten years.

As an aspect of social dynamics, natal emigration is a particularly striking phenomenon because it entails a radical restructuring of all the emigrant's social relationships. To what, then, is this restructuring due? Some authors have approached this question by investigating the ultimate causes of natal emigration (Cheney, 1983; Greenwood, 1980; Harcourt, 1978; Marsh, 1979; Packer, 1979). However, the focus in this chapter will be on proximate causes. Consideration of this issue raises the following questions. Is natal emigration due simply to a decreased interest in all relationships in the natal group, accompanied by an increased interest in potential relationships in a neighboring group? Or, alternatively, does it arise in response to *particular* cues, with the result that most bonds are broken and then rebuilt elsewhere because of some overriding but unrelated propensity? If the latter were found to be true, then it is necessary to ask whether the particular cues involved are related to a single dyadic relationship or to general aspects of the social situation and, further, whether they are related to intra or to intergroup factors, or to some combination of these. Consideration of potential intra and intergroup factors suggests that emigration should on the one hand be stimulated by negative intragroup factors and/or by positive intergroup factors, but on the other hand be delayed or inhibited by positive intragroup factors and/or by negative intergroup factors.

One method of determining which social factors are associated with male emigration in primates is to compare the social relationships of immature males, who do emigrate, with those of immature females, who do not (Seyfarth et al., 1978). A more precise method is to take advantage of individual variation among males themselves in the age of emigration. Because males emigrate at a variety of ages through adolescence and early adulthood (Cheney, 1983; Drickamer and Vessey, 1973; Kawanaka, 1977; Packer, 1977; Sugiyama, 1976), the social situations of males who stay in their troop can be compared with those who leave, at any particular age.

This latter method is used here, drawing upon detailed data from a field study of the natal emigration of immature male rhesus monkeys on Cayo Santiago (Colvin, 1982) as well as population data from this colony. In this chapter, an earlier analysis of these data (Colvin, 1983b) is updated within a comprehensive analysis of these and more recent data from the same colony, in order to elucidate differences between those males who stay in their natal troop and those who emigrate at 3-3½ years, or around the age of puberty.

METHODS

Subjects and Study Troops

Two types of data are reported here, based upon two subject pools. The first type of data is based on the life histories of all males born into the Cayo Santiago population between 1970 and 1977 and concerns their troop memberships over the ten-year period 1973-82. The migration histories of 218 males are considered here.

At the start of the study period in June 1973, these males were part of a population which comprised five troops—Groups F, I, J, L, and M— totalling 275 individuals; by June 1982 the population comprised six troops (in addition, Group O) totalling 1,030 individuals. Each troop contained between one and five matrilineages. Group F underwent fission over the two-year period 1971-73 to form Groups F and M and then divided a second time in 1976 into Groups F and O. The first division of Group F was complex and resulted at various times in four of the six original Group F lineages becoming split between Groups F and M (Chepko-Sade and Sade, 1979). After the final division, Group F contained five lineages and Group M contained three. In the division of Group O from Group F the middle-ranking lineage divided from the remaining four and there was no intra-genealogical splitting. The other three troops had split from parent troops in 1960 (Group I), 1964 (Group J) and 1969 (Group L) and contained three, two, and four lineages repectively. In 1973, the largest troop contained 70 individuals (Group J) and the smallest 35 individuals (Group M); by 1982 Group J had grown to 255 individuals while Group O was the smallest troop with 90 individuals.

Given that there is a distinct birth season on Cayo Santiago, males can be divided into distinct age-cohorts according to year of birth. For the age-cohorts 1970 through 1977 the median cohort size was 7 in Group F (range 3 to 10), 5 in Group I (range 2 to 9), 7½ in Group J (range 1 to 12), 4 in Group L (range 1 to 9), 4½ in Group M (range 1 to 8), and 3 in Group O (range 1 to 4).

The second type of data reported in this chapter is based on much more detailed studies of the social relationships of particular cohorts of males in two troops, Groups I and L. The subjects of these studies were the five males of the 1974 cohort (Group I), the nine males of the 1975 cohort (Group I), and the six males of the 1978 cohort (Group L).

Data Collection

During the ten-year period 1973-82 the population was censused daily, chiefly by Angel Figueroa, so that dates of birth, natal emigration, and death are known accurately to the exact month in almost all cases. From the year 1973 through 1976 these census data were collected under the direction of Donald Sade; from the years 1977 through 1981 under the direction of Richard Rawlins; and during 1982 under the supervision of Matt Kessler.

Observations comprising the more detailed studies of the 20 males were carried out by myself between March 1977 and early October 1978 and by Gerry Tissier during August and September 1981. The 1978 cohort was observed during the summer of 1981 only. The 1975 cohort was observed during the spring and summer of both 1977 and 1978, while the 1974 cohort was observed during the spring and summer of 1977 only.

Focal-animal sampling techniques (Altmann, 1974) were used to record the social interactions of focal males with each other and with other members of either their own or other troops. Each male was observed individually over periods of either eight weeks (summer 1977, spring 1978, summer 1981) or 12 weeks (spring 1977, summer 1978), for an average of 55 minutes per week (122 minutes per week in summer 1981). All interactions involving the male were listed on checksheets chronologically, along with the identities of the initiator and recipient of the interaction and the time they occurred. At 30-second intervals, point (instantaneous) time samples were taken which identified the current behavior of the male, all individuals within 2 meters of the male and the nearest neighbor beyond 2 meters of the male.

In addition *ad libitum* samples were taken whenever any focal male was observed in agonistic interaction, or grooming. All observations were made between 0700 and 1800 hours and all males were observed equally in the five segments into which each morning or afternoon was divided. The order in which males were observed was determined according to an "optimum sampling schedule" (described in Colvin, 1982) which followed as closely as possible a randomized order.

Analysis of population data

From the daily census of the population, monthly summary sheets were constructed listing the current troop status of each individual in the population. In some cases the troop membership of particular males was uncertain because they had rarely been seen or because they were temporarily affiliating with more than one troop. In the majority of cases, however, it was possible to determine accurately from these summary sheets the exact month in which each male emigrated from his natal troop.

Ages of natal emigration were then grouped into categories covering six month intervals (e.g., '3 years' covers 3 yrs. 0 mo. to 3 yrs. 5 mo. and '3½ years' covers 3 yrs. 6 mo. to 3 yrs. 11 mo.), up to the age of 8+ years (i.e., a male still in his natal troop after 8 years of age is scored as emigrating at 8+ years). Males born in 1976 who had not emigrated by the end of the 1982 mating season (census for January 1983) were scored conservatively as emigrating at 7 or 7½ years and males born in 1977 who were still with their natal troops in January 1983 were scored conservatively as emigrating at 6 or 6½ years (the ratios of these estimates—6 versus 6½ yrs., or 7 versus 7½ yrs.—follows the actual ratios observed for males born 1970 through 1975).

For the analysis of the population data, eight independent variables were considered in relation to the dependent measure, age of natal emigration. The first two variables concern a male's natal troop, and the troop to which he first emigrated (his *transfer troop*). Whereas the category of natal troop refers to the six troops already described, the category of transfer troop includes an additional member, *extra-troop*. This refers to those males who became solitary after natal emigration or who associated with subgroups of extra-troop males which themselves did not associate consistently with any of the six main troops. The particular transfer troop to which a male was considered to have emigrated was that in which he was first censused for at least two consecutive months, irrespective of whether he subsequently transferred to other troops within the same season, as did a small number of males.

The next four independent variables concerned measures of male rank within the natal troop. These are: lineage rank, relative rank in cohort at one and three years, and relative rank within cohort at age of natal emigration. The derivation of the first of these measures, lineage rank, was as follows: within troops comprising three or more lineages (Groups F, I, L, and M) these lineages were classified as either high or low-ranking, with intermediate-ranking lineages grouped into a third, medium-ranking class. Males in Group J could be assigned only to a high or to a low-ranking lineage, while males in Group O, the single-lineage troop, could not be ranked in this way. Lineage ranks within troops were assigned on the basis of *ad libitum* observations of interlineage agonistic observations, primarily between the mature females comprising these lineages. These observations come from a number of sources: from observations by Schulman (1980) on Group L in 1977 and 1978, by Chapais (1981) on Group F in 1978, by Datta (1981) on Group J in 1975 and 1976, by Berman (1978) on Group I in 1974 and 1975, and by myself in 1977 and 1978, and by Scanlon in 1979, 1981, and 1982 (Scanlon et al., 1979; Scanlon, in preparation), and by Brereton (personal communication, 1982) on Group M and by myself on Group O in 1977. The stability of the ranks of Group F lineages receives confirmation from observations in 1971, 1973, and 1976 taken from Chepko-Sade and Sade (1979) who also give the lineage ranks of Group M for 1971 and 1973. While the stability of these lineage rankings receives more reliable confirmation in cases such as Group F and Group I where long-term observations have been maintained, the *ad libitum* observations of long-term census workers provide no indication of any instability of lineage ranks in the other cases (Rawlins, personal communication, 1982).

Since the ranks of immature males follow those of their mothers until puberty (Loy and Loy, 1974; Sade, 1967), maternal lineage rank can be taken as a reliable indicator of male lineage rank up until this age. Considering males at one year of age, an average of 36 percent were found within the high-ranking lineage of their troop, 25 percent within the medium-ranking lineages, and 39 percent within the low-ranking lineage (data taken only from males born 1970 through 1975; Group O males excluded from this analysis). Relative lineage sizes within each cohort remained roughly constant within each troop between 1971 and 1976 (Colvin, 1982).

The calculation of relative rank in cohort at one and three years of age (relative rank in cohort at one year of age: RRC1; relative rank in cohort at three years of age: RRC3) was based on estimates of the individual dominance ranks of males at these ages. These estimates were based on measures of maternal dominance rank, taken from the sources already described. As noted above, maternal dominance rank can be taken as a reliable indicator of male dominance rank at one year, given that the ranks of prepubertal immature males follow those of their mothers. A male's relative rank within his cohort was then calculated as the percentage of other males within his cohort to whom he was dominant. For example, a male from a cohort of nine males who ranked fourth within the cohort at one year of age could be assigned an RRC1 of 63 percent (5/8 x 100).

The calculation of relative rank in cohort at natal emigration (RRCE) was based on the assumption that individual dominance ranks at one year of age remained stable as long as males remained within the natal group. Strictly speaking, this assumption does not always hold true—rank reversals between natal males may occur from 3½ years onward (Sade, 1967; Colvin, 1982)—but it was considered safe as a general assumption. A male's relative rank within his cohort at natal emigration was therefore calculated as the percentage of males remaining within his cohort at the beginning of the season of his natal emigration to whom he was dominant.

The final two independent variables concern the troop status of other males emigrating either in the same or in previous years to that of each male. The first measure therefore concerns the identity (in terms of age and relatedness) of *companions* who emigrate to the same transfer troop in the same season as the male concerned. The second measure concerns the identity of males already residing in the transfer troop of a particular male, but who had emigrated from his natal troop in a previous year and could therefore be classified as *potential previous acquaintances.*

Analysis of Focal Data

The analysis of focal data provided a much more detailed set of measures describing the nature of an immature male's relationships with others in his troop. These measures concerned both direct and relative measure of behavioral frequency and duration as well as derived indexes, and are defined as follows:

1. *Time in proximity*—the proportion of time that other individuals were in proximity was estimated from the number of 30-s point time samples such individuals were within 2m of or nearest neighbor to the focal male.
2. *Play*—from the focal and point time data it was possible to subdivide the continuous focal record of behavior into bouts of activity. Whenever a playful interaction was recorded, the following point time sample was recorded as 'in play' and any further playful interaction within the next three 30-s intervals was defined as part of the same play 'bout' and the point time samples included with this period scored as 'time in play'. The proportion of time spent playing was estimated from the number of point time samples scored as 'in play'.
3. *Grooming*—grooming bouts were defined in a similar way, with a

minimum interbout interval of two minutes. However, for grooming bouts this interval could be estimated more accurately than for play bouts, since the initiation and termination of grooming was recorded to the nearest five seconds as part of the focal sampling procedure. The proportion of time spent grooming/being groomed was estimated from the number of point time samples which occurred during grooming bouts. These focal data on grooming also provided a measure of the frequency of grooming bouts.

4. *Observation*—'observation' was an activity directed by immature males primarily toward receptive females during the breeding season; immature males were scored as 'in observation' when they sat observing, often at very close range, or followed receptive females in consortship with a particular adult male, or associating with several adult males (Colvin, 1985). The proportion of time spent in observation was estimated from the number of point time samples scored as 'in observation'.
5. *Copulation*—series mounts with or without ejaculation by the male. Only raw counts were used, since these were based upon both focal and *ad libitum* sampling.
6. *Responsibility for proximity*—from data on approaches (from more than to less than 2m between partners) and leavings (from less than to more than 2m between partners) a 'close proximity index' was calculated for each male in each of his male peer relationships. This index is used to assess the relative roles of two partners in the maintenance of proximity in a relationship and is defined as the 'percentage of approaches' (%*Ap*) made toward a particular partner minus the 'percentage of leavings' (%*L*) made from that partner. If the index is positive, the male is primarily responsible for the maintenance of proximity but if it is negative, the partner is (see also Hinde and Atkinson, 1970).
7. *Interference*—if a male approaches simultaneously more than one partner, then the consequence of this approach may be to 'interfere' with the previous interaction of those partners. In analyzing triadic relationships among immature males, interference rates were calculated as the rate at which a male approached two other males, relative to the time these males spent in close proximity (i.e. within 2m only) to each other.
8. *Aiding*—a more active type of intervention occurs when a third party forms an alliance with the victim of aggression against the aggressor. This is known as 'aiding' (Cheney, 1977; Ali, 1981). Combined data from focal and *ad libitum* sampling provide raw counts for this type of interaction.
9. *Aggression*—'agonistic interactions' were scored whenever a gesture of threat (as defined by Sade, 1967) was followed by a gesture of submission, or a submissive gesture (cower and/or grimace)

appeared alone. Using focal data, frequencies of aggression received in interactions of this type with all individuals other than adult females were expressed relative to the time spent in proximity to the aggressors concerned. In the case of agonistic interactions with adult females (apart from the mother), focal and *ad libitum* data were pooled and then divided by the number of weeks of sampling to give estimates of rates of agonistic interaction. In addition to rates of agonistic interaction, rates of displacement (supplants and avoids) were also calculated on a similar basis.

RESULTS AND DISCUSSION

Intertroop Factors

The proximate causes of natal emigration have been studied in several species of birds and mammals and the comparative evidence so far available indicates that, under certain ecological conditions, intertroop factors may play an important role. For example, in the cooperatively breeding acorn woodpecker, *Melanerpes formicivorus,* in which young remain as helpers in their natal group for up to several years, emigration is delayed because the habitat is saturated and there are no spaces for such birds to set up a breeding territory (Koenig and Pitelka, 1981; Stacey, 1979). Emigration occurs as soon as alternative space becomes available. However, since there is intense competition to occupy such space (Koenig, 1981), dispersal in the more cooperative males typical of this species is delayed longer than in the more competitive females (Koenig et al., 1983).

In some primate species also, there is good evidence that negative intertroop factors may delay natal emigration while positive intertroop factors may stimulate it. A number of authors have noted an attraction during intertroop encounters between young males and unfamiliar females, especially young females in a sexually receptive state (Cheney and Seyfarth, 1977, 1982; Enomoto, 1974; Hamilton et al., 1975; Hrdy, 1977; Marsden, 1968; Packer, 1979) and have suggested that the immediate motivation to transfer may be a strong sexual attraction to unfamiliar individuals of the opposite sex (Packer, 1979; Pusey, 1980). Affiliative interactions with subadult males from other troops may also facilitate transfer into these troops (Boelkins and Wilson, 1972; Cheney and Seyfarth, 1977; Packer, 1979), particularly for those younger males who initially integrate into peripheral all-male subgroups. However, such potentially stimulating factors to natal emigration must be set against negative factors, in particular aggression against young males during intertroop encounters. In some populations of vervets and in Japanese macaques, female aggression during intertroop encounters may be a primary factor to limit male movement (Cheney, 1981; Packer and Pusey,

1979), in part because males who are not yet fully grown are particularly vulnerable to attack and injury by adult females of these species (Cheney, 1983). In baboons and in other populations of vervet monkeys, competition with the resident males of the opposing troop may exert the major inhibitory influence, since female involvement in intertroop encounters is very much reduced (Harrison, 1983; Packer, 1979; Strum, 1982). Whereas these differences may be in part accounted for by species differences in sexual dimorphism in the case of baboons (Packer and Pusey, 1979), they would appear to be mainly due to population differences in patterns of ranging and defense of food resources (Cheney, 1983; Harrison, 1983).

In some populations, however, intertroop encounters are either rare and nonaggressive (Anderson, 1981) or, as in the present case, frequent yet seldom aggressive. As a result, those transfers which have been observed suggest that these are accomplished with little or no opposition (Anderson, 1981). Thus, in some populations it appears that negative intertroop factors do not influence emigration. The absence of any data relating individual differences in age of emigration to individual differences in involvement in aggressive intertroop encounters supports this interpretation. In addition, in view of the fact that Cheney (1981) found no strong individual differences in rates of aggression received during intertroop encounters in vervets, this argument could apply also to those populations and/or species in which encounters are both frequent and aggressive. In these cases, aggression from females appears to play a role in determining to *which* troops a young male transfers.

In a similar fashion, the positive intertroop factors of attraction to unfamiliar individuals and affiliative interactions between strangers probably influence the emigrating male's choice of troop but not his age of emigration. For attraction to unfamiliar but proceptive females or familiarity with subadult males is likely to offset the effects of aggressive encounters with these individuals' troops. As a result, the pattern of male natal intertroop movements may not be random, but affected by the distribution of previous transfers. There is some evidence to support this idea in vervet and rhesus monkeys. In the former case, of 15 migrants who were either known to be natal males or not yet fully grown, 14 (93 percent) transferred to groups where others of their previous group had migrated (Cheney, 1983). In the latter case, it was found that the predominant tendency was for males to transfer to troops of adjacent but higher rank (Figure 6.1). In so doing, they joined potential previous acquaintances from their natal troops (that is, either brothers or nonfraternal relatives) significantly more often than expected by chance (see also Meikle and Vessey, 1981).

The fact that intertroop factors are unlikely to influence directly the age of emigration suggests that *intra*troop conditions might be of

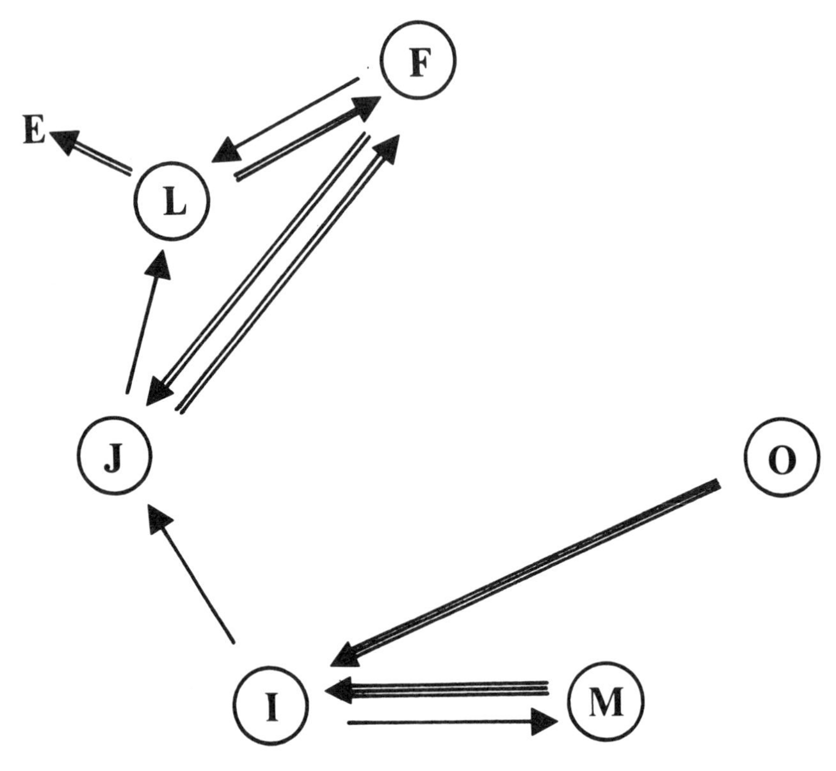

Figure 6.1 Intertroop preferences based on male transfers among the six Cayo Santiago troops, arranged in order of intertroop dominance (F[highest] to O[lowest]; E=extra-troop transfers). The percentage of emigrations (including males who remained with their natal troop) are as follows: 20-29% (single arrow); 30-39% (double arrow); 40-49% (triple arrow).

primary importance. Such conditions, to which the male would be exposed more or less continuously, could determine both the onset and/or the strengthening of a propensity to emigrate, with intertroop factors acting only as the final stimulus to emigrate. As argued above, an overall balance of positive intratroop factors should delay emigration, whereas an overall balance of negative intratroop factors should stimulate emigration.

Intratroop Factors

A major clue to the nature of these factors comes from the analysis of population data in the present study. A number of studies have looked at the effect of maternal rank on the age of male natal emigration (Cheney, 1978; Drickamer and Vessey, 1973; Itoigawa, 1975; Kawanaka, 1977; Sugiyama, 1976). None of these studies has demonstrated particularly striking rank effects, although reanalysis of Kawanaka's (1977) data, taken from observations of a single troop of Japanese macaques, reveals a significantly positive correlation between lineage rank and the mean age of male emigration. Seventeen out of twenty males were from medium or low-ranking lineages, and all but one of them had emigrated by six years of age. The three males from the high-ranking lineage, on the other hand, emigrated at five, eight, and eleven years respectively. This supports the hypothesis of Sade (1980) that males from high-ranking lineages tend to remain longer with their natal troops than do males from low-ranking lineages.

For the Cayo Santiago population 1973-82, the overall pattern was for most males to emigrate between 3 and 5½ years of age, with a median age of natal emigration of 4½ years. Occasionally, males emigrated while still juvenile (2½ years old), while 43 out of 218 (20 percent) remained in their natal troops into adulthood (6 years old) (Figure 6.2). Of the latter, one was in his twelfth year as a natal male and another in his fifteenth year as of 1983, and both are alpha males of their respective troops. Such findings concerning the range of emigration ages parallel those from Cheney's (1983) study of emigration in vervet monkeys. In this study the movements of 25 males into and from three study groups were documented over a five-year period. Of these males, none whose ages were known remained in their natal group beyond seven years of age, while the youngest male known to have transferred was 2½ years of age.

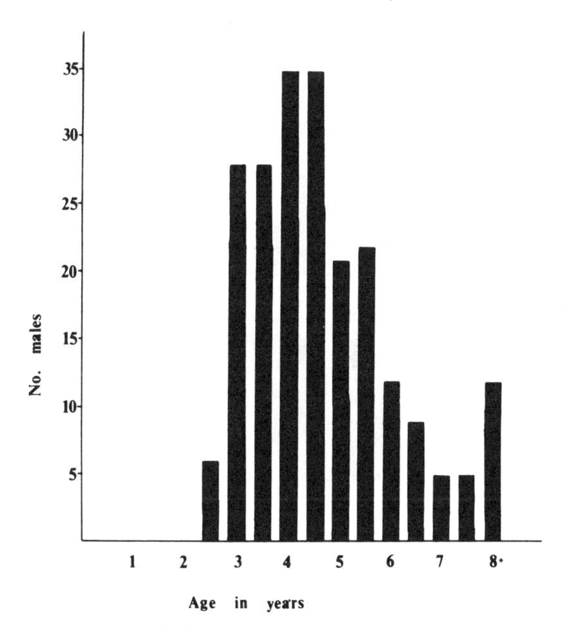

Figure 6.2 Frequency distribution of natal emigration ages for males born into all troops (including Group O) between 1970 and 1977 and surviving until emigration. The median age of natal emigration is 4½ years old.

Beneath this general pattern, in the present study the effects of lineage rank showed much more strongly than in the studies mentioned above. Within the five troops whose lineages could be ranked, there were significant differences between high, medium, and low-ranking lineages in the age of natal emigration (F 2,201 = 44.85, $p < 0.001$). The median ages of natal emigration from high, medium, and low-ranking lineages were 5½, 4½ and 3½ years of age, respectively. All but one of the 15 males who emigrated at 7 years or later were from high-ranking lineages, whereas it was only low-ranking males who emigrated as early as 2½ years (Figure 6.3). Males from the single-lineage troop, Group O, behaved even more extremely than the high-ranking males of the other troops, delaying emigration by a median of seven years (Figure 6.3).

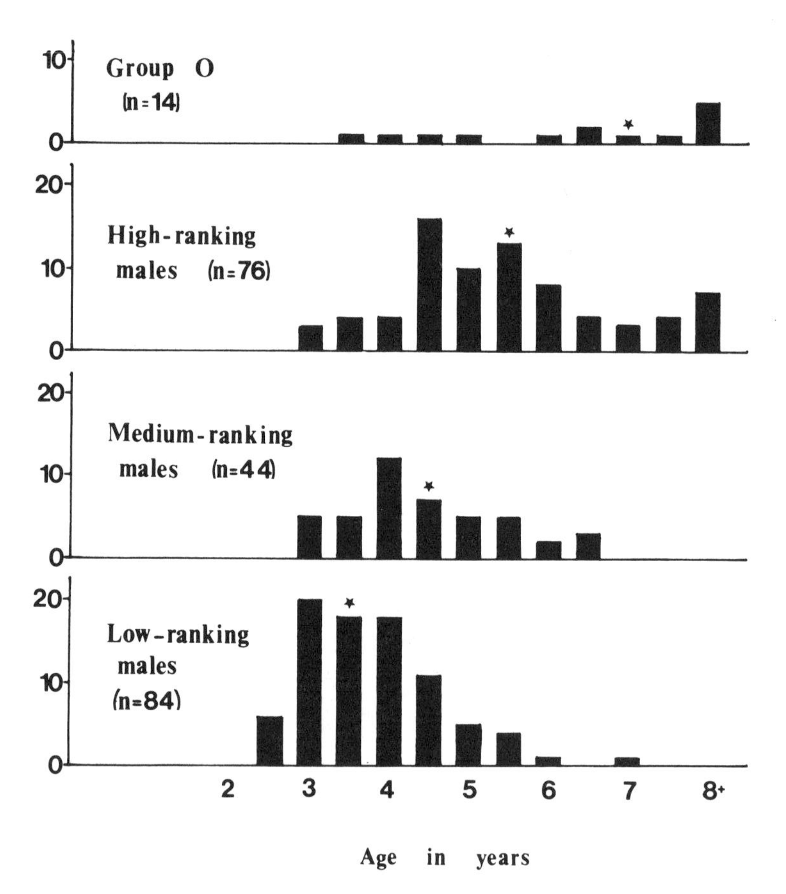

Figure 6.3 Frequency distribution of natal emigration ages in relation to lineage rank. Asterisks mark the median age of natal emigration, which increases with lineage rank. Males from the single-lineage troop, Group O, are shown separately.

The effect of lineage rank on the age of natal emigration could be transmitted through any set of interactions or relationships whose patterning is influenced by rank. At one extreme, this could involve a single, key relationship of a male. At the other extreme, it could involve general aspects of a male's interactions with all troop members; for example, overall time spent in proximity to others or overall rates of aggression received from others. Between these two extremes a male could be affected by various subsets of his relationships, of which three will be discussed here: the subset of a male's relationships with mature females, the subset of his relationship with adult males, and the subset of his relationships with his immature male peers.

General Aspects of the Social Situation

From the study of focal males in Group I there was little evidence to support the notion that rank-related effects might be transmitted through the most general aspects of a male's interactions. Thus low-ranking males between 2 and 3½ years of age spent no less time than did high-ranking males in proximity to others or being groomed by others, nor did they receive aggression from others at a higher rate. However, low-ranking males were less likely to be related to their aggressors than were high-ranking males, and this may have affected the psychological value of the aggression received.

Relationship with Adult Females

Considering the subset of an immature male's relationships with adult females, there is no evidence in rhesus monkeys that low-ranking immature males receive aggression from adult females at higher rates than do high-ranking immature males (Colvin, 1985), as has been reported among captive bonnet macaques (Silk et al., 1981). There may, on the other hand, be advantages to be gained by high-ranking males who delay emigration, which are not available to low-ranking males. In particular, the focal rhesus study shows that, while all males who remained with their natal troop at 3½ years spent considerable time during the breeding season observing at close range consorting females, high-ranking males devoted more time to this activity than did low-ranking males (Colvin, 1985). The benefits to be gained from this could include the observational learning of a wide range of social skills relevant to mating. In addition, at both 3½ and 4½ years, only high-ranking immature males copulated with receptive females (Colvin, 1982). If, as a result, these males were to father offspring in the natal troop, this might then delay their emigration even further. For in chacma baboons (*Papio ursinus*) it is known that males having infant offspring in the troop emigrate significantly less than expected if paternity were not a factor (Hamilton and Busse, 1980).

By contrast, lack of mating success in the natal troop, considered by a number of authors as a key factor in causing emigration (Altmann and Altmann, 1970; Marsh, 1979; Packer, 1979), is likely to take on increasing importance as males grow older. Males may be affected either because access to receptive females is frustrated or, in nonseasonal breeders, by finding most females impregnable; this latter factor is known to stimulate breeding emigration by adult males in yellow baboons (Rasmussen, 1979). In addition, lack of mating success may lead to increased intrasexual competition, in turn causing emigration (Dunbar, 1979).

Relationships with Adult Males

The effects of relationships with adult males on natal emigration have been stressed by Dittus (1979). In langurs and patas monkeys, immature males are rejected by the adult male from the natal troop at an early age (Hall, 1967; Poirier, 1969; but see Gutstein, 1978). In other species, however, it is the consequences of weak or inhibited relationships with adult males which appear to be more important. Harcourt and Stewart (1981) have shown that for immature male gorillas nearing maturity, lack of a close relationship with the current leading male when they were younger leads to gradual peripheralisation and dispersal, rather than to continued integration and potential takeover of group leadership. In rare cases in bonnet macaques, lack of strong bonds with adult males may also induce peripheralization and emigration in subadult males (Simonds, 1973).

For the rhesus macaques on Cayo Santiago, relationships with adult males may have less importance than relationships with male peers, at least in relation to emigration at younger ages. Thus, on the one hand, all immature males regardless of rank had at least one strong bond with an adult male, who groomed them frequently and on occasion might provide aid. On the other hand, while immature males received the most aggression from peer and older males, there were no rank differences among the immature males in the amounts of aggression received. Furthermore, for males emigrating at puberty (3-3½ years) there was no direct competition with adult males for access to receptive females; indeed, immature males do not begin seriously to establish a position for themselves in the adult male hierarchy until they reach the age of four years (Colvin, 1982), although when they do, previous relationships with adult males may contribute to the new rank they achieve. It is here that high-ranking males might be at an advantage, because there is a significant positive correlation between immature male rank and the rank of the 'central' adult males with whom strong bonds are formed.

Relationships with Male Peers

In certain species of birds and mammals, it has been observed that competition among peers is associated with early dispersal, whereas social cohesion among peers is associated with delayed dispersal. In the acorn woodpecker it is the more competitive female sex which disperses earlier than the more cooperative male sex, as described above (Koenig et al., 1983). In the comparison of three species of the mammalian genus *Marmota*, there is an association between the age of natal dispersal and the ontogeny of aggressive interactions (Barash, 1974). Thus the highly social Olympic marmot shows a delay in the ontogeny of aggressive behavior as compared to the less social yellow-bellied marmot, which in turn shows a delay in comparison to the solitary woodchuck. Olympic marmots typically disperse at approximately two years of age, yellow-bellied marmots at approximately one year of age, and woodchucks at weaning. A strikingly similar relation is found in certain canids, where delayed dispersal is associated not only with lower levels of aggressive interaction among juveniles but also with heightened sociality. Thus the more social canids, such as wolves, play more with their litter mates as well as fight less very early in life than do the less social canids, such as coyotes and foxes (Bekoff, 1977).

In one study, a more subtle distinction than that between competition and cooperation has been associated with variation in the probability of emigration. Thus Russell (1983) has provided evidence that it is the patterning of affiliation, in the form of grooming reciprocity, that may influence emigration. In a study of peer relationships among adult female coatis (*Nasua narica*), he found that the maintenace of these relationships was dependent upon an equitable patterning, with inequitable patterning giving rise to emigration.

In primate studies consideration of the relation between peer relationships and dispersal has focused on the role of peers at the time of emigration itself rather than on the quality of peer relationships prior to this (Boelkins and Wilson, 1972; Cheney and Seyfarth, 1977; Drickamer and Vessey, 1973; Itoigawa, 1975; Sugiyama, 1976). Yet the comparative evidence would suggest that such relationships could play an important part in influencing the age of natal emigration and, in particular, that affiliative relationships with peers early in development could act to delay emigration. In the present study this leads us to a consideration of the relation between a male's relative rank in his cohort (RRC) and his age of natal emigration. Although a male's peers need not necessarily be restricted to those of his age cohort, but, depending on demographic factors, could cover a varying age range (Altmann and Altmann, 1979), this restriction was applied in the present case in view of the large size of the cohorts concerned.

Up until the age of four years, most emigrating males (71 out of 97, or 73 percent) had an RRCE of less than 50 percent. The RRCE of most males (56 out of 78, or 72 percent) emigrating between 4½ and 5½ years was less than 99 percent. From six years of age and older most males (31 out of 43, or 72 percent) who emigrated were single or had an RRCE of 100 percent (Figure 6.4). This suggests that up until four years of age natal emigration is associated with overall subordinacy in peer relationship (RRC < 50 percent), but that at older ages this relation no longer holds.

The importance of the relation between overall subordinacy in peer relationships and early emigration also emerges in the strong relation which pertains between age of natal emigration and RRC at both one and three years of age. Not only is it possible to predict at what age a male will emigrate from his RRC at three years, but an even stronger prediction is possible from a knowledge of his RRC at one year (Table 6.1). Thus the median age of natal emigration for males with an RRC at one or three years of 100 percent was six years. Such males delayed emigration for considerably longer than those with an RRC1 or RRC3 less than 100 percent but greater than or equal to 50 percent, their median age of natal emigration being 4½ years. Males with an RRC1 or RRC3 less than 50 percent but greater than 0 percent delayed natal emigration by a median of four years, whereas males with an RRC1 or RRC3 of 0 percent delayed the least, their median age of natal emigration being 3½ years (see also Figure 6.5). This effect is both clearer than that described for lineage rank alone and also more precise because it is related specifically to relationships within the peer network.

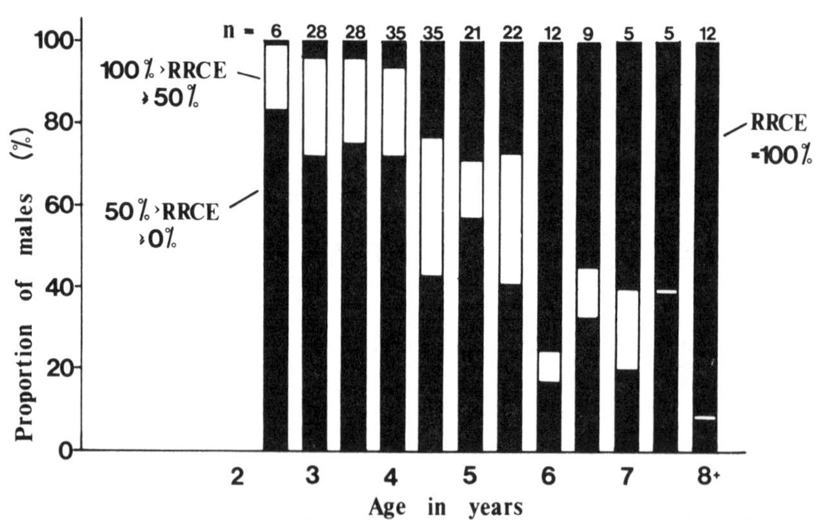

Figure 6.4 Frequency distribution of relative ranks in the cohort at emigration (RRCE) in relation to age of natal emigration. For those males emigrating at each age (sample size = n), the relative proportions of three categories of RRCE are shown: RRCE=100% (+singletons) (upper solid bar); 100% > RRCE ≥ 50% (open bar); 50% > RRCE ≥ 0% (lower solid bar).

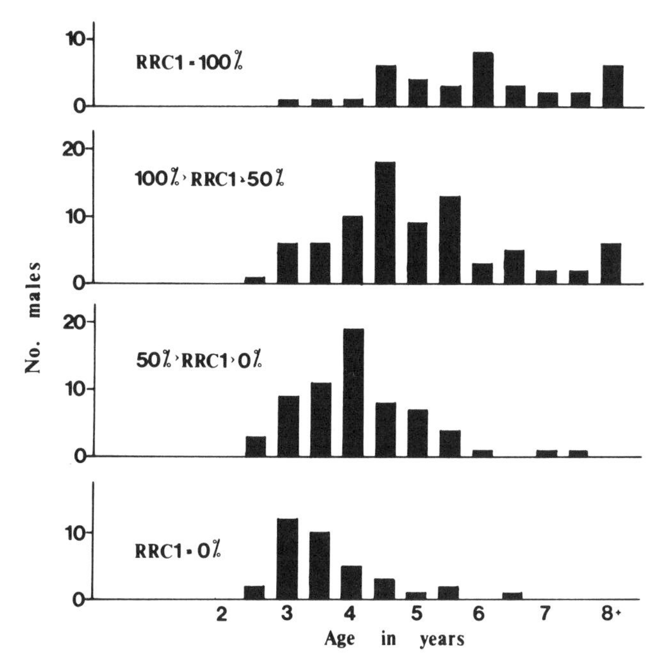

Figure 6.5 Frequency distribution of male natal emigration ages in relation to their relative rank in the cohort at one year of age (RRC1). Males with the lowest RRC1 emigrate at the youngest ages and the age of emigration increases with the RRC1.

Table 6.1. Relation between Relative Rank in Cohort and Median Age of Natal Emigration for Males at the Age of One Year, Three Years and at Emigration.

	Relative rank in cohort (RRC)			
Type of RRC	100%	99-50%	49-1%	0%
RRC1	6 z=3.24** (n=37,81)	4½ z=4.30*** (n=81,64)	4 z=2.51* (n=64,36)	3½
RRC3	6 z=3.90*** (n=43,73)	4½ z=3.69*** (n=73,57)	4 z=1.84+ (n=57,39)	3½
RRCE	6 z=5.80*** (n=55,47)	4½ z=1.85+ (n=47,48)	4 z=1.42 (n=48,68)	4

Singletons (males with no peers in their cohort) are included within the category RRC=100%. Differences between medians are compared using the Mann-Whitney U test, with the significance of U being calculated from the value of z. Sample sizes are given in brackets beneath the z-value. ***=p< 0.001; **=p< 0.002; *=p< 0.02; +=p< 0.10.

How might relative overall dominance or subordinancy in peer relationships influence emigration? The study of the focal males in Groups I and L points to at least four factors which might be involved:

1. Considering only those peer relationships with an aggressive component, at both 3 and 3½ years low-ranking males received higher rates of aggression from their peers than did high-ranking males. Thus, as shown in Figure 6.6, the rate of aggression received increases as relative rank decreases (data are shown for two cohorts only).
2. Considering all peer relationships within a network, low-ranking males at three years showed the least responsibility for maintaining close proximity to their peers out of deference to their higher-ranking partners. Thus the index of responsibility for close proximity moves from positive median values, indicating greater responsibility than the partner, to negative median values, indicating less responsibility than the partner, as relative rank decreases (Figure 6.7).
3. Related to this, low-ranking males were the most inhibited from interfering between males of other peer dyads and thus least able

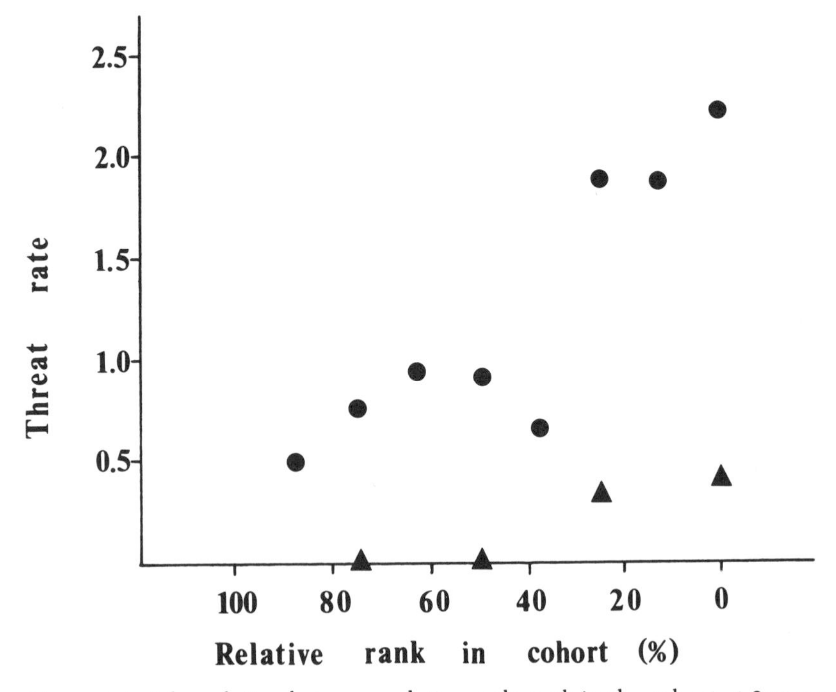

Figure 6.6 The relation between relative male rank in the cohort at 3 years (RRC3) and aggression received in male peer relationships at 3 (solid triangle, 1974 cohort; solid circle, 1975 cohort) and 3½ years (solid square, 1975 cohort). The rate of aggression received increases as the RRC3 decreases.

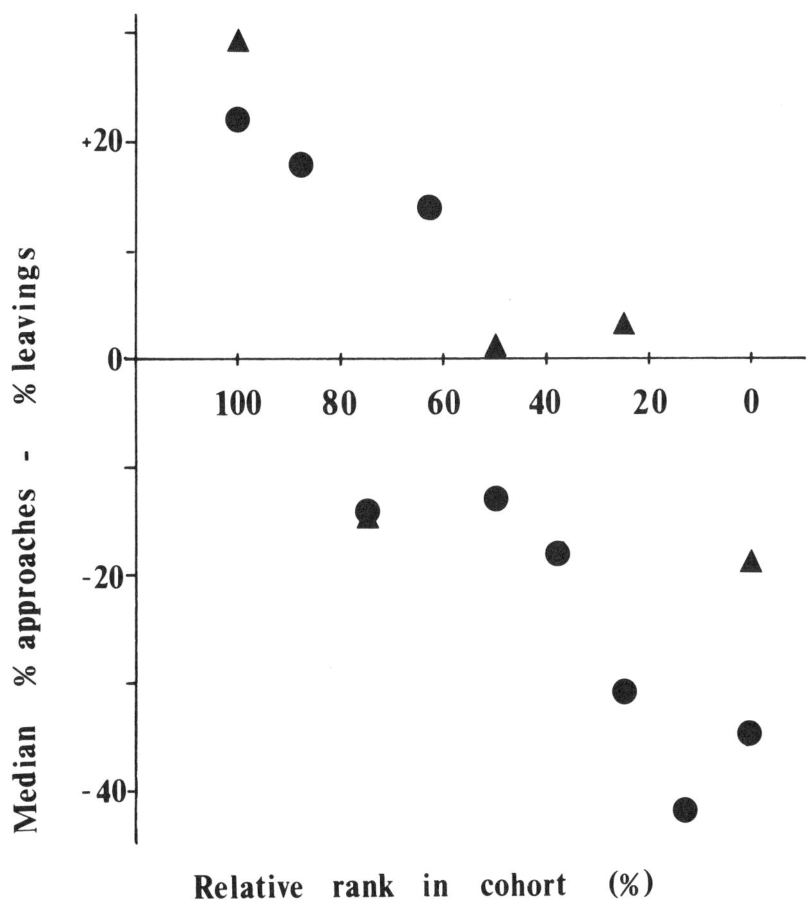

Figure 6.7 The relation between relative male rank in the cohort at 3 years (RRC3) and responsibility for close proximity in male peer relationships at 3 years (solid triangle, 1974 cohort; solid circle, 1975 cohort). Responsibility for close proximity moves from positive to negative median values as the RRC3 decreases.

to protect their most valued peer relationships. As shown in Figure 6.8, data from the 1975 cohort, but not the 1974 cohort, are supportive of this observation.

4. Finally, taking a role less often as a dominant than as a subordinate partner in the most important peer relationships may also be associated with early emigration. Thus, low-ranking males played a proportionately lower role as a dominant than as a subordinate partner in their key peer relationships, whereas high-ranking males played a proportionally higher role. Here these findings are supported by data from an additional cohort, the 1978 cohort from Group L on Cayo Santiago (Figure 6.9).

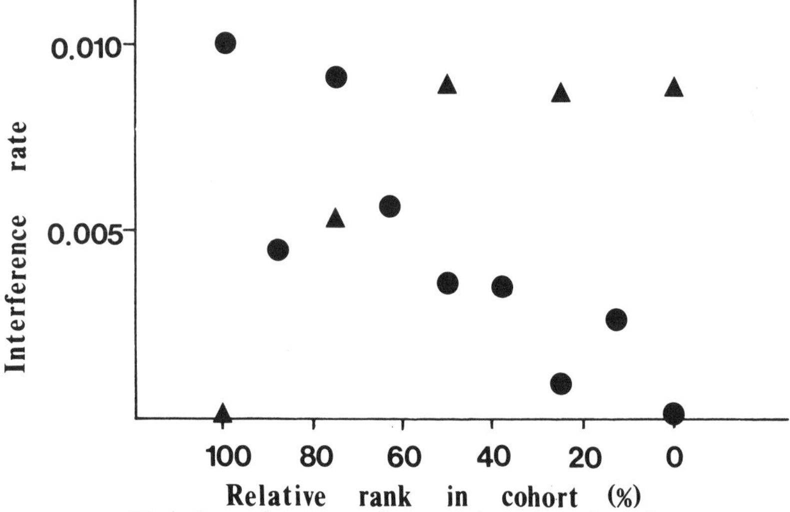

Figure 6.8 The relation between relative male rank in the cohort at 3 years (RRC3) and the rate at which males interfered between males of other peer dyads (solid triangle, 1974 cohort; solid circle, 1975 cohort). The rate of interference decreases as the RRC3 decreases, in the 1975 cohort but not in the 1974 cohort.

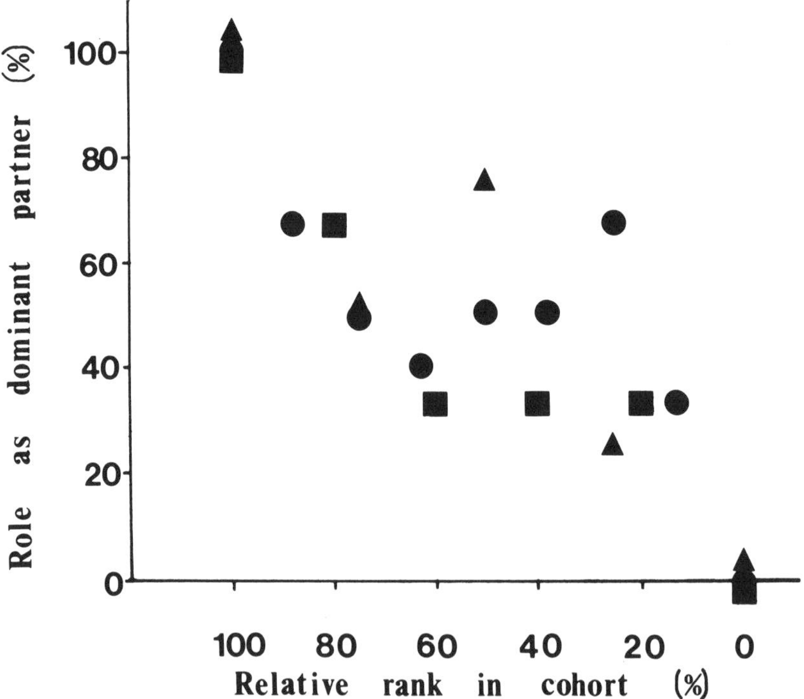

Figure 6.9 The relation between relative male rank in the cohort at 3 years (RRC3) and the relative role as dominant or subordinate partner in key male peer relationships at 3 (solid triangle, 1974 cohort; solid circle, 1975 cohort) and 3½ years (solid square, 1978 cohort). Relative role as a dominant partner decreases as the RRC3 decreases.

Taken together, these four factors might create in the low-ranking male a weakened sense of affiliation toward his peers, encouraging him to emigrate early. Conversely, high-ranking males might be expected to delay emigration due to the absence of weakening or inhibiting influences in their peer relationships.

Single Relationships

Finally, we turn to the possibility that the effects of lineage rank on age of male emigration could be transmitted through a single, key relationship or, alternatively, that such a relationship could influence natal emigration in other ways not directly related to lineage rank. Here two key relationships—the sibling relationship and the mother-son relationship—will be considered briefly.

A number of authors have suggested that males may emigrate in the company of their brothers, if they have them (Drickamer and Vessey, 1973; Meikle and Vessey, 1981). However, there is little evidence for any relation between the quality of the sibling relationship and age of natal emigration. On Cayo Santiago, all immature males, irrespective of rank, who have brothers have strong friendships with at least one of these (Colvin, 1983a). Yet when they emigrate there is no evidence that they do so in the company of their brothers. Thus only 19 percent of males who could have accompanied older or younger brothers did so, while almost half the males who emigrated alone had brothers who were potential companions.

Finally, the relation between the nature of the mother-son relationship and emigration must be considered. Data on orphaning in vervet monkeys suggest that this may lead to early emigration (Cheney et al., 1981), although there is no evidence to support this in Japanese monkeys (Hasegawa and Hiraiwa, 1980) or in rhesus monkeys (Drickamer and Vessey, 1973). On the other hand, Itoigawa (1975) found that often it was possible to predict when young male Japanese monkeys would leave their natal troop on the basis of the degree of association between mother and son at one year of age and changes in association thereafter.

In the study of focal rhesus males, a similar relation was found (Colvin, 1982). Although based on very small samples, data from this study suggest that the nature of the changes in the mother-son relationship between 2 and 3 years may be predictive of emigration at 3½ years. Furthermore, while at later ages differences in the relationship with the mother between males who stay and those who emigrate are less clear since the expression of mother-son bonding declines from four years onward, some very high-ranking females may continue to influence their sons' social integration by assisting their integration into the adult male hierarchy (Koford, 1963; Chapais, 1983).

Summary

It is clearly not the case that a uniform decrease of interest across all relationships in the natal troop precedes male emigration. Rather, it would appear that certain key social factors influence emigration, although we are not yet able to specify precise casual links, or to say whether emigration is preceded by psychological changes only, or by physiological or physical intermediaries, such as stress or weight loss. These key factors are associated with the young male's social situation vis-a-vis his male peer network. Greater overall dominance within this network delays emigration whereas greater overall subordinacy encourages emigration. In addition, various other aspects of troop structure may influence emigration at puberty. These include aspects of the subset of relationships with mature females and in particular aspects of the mother-son relationship before puberty. In contrast, general aspects of the social situation, relationships with adult males, and sibling relationships appear not to influence emigration at puberty, although certain of these relationships may have greater influence at later ages. Finally, whereas it is primarily *intra*troop factors which influence the age of emigration, the available evidence suggests that *inter*troop factors may influence the emigrating male's choice of troop.

CONCLUSIONS

1. At the Cayo Santiago colony, there is considerable variation in the age of natal emigration, with most males leaving in their fourth, fifth, or sixth years.

2. The key social factors influencing emigration at puberty are those associated with a male's position in his male peer network. Greater overall dominance in this network delays emigration whereas greater overall subordinacy encourages emigration.

3. Other aspects of troop structure which may influence emigration at puberty include a male's relationships with mature females and in particular with his mother.

4. General aspects of the social situation, a male's relationships with adult males and sibling relationships do not appear to influence emigration at puberty, although certain of these relationships may have greater influence at later ages.

5. Whereas it is primarily intratroop factors which influence the age of emigration, the available evidence suggests that *inter*troop factors may influence the emigrating male's choice of transfer troop.

ACKNOWLEDGEMENTS

I am grateful to the Directors of the Caribbean Primate Research Center for permission to collect focal data on Cayo Santiago, to Donald Sade and Richard Rawlins to analyze the population data, and especially to Richard Rawlins for advice and hospitality during the field project. I thank Gerry Tissier for assistance with data collection, Donald Sade for discussion, and Robert Hinde, Jill Hooley, and Gerry Tissier for comments on earlier drafts of this paper. This research was supported by grants from the Medical Research Council and the Nuffield Foundation of Great Britain. Collection of census data was supported in part by grants from the National Science Foundation to D.S. Sade through Northwestern University and the University of Puerto Rico. National Institutes of Health contract RR-7-2115 and grant RR-01293 to the University of Puerto Rico funded Cayo Santiago for the duration of this study.

REFERENCES

Ali, R. The ecology and social behaviour of the Agastyamalai bonnet macaque (*Macaca radiata diluta*). Ph.D. Thesis, University of Bristol, 1981.

Altmann, J. Observational study of behaviour: sampling methods. *Behaviour* 49:227-265, 1974.

Altmann, S.A.; Altmann, J:. *Baboon Ecology: African Field Research.* Chicago, University of Chicago Press, 1970.

Altmann, S.A.; Altmann, J. Demographic constraints on behavior and social organization. pp. 47-63 in *Primate Ecology and Human Origins.* I.S. Bernstein; E.O. Smith, eds. New York, Garland Publishing, 1979.

Anderson, C.M. Intertroop relations of Chacma baboons (*Papio ursinus*). *International Journal of Primatology* 2:285-310, 1981.

Baker, R.R. *The Evolutionary Ecology of Animal Migration.* London, Hodder & Stoughton, 1978.

Barash, D.P. The evolution of marmot societies: a general theory. *Science* 185:415-420, 1974.

Bekoff, M. Mammalian dispersal and the ontogeny of individual behavioral phenotypes. *American Naturalist* 111:715-732, 1977.

Berman, C.M. Social relationships among free-ranging infant rhesus monkeys. Ph.D. Thesis, University of Cambridge, 1978.

Boelkins, R.C.; Wilson, A.P. Intergroup social dynamics of the Cayo Santiago rhesus with special reference to changes in group membership by males. *Primates* 13:125-140, 1972.

Chapais, B. The adaptiveness of social relationships among adult rhesus monkeys. Ph.D. Thesis, University of Cambridge, 1981.

Chapais, B. Matriline membership and male rhesus reaching high ranks in their natal troops. pp. 171-175 in *Primate Social Relationships: An Integrated Approach.* R.A. Hinde, ed. Oxford, Blackwell Scientific Publications, 1983.

Cheney, D.L. The acquisition of rank and the development of reciprocal alliances among free-ranging immature baboons. *Behavioral Ecology and Sociobiology* 2:303-318, 1977.
Cheney, D.L. Interactions of immature male and female baboons with adult females. *Animal Behaviour* 26:389-408, 1978.
Cheney, D.L. Intergroup encounters among free-ranging vervet monkeys. *Folia Primatologica* 35:124-146, 1981.
Cheney, D.L. Proximate and ultimate factors related to the distribution of male migration. pp. 241-249 in *Primate Social Relationships: An Integrated Approach.* R.A. Hinde, ed. Oxford, Blackwell Scientific Publications, 1983.
Cheney, D.L.; Lee, P.C.; Seyfarth, R.M. Behavioral correlates of non-random mortality among free-ranging female vervet monkeys. *Behavioral Ecology and Sociobiology* 9:153-161, 1981.
Cheney, D.L.; Seyfarth, R.M. Behaviour of adult and immature male baboons during inter-group encounters. *Nature* 269:404-406, 1977.
Cheney, D.L.; Seyfarth, R.M. Recognition of individuals within and between free-ranging groups of vervet monkeys. *American Zoologist* 22:519-529, 1982.
Chepko-Sade, B.D.; Sade, D.S. Patterns of group splitting within matrilineal kinship groups. *Behavioral Ecology and Sociobiology* 5:67-86, 1979.
Colvin, J.D. Social integration and emigration of immature male rhesus monkeys. Ph.D. Thesis, University of Cambridge, 1982.
Colvin, J.D. Description of sibling and peer relationships among immature male rhesus monkeys. pp. 20-27 in *Primate Social Relationships: An Integrated Approach.* R.A. Hinde, ed. Oxford, Blackwell Scientific Publications, 1983a.
Colvin, J.D. Influences of the social situation on male migration. pp. 160-170 in *Primate Social Relationships: An Integrated Approach.* R.A. Hinde, ed. Oxford, Blackwell Scientific Publications, 1983b.
Colvin, J.D. Breeding season relationships of immature male rhesus monkeys with females I. Individual differences and constraints on partner choice. *International Journal of Primatology* 6:261-287, 1985.
Datta, S.B. Dynamics of dominance among rhesus females. Ph.D. Thesis, University of Cambridge, 1981.
Dittus, W.P.J. The evolution of behaviors regulating density and age-specific sex ratios in a primate population. *Behaviour* 59:265-302, 1979.
Drickamer, L.C.; Vessey, S.H. Group changing in free-ranging male rhesus monkeys. *Primates* 14:359-368, 1973.
Dunbar, R.I.M. Population demography, social organization, and mating strategies. pp. 65-88 in *Primate Ecology and Human Origins.* I.S. Bernstein; E.O. Smith, eds. New York, Garland Publishing, 1979.
Enomoto, T. The sexual behavior of Japanese monkeys. *Journal of Human Evolution* 3:351-372, 1974.
Greenwood, P.J. Mating systems, philopatry and dispersal in birds and mammals. *Animal Behaviour* 28:1140-1162, 1980.
Greenwood, P.J. Mating systems and the evolutionary consequences of dispersal. pp. 115-131 in *The Ecology of Animal Movement.* I.R. Swingland; P.J. Greenwood, eds. Oxford, Clarendon Press, 1983.

Greenwood, P.J.: Harvey, P.H. The natal and breeding dispersal of birds. *Annual Review of Ecology and Systematics* 13:1-21, 1982.

Greenwood, P.J.; Harvey, P.H.; Perrins, C.M. The role of dispersal in the great tit (*Parus major*): the causes, consequences and heritability of natal dispersal. *Journal of Animal Ecology* 48:123-142, 1979.

Gutstein, J. Behavioural correlates of male dispersal in patas monkeys. pp. 79-82 in *Recent Advances in Primatology, Vol. I.* D.J. Chivers; J. Herbert, eds. London, Academic Press, 1978.

Hall, K.R.L. Social interactions of the adult male and adult females of a patas monkey group. pp. 261-280 in *Social Communication Among Primates.* S.A. Altmann, ed. Chicago, University of Chicago Press, 1967.

Hamilton, W.J.; Buskirk, R.E.; Buskirk, W.H. Chacma baboon tactics during inter-troop encounters. *Journal of Mammalogy* 56:857-870, 1975.

Hamilton, W.J.; Busse, C. Male transfer and offspring protection in chacma baboons. *Anthropologia Contemporanea* 3:207, 1980.

Harcourt, A.H. Strategies of emigration and transfer by primates, with particular reference to gorillas. *Zeitschrift Fur Tierpsychologie* 48:401-420, 1978.

Harcourt, A.H.; Stewart, K.J. Gorilla male relationships: can differences during immaturity lead to contrasting reproductive tactics in adulthood? *Animal Behaviour* 29:206-210, 1981.

Harrison, M.J. Territorial behaviour in the green monkey, *Cercopithecus sabaeus:* seasonal defense of local food supplies. *Behavioral Ecology and Sociobiology* 12:85-94, 1983.

Hasegawa, T.; Hiraiwa, M. Social interactions of orphans observed in a free-ranging troop of Japanese monkeys. *Folia Primatologica* 33:129-158, 1980.

Hinde, R.A.; Atkinson, S. Assessing the roles of social partners in maintaining mutual proximity as exemplified by mother-infant relations in rhesus monkeys. *Animal Behaviour* 18:169-176, 1970.

Hrdy, S.B. *Langurs of Abu: Female and Male Strategies of Reproduction.* Cambridge, Massachusetts, Harvard University Press, 1977.

Itoigawa, N. Variables in male leaving a group of Japanese macaques. pp. 233-245 in *Proceeding from the Symposia of the Fifth Congress of the International Primatological Society.* S. Kondo; M. Kawai; A. Ehara; S. Kawamura, eds. Tokyo, Japan Science Press, 1975.

Kawanaka, K. Division of males in a Japanese monkey troop on the basis of numerical data. *Bulletin of the Hiruzen Research Institute* 3:11-44, 1977.

Koenig, W.D. Space competition in the acorn woodpecker: power struggles in a cooperative breeder. *Animal Behaviour* 29:396-409, 1981.

Koenig, W.D.; Mumme, R.L.; Pitelka, F.A. Female roles in cooperatively breeding acorn woodpeckers. pp. 235-262 in *Social Behavior of Female Vertebrates.* S.K. Wasser, ed. New York, Academic Press, 1983.

Koenig, W.D.; Pitelka, F.A. Ecological factors and kin selection in the evolution of cooperative breeding in birds. pp. 261-280 in *Natural Selection and Social Behavior: Recent Research and New Theory.* R.D. Alexander, D.W. Tinkle, eds. New York, Chiron Press, 1981.

Koford, C.B. Rank of mothers and sons in bands of rhesus monkeys. *Science* 141:356-357, 1963.

Loy, J.; Loy, K. Behavior of an all-juvenile group of rhesus monkeys. *American Journal of Physical Anthropology* 40:83-95, 1974.

Marsden, H.M. Behavior between two social groups of rhesus monkeys within two tunnel-connected enclosures. *Folia Primatologica* 8:240-246, 1968.

Marsh, C.W. Comparative aspects of social organization in the Tana River red colobus, *Colobus badius rufomitratus. Zeitschrift Fur Tierpsychologie* 51:337-362, 1979.

Meikle, D.B.; Vessey, S.H. Nepotism among rhesus monkey brothers. *Nature* 294:160-161, 1981.

Packer, C. Inter-troop transfer and inbreeding avoidance in *Papio anubis* in Tanzania. Ph.D. Thesis, University of Sussex, 1977.

Packer, C. Inter-troop transfer and inbreeding avoidance in *Papio anubis. Animal Behaviour* 27:1-36, 1979.

Packer, C.; Pusey, A.E. Female aggression and male membership in troops of Japanese macaques and olive baboons. *Folia Primatologica* 31:212-218, 1979.

Poirier, F.E. The Nilgiri langur troop: its composition, structure, function, and change. *Folia Primatologica* 10:20-47, 1969.

Pusey, A.E. Inbreeding avoidance in chimpanzees. *Animal Behaviour* 28:543-552, 1980.

Rasmussen, D.R. Correlates of patterns of range use of a troop of yellow baboons (*Papio cynocephalus*) I. Sleeping sites, impregnable females, births and male emigrations and immigrations. *Animal Behaviour* 27:1098-1112, 1979.

Russell, J.K. Altruism in Coati bands: nepotism or reciprocity? pp. 263-290 in *Social Behavior of Female Vertebrates.* S.K. Wasser, ed. London, Academic Press, 1983.

Sade, D.S. Determinants of dominance in a group of free-ranging rhesus monkeys. pp. 99-114 in *Social Communication Among Primates.* S.A. Altmann, ed. Chicago, University of Chicago Press, 1967.

Sade, D.S. Population biology of free-ranging monkeys on Cayo Santiago, Puerto Rico. pp. 171-180 in *Biosocial Mechanisms of Population Regulation.* M.N. Cohen, R.S. Malpass, G. Klein, eds. London, Yale University Press, 1980.

Scanlon, K.; Dewhurst, A.; Gimlette, P. Some aspects of the behaviour of young male rhesus monkeys in a free-ranging band on Cayo Santiago. Unpublished report of the Cambridge Primatological Expedition to Cayo Santiago, 1979.

Scanlon, K. Development of social skills in free-ranging infant rhesus monkeys. Ph.D. Thesis, The Open University (In preparation)

Schulman, S.R. Intragroup spacing and multiple social networks in *Macaca mulatta.* Ph.D. Thesis, Yale University, 1980.

Seyfarth, R.M.; Cheney, D.L.; Hinde, R.A. Some principles relating social interactions and social structure among primates. pp. 39-52 in *Recent Advances in Primatology, Vol. I.* D.J. Chivers; J. Herbert, eds. London, Academic Press, 1978.

Silk, J.B.; Clark-Wheatley, C.B.; Rodman, P.S.; Samuels, A. Differential reproductive success and facultative adjustment of sex ratios among captive female bonnet macaques (*Macaca radiata*). *Animal Behaviour* 29:1106-1120, 1981.

Simonds, P.E. Outcast males and social structure among bonnet macaques, (*Macaca radiata*).*American Journal of Physical Anthropology*38:599-604,1973.
Stacey, P.B. Habitat saturation and communal breeding in the acorn woodpecker. *Animal Behaviour* 27:1153-1166, 1979.
Strum, S.C. Agonistic dominance in male baboons: an alternative view. *International Journal of Primatology* 3:175-202, 1982.
Sugiyama, Y. The life history of male Japanese monkeys *Macaca fuscata*. *Advances in the Study of Behaviour* 7:255-284, 1976.
Wrangham, R.W. An ecological model of female-bonded primate groups. *Behaviour* 75:262-300, 1980.

Illustration 7.1 An adult male during the breeding season.

CHAPTER SEVEN

Seasonal Differences in the Spatial Relations of Adult Male Rhesus Macaques

DAVID A. HILL

INTRODUCTION

The multimale group typical of macaque and baboon species consists of a stable core of related adult females, their immature offspring, and a number of relatively transient adult males. As they normally leave their natal group and transfer into another upon reaching sexual maturity, most adult males will not be related to the females in their group. Therefore, whereas kinship may be postulated as a major source of group cohesion amongst the females, this cannot be true for most adult males.

Zuckerman (1932) maintained that sexual attraction was the basis of cohesion in all primate groups. This assertion was founded on the belief that all primates had sexual relationships throughout the year. Subsequent field studies revealed that most primate species have either a birth peak or discrete birth and mating seasons (Lancaster and Lee, 1965). It became clear that current sexual relationships could not be the only social bond underlying group cohesion.

Several recent studies of savannah baboons have examined aspects of social relationships between adult males and anoestrous females. In a troop of chacma baboons Saayman (1971) found that one male acted as a focal point for lactating females and their offspring. In another troop of the same species Seyfarth (1975) described two types of long-term affiliative bonds which persisted irrespective of the female's reproductive state. Both of these studies were of small troops with very few adult males. Similar relationships have been found in much larger troops of olive baboons (Packer, 1979; Smuts, 1983) and yellow baboons (Altmann, 1980; Rasmussen, 1983).

Most macaque species for which there are adequate data have clearly defined birth and mating seasons (Vandenbergh, 1973; Roonwal and Mohnot, 1977). For more than half of the year there are no females in oestrous. This provides an ideal situation in which to investigate the importance of sexual attraction in group cohesion. If sexual relationships were the main social bonds linking adult males to the group one would expect the attachment of males to the group to be stronger during the mating season.

Among free-ranging rhesus monkeys, Kaufmann (1967) found that adult males tend to be more central during the mating season and to direct a greater proportion of their grooming to adult females. However, levels of aggression are also higher during the mating season, both in the Cayo Santiago colony (Kaufmann, 1967; Wilson and Boelkins, 1970) and in feral rhesus monkeys in Nepal (Teas, Taylor, and Richie, 1978). In addition most changes in male group membership occur during the mating season *(M. mulatta,* Lindburg, 1969 (feral); Boelkins and Wilson, 1972 (free-ranging); *M. sinica,* Dittus, 1975; *M. Fuscata,* Sugiyama, 1976), suggesting that this is, in fact, a time of relative instability for the males.

Two studies have examined the birth season relationships of adult macaques in some detail. In a two-and-a-half year study of feral Japanese macaques, Takahata (1982) described "peculiar proximate relations" between adult males and females during the birth seasons. These were characterized by persistent proximity within or across birth season and by high levels of grooming and following. The number of such relationships that each male had was correlated with dominance rank. Similarly Chapais (1981, 1983) found that long-term, high-frequency grooming relations were present among free-ranging rhesus monkeys during the birth season. He attributed the social structure observed to two main features: the attraction of adult females to the high-ranking males and the attractiveness of the alpha female to the high-ranking males.

This paper examines seasonal changes in the spatial relations of adult male and female rhesus macaques. It is a preliminary report from a study of the social relationships of the adult males.

METHODS

The data presented here were collected from Group I, the second largest group on Cayo Santiago, which grew from 180 to 230 members during the study period. Initially these included 25 adult males (7 years or older) and 44 adult females (4 years or older). The three-year-old females, who reached sexual maturity during the study period, were not included in the analyses. Sixteen adult males were chosen as focal animals (Table 7.1). Among them were the six top-ranking males who maintained their ranks throughout the study and 10 mid- and low-ranking males. Two mid-ranking males left the group. K3 left at an early stage and was not included in the following analyses. Another four males were involved in the formation of a subgroup during the first half of the study. Although often spatially distinct, it was clear that this was a subdivision of Group I rather than a group in its own right

Table 7.1 Length of tenure at the beginning of the study and dominance rank of the focal males (↓ = dropped in rank; ↑ = climbed in rank).

ID	Tenure Months	Initial Rank	Final Rank	Changes
9L	NATAL	1	1	—
376	90	2	2	—
1C	87	3	3	—
435	74	4	4	—
504	68	5	5	—
539	63	6	6	—
628	44	7	—	
538	35	8	28	↓
596	39	9	7	—
599	32	12	23	Sub Gp.
K3	107	14	—	Left
515	53	18	25	Sub Gp.
552	5	21	26	Sub Gp.
513	6	22	12	↑
512	6	24	29	Sub Gp.
537	0	26	27	—

and so these males were included in the analyses. Relative dominance was assessed using the direction of dyadic agonistic interactions involving a submissive gesture. These gave a clearly defined linear dominance hierarchy with few reversals. The subgroup males and two others underwent changes in dominance rank during the study. Table 7.1 shows the ranks of the males at the beginning and the end of data collection. During the last three-fifths of the study there were no changes in the relative ranks of the focal males. The rank order from this period was used in the analyses below unless otherwise stated.

Details of interactions and proximity data were recorded in 20-minute samples. These were organized into fifteen Blocks, each of which contained four hours of data per focal male. The data covered a 14-month period including parts of two birth seasons and the intermediate mating season. The mating season was defined as that period during which the total frequency of full mounts observed in focal samples exceeded five per Block. The distribution of data across the season was uneven with four Blocks in the first birth season, six Blocks in the mating season, and five Blocks in the second birth season.

RESULTS

Spatial relations of males and females in general

Two methods were used to measure seasonal differences in the spatial relations of the males and adult females in general.

The percentages of time that a focal male had one or more females within 1 meter, and within 5 meters, of him were calculated for each season. The plots in Figure 7.1 give the results for each focal male. They are arranged in rows, in order of dominance rank, with the alpha male in the top right corner. There was a tendency for high-ranking males to be close to one or more females more frequently than lower-ranking males. This correlation was statistically significant for both distance categories in all three seasons using the relevant dominance rank order for each season (six Spearman's tests—all $p < 0.01$). Of the 14 males with data for all three seasons, eight showed a peak in the mating season in the 5 meter data, and nine showed a peak in the 1 meter data. 628, who left the group at the end of the mating season, showed a similar trend at both distances. Five males did not show this trend in either the 1 meter or the 5 meter data. Four of these were involved in the formation of the subgroup and the other (538) dropped from rank eight to rank 24 in the male dominance hierarchy. Both of these events occurred shortly after the onset of the mating season.

The seasonal trends shown in Figure 7.1 could be due to more females being close to the males during the mating season, or to the same number of females being close more often. In order to examine this point the following procedure was used. For each Block only females who were within 1m for at least 5 percent of a male's observation time

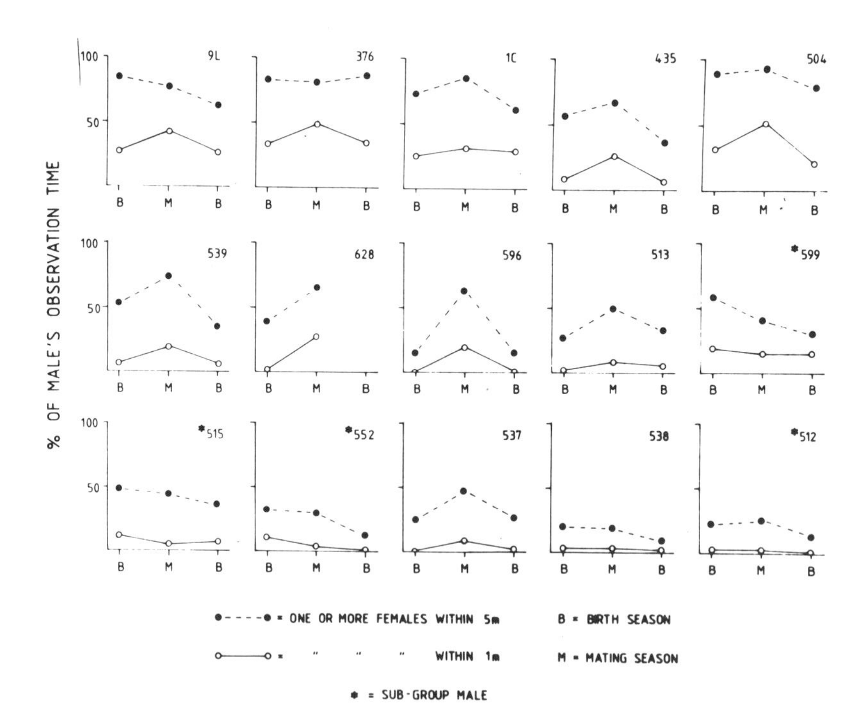

Figure 7.1 Seasonal changes in the percentage of time with females within 1m and 5m of the focal males. Each plot shows data for a single focal male.

were considered. These will be referred to as having reached the proximity criterion. All but three of the 44 adult females reached the criterion with a focal male at some point in the study.

Figure 7.2 shows the total number of females reaching the proximity criterion plotted against time. The shaded bars represent the mating season as defined in the methods section and the open bars represent the birth seasons. There was a clear seasonal difference with more females reaching criterion in the mating season than in either birth season. If the subgroup males were omitted t-tests on the seasonal means for the remaining 10 males showed a significant difference between the mating season and each of the birth seasons ($p < 0.005$ and $p < 0.01$) but no significant difference between the two birth season periods ($p > 0.50$). That this trend actually begins in Block 4, before the observation of a high frequency of full mounts, is largely due to the early mating activity of the alpha male. For this reason it may be better to consider Block 4 as a transition period between seasons.

Summing the data for all focal males in this way gave no indication of the persistence of individual male-female dyads. The next section examines this question.

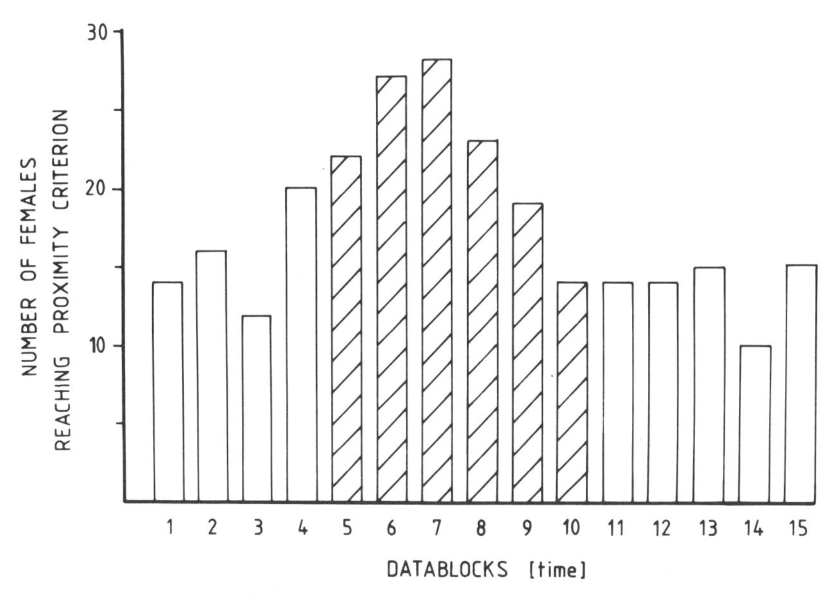

Figure 7.2 Seasonal distribution of the total number of females reaching the proximity criterion in each Block. (Shaded bars = mating season; open bars = birth seasons.)

Persistence of relationships within seasons

The four categories listed below give a crude measure of the persistence of spatial relations. They refer to the number of Blocks within a season that a female reached the proximity criterion with a particular male.

Female reaches criterion for:

Category A	Three Blocks
Category B	Two consecutive Blocks
Category C	Two nonconsecutive Blocks
Category D	One Block

All male-female dyads that reached the criterion were classified according to these categories. The number of dyads in each category was then calculated for each season, for each focal male. The results are given in Table 7.2 for the 14 males with data for all three seasons.

The high proportion of low and zero values made the results unsuitable for correlation analyses. A clear visual impression of the distribution of categories can be gained by comparing the seven top-ranking males (9L through 596) with the seven lower-ranking males (513 through 512). Very few lower-ranking males were involved in any category A, B, or C dyads. Eighty-four percent of these dyads involved the top-ranking males. Although all males scored at least two category D dyads, the top-ranking males accounted for 77 percent of those recorded.

Table 7.2 Distribution of persistent proximity categories A-D (explained in text) among the focal males during the birth (BS) and mating (MS) seasons (* indicates sub-group male).

	Persistence Categories											
	A			B			C			D		
	BS	MS	BS	BS	MS	BS	BS	MS	BS	BS	MS	BS
9L		2	2		2		1	1	1	8	9	3
376	1	4	1	1	1		1	2	2	5	6	4
1C	1	1	2		2			1	1	6	8	3
435								1		2	5	4
504	1	1	1		3			2		3	9	3
539				1	1				1		6	
596		1									6	
513										1	3	3
*599	2	2	1			1				1	1	2
*515						1				4	1	
*552				1						2	2	
537											2	1
538											1	1
*512										1	1	

The individual scores were too low to examine seasonal differences and so they were combined for the 14 focals with complete data. As there was a different amount of data collected in each season it was necessary to correct these values by calculating the mean frequency of each category per Block. The raw frequencies and corrected values are given in Table 7.3. Category D dyads were the most frequent in each of the three seasons. All four categories were more frequent in the mating season than in either birth season. The figure in parentheses in Table 7.3 is the mean for category D in the first birth season if transitional Block 4 data are omitted. This clearly enhances the seasonal difference.

Having established that there were persistent spatial relationships between adult males and females, and that the frequency of these varied seasonally, the next step was to ask which of the partners was responsible for maintaining proximity. Two sets of data were relevant to this problem: follows and approach/leave data.

Approaches and leaves were recorded over a 1 meter boundary around the focal animal. For those dyads with sufficient data the approach/leave index developed by Hinde and coworkers (e.g., Hinde and Atkinson, 1970) was calculated. Responsibility was attributed to the male if the index was ≥ 10 and to the female if it was ≤ 10. If the index fell between these values responsibility was said to be equal.

Follows were recorded using a point sampling technique with an interval of 30 seconds. Follows that were continuous for more than one point sample were scored only once in this analysis. For those dyads with sufficient data responsibility was attributed to an animal if it had scored at least twice as many follows as its partner.

Figure 7.3 shows the number of dyads in which the males (solid bars) or females (hatched bars) were more responsible for maintaining proximity. Responsibility as calculated from approach/leave data is shown on the left of the figure and follows on the right. For both approach/leaves and follows there were very few dyads with sufficient data in either birth season compared to the mating season.

Table 7.3 Seasonal distribution of persistent proximity categories A-D (Abbreviations as for Table 7.2).

	Raw Frequency			Mean Frequency per Block		
	BS	MS	BS	BS	MS	BS
A	5	11	7	1.25	1.83	1.4
B	3	9	2	0.75	1.5	0.4
C	2	7	5	0.5	1.2	1.0
D	33	60	24	8.25	10.0	4.8

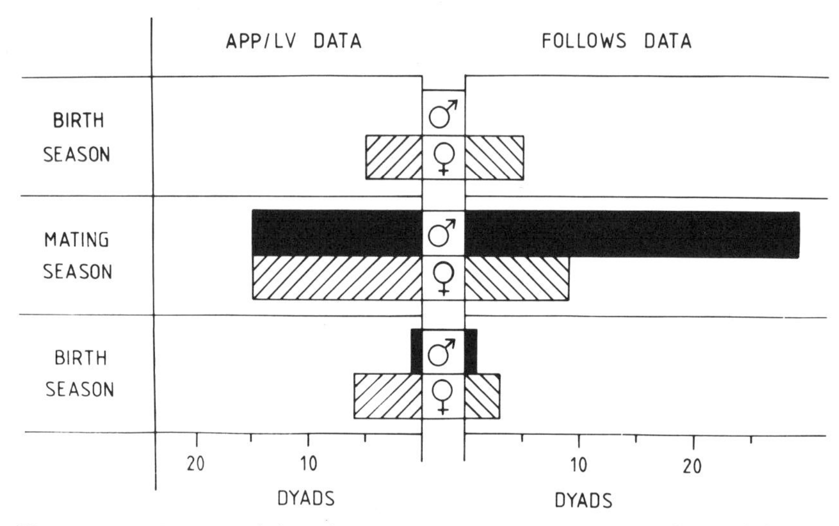

Figure 7.3 Responsibility for maintaining proximity as derived from approach/leave and follows data.

Of the 12 dyads for which there was sufficient approach/leave data in the birth seasons, the females were responsible for maintaining proximity in all but one. By contrast, in the mating season males were responsible in as many dyads as females. The follows data showed a similar pattern in the birth seasons with the male being responsible in only one of the nine dyads. In the mating season, the males were responsible in more than three times as many dyads as the females. There is, therefore, a general pattern of the females being responsible for maintaining proximity in the birth seasons and both males and females being responsible in the mating season.

Persistence of relationships across seasons

There were seven male-female dyads that were in a persistent proximity category (A or B) for more than one season and they are listed in Table 7.4. They included five males and seven of the 44 females in the group. All of these males, except for 599, were high-ranking. 599 was a mid-ranking male at the beginning of the study, but was involved in the formation of the subgroup in Block 6 and subsequently dropped in rank. He was, however, the most dominant of the five males in the subgroup.

There was no apparent common factor linking the seven females except that they were all sexually mature. They included both parous and nonparous individuals and members of all three genealogies. They ranged in age from four to 19 years and in dominance rank from the alpha female to the 42nd-ranking female (Table 7.4).

In all persistent proximity dyads (PPD's) for which there were sufficient data, responsibility for maintaining proximity in the birth season lay with the female. The pattern in the mating season compared with that of other dyads is shown in Figure 7.4. From approach/leave data the PPD females were responsible in five out of seven dyads. In other dyads the sexes were approximately equally responsible (no significant difference, $p = 0.19$, Fisher's Exact Test). From the follows data the PPD females were responsible in five out of six dyads whereas other females were responsible in only four out of 31 dyads ($p < 0.002$, Fisher's Exact Test).

Table 7.4 Dyads with a persistent proximity relationship in more than one season (Abbreviations as in Table 7.2).

MALE	FEMALE			SEASON		
ID	ID	AGE (yrs)	RANK	BS	MS	BS
9L	WE	17	1		A	A
376	777	5	30	A	A	A
1C	605	7	20	A	B	A
1C	NU	19	31		A	A
504	803	4	42	A	A	A
599	808	4	16	A	A	A
599	528	8	18	A	A	A

RESPONSIBILITY

APPROACH/LEAVES		♂	♀
	PP dyads	2	5
	Others	13	10

NO SIG. DIFF.

FOLLOWS		♂	♀
	PP dyads	1	5
	Others	27	4

p<0·002
Fisher's exact test

Figure 7.4 Comparison of the number of dyads in which males and females were responsible for maintaining proximity in the mating season in persistent proximity (PP) dyads and others.

Although these data have shown that affiliative bonds between adults can persist across reproductive seasons, it is important to put them into perspective. Of the 14 males with data from the whole study, only five were involved in these relationships. Moreover, they involved only seven of the 44 adult females in the group. *Ad libitum* data indicated that two females may have had similar relationships with non-focal males. Even if these were included, persistent proximity relationships involved a small proportion of the group members.

Discussion

The analyses clearly indicate that both reproductive season and male dominance rank are important factors in the control of spatial relations between adult male and female rhesus monkeys. Although both variables act simultaneously each has a clear effect which is not confounded by the other.

Kaufmann (1967) reported seasonal differences in the "social spacing" of the adult males. He found a tendency for most males to be more central during the mating season than in the birth season. The size of Group I precluded the use of Kaufmann's measure of centrality in the present study. Nevertheless, it is clear that there were seasonal differences in the spatial relations of the focal males with adult females. Most males were close to one or more females more frequently in the mating season than in either birth season. The main exceptions to this were the subgroup males and 538. The four focal males who were involved in the formation of the subgroup experienced a concurrent drop in rank. 538 also dropped in rank at about the same time. So the lack of a clear seasonal pattern in these males appeared to be related to their decline in dominance rank. This is further supported by the fact that dominance rank was significantly correlated with the frequency of being close to females.

The mating season also saw an increase in the number of females within 1 meter of a male for 5 percent of his time or more. This trend was apparent both in the combined data for 14 focal males and in comparisons of the seasonal means for the 10 males who were not involved in the subgroup. Although these data are not directly comparable to those of Kaufmann, it seems likely that the seasonal trends shown here, and his finding that males were more central during the mating season, represent the same phenomenon. Kaufmann also found that the tendency to remain in the central group was lower among males of lower rank. As he defined the central group by the distribution of the females, the same tendency is probably manifest in the correlation between dominance rank and time spent in proximity to females found in this study. Male dominance rank was also related to the persistence of proximity relationships with females. The higher-ranking males accounted for almost all of the dyads in categories A, B, and C in all three seasons. The only major exception to this pattern was 599 who was also the highest-ranking male in the subgroup.

If it is assumed that persistent proximity between a male and female represents an affiliative relationship, then these were most frequent during the mating season. At this time the males were responsible for maintaining proximity by following in more cases than the females. However, both sexes were responsible for maintaining proximity within 1 meter in equal numbers of dyads. It is impossible to tell from spatial data alone if all of the dyads that reached the proximity criterion during the mating season were sexual relationships. There is, however, evidence of affiliative relationships outside the mating season which cannot have been sexual. In 18 birth season dyads, mainly categories A and B, it was possible to attribute responsibility for maintaining proximity to one of the partners. In all but two of these, proximity was due to the females.

Many early studies mentioned affiliative relationships between adult male and anoestrous female rhesus monkeys both in laboratory conditions (Bernstein, 1963; Reynolds, 1970) and in free-ranging colonies during the birth season (Altmann, 1962; Kaufmann, 1967; Loy, 1971). More recently, Chapais (1981; 1983) provided a quantitative analysis of the birth season relationships between adult males and females of Group F on Cayo Santiago. The highest-ranking males were found to be the most central and had long-term, high-frequency grooming relations with females. Approach/leave data indicated that their centrality resulted from: 1) four males being attracted to the alpha female; and 2) several females being attracted to each of three high-ranking males. The main differences in the data presented here are that no female acted as a center of attraction for males and that no male had persistent proximity relationships with more than two females during either birth season.

In the current data, all five of the category A relationships recorded in the first birth season persisted throughout the study. Another two dyads were category A in the mating season and in the second birth season. In these persistent proximity dyads, there was a strong tendency for the females to maintain proximity during the mating season as well as in the birth seasons. In the follows data, this trend was significantly different from that of the other dyads. Chapais (1981) did not find relationships that were persistent across seasons. However, in the troop of Japanese macaques where Takahata (1982) found "peculiar proximity relationships" between males and females many of these persisted from one birth season to the next. Unfortunately, Takahata did not analyze approach/leave data for individual dyads so there is no indication of which partner contributed the most to maintaining proximity.

The existence of long-term relationships during the birth season indicates that bonds, other than immediate sexual attraction, occurred between the adult males and females. In baboons, affiliative relationships between males and anoestrous females may influence subsequent sexual relationships. Packer (1979) found that for some males these affiliative bonds were significantly related to the pattern of consort formation. In addition, Rasmussen (1983) found that mutual preference outside the sexual consort may be related to greater coordination within it. However, sexual relationships do not always follow affiliative relationships between adult males and anoestrous females. In Japanese macaques Takahata (1982) found that mating was significantly less likely between pairs which had had a persistent birth season relationship. Also, Chapais (1981) found no correlation between male-female relationships in the birth season and the subsequent pattern of mating. He concluded that the highest-ranking males and the adult females benefit from each other's support in agonistic encounters and so "enhance their competitive ability in relation to lower-ranking males". The interaction data from the present study are currently under analysis. When they are available it will be possible to assess the relation, if any, between the pattern of mating observed and affiliative relationships in the preceding and subsequent birth seasons.

CONCLUSIONS

1. Reproductive season and male dominance rank were found to be major factors affecting spatial relations between adult males and females.
2. Persistent proximity relationships between males and females were most frequent in the mating season, but also occurred in the birth seasons.
3. Females were responsible for the proximity patterns observed in the birth seasons. Both sexes were responsible for the proximity patterns observed in the mating season.
4. Some proximity relationships persisted for more than one season. In most of these, the female was responsible for maintaining proximity throughout the year.
5. The existence of long-term relationships between males and ʻ males during the birth season strongly suggests that factors, other than immediate sexual attraction, are important in the cohesion of males to the group.

ACKNOWLEDGEMENTS

This study was funded by a research studentship awarded by the Science and Engineering Research Council. I thank S. Datta, R.A. Hinde, A.E. Main, C.E. Scanlon, and M.J.A. Simpson for their helpful comments on earlier drafts.

REFERENCES

Altmann, J. *Baboon Mothers and Infants.* Cambridge, Massachusetts, Harvard University Press, 1980.

Altmann, S.A. A field study of the sociobiology of rhesus monkeys, *Macaca mulatta. Annals of the New York Academy of Science* 102:338-435, 1962.

Bernstein, I.S. Social activities related to rhesus monkey consort behavior. *Psychological Reports* 13:375-379, 1963.

Boelkins, R.C.; Wilson, A.P. Intergroup social dynamics of the Cayo Santiago rhesus (*Macaca mulatta*) with special reference to changes in group membership by males. *Primates* 13:125-140, 1972.

Chapais, B. *The Adaptiveness of Social Relationships Amongst Adult Rhesus Monkeys.* Ph.D. thesis. University of Cambridge, 1981.

Chapais, B. Structure of the birth season relationship among adult male and female rhesus monkeys, pp. 200-208 in *Primate Social Relationships.* R.H. Hinde ed. Oxford, Blackwell, 1983.

Dittus, W.P.J. The dynamics of migration in *Macaca sinica* and its relation to male dominance, reproduction and mortality. *American Journal of Physical Anthropology* 42:298, 1975.

Hinde, R.A.; Atkinson, S. On assessing the roles of social partners in maintaining mutual proximity, as exemplified by mother-infant relations in rhesus monkeys. *Animal Behaviour* 18:169-176, 1970.

Kaufmann, J.H. Social relations of adult males in a free-ranging band of rhesus monkeys, pp. 73-98 in *Social Communication Among Primates.* S.A. Altmann ed. Chicago, University of Chicago Press, 1967.

Lancaster, J.B.; Lee, R.B. The annual reproductive cycle in monkeys and apes. pp. 486-513 in *Primate Behavior: Field Studies of Monkeys and Apes.* I.H. De Vore ed. New York, Holt, Rhinehart and Winston, 1965.

Lindburg, D.G. Rhesus monkeys: mating season mobility of adult males. *Science* 166:1176-1178, 1969.

Loy, J. Estrous behavior of free-ranging rhesus monkeys (*Macaca mulatta*). *Primates* 12:1-31, 1971.

Packer, C. Male dominance and reproductive activity in *Papio anubis. Animal Behaviour* 27:37-45, 1979.

Rasmussen, K.L.R. Influence of affiliative preferences upon the behavior of male and female baboons during sexual consortships, pp. 116-120 in *Primate Social Relationships.* R.A. Hinde ed. Oxford, Blackwell, 1983.

Reynolds, V. Roles and role changes in monkey society: the consort relationship of rhesus monkeys. *Man* (New Series) 5:449-465, 1970.

Roonwal, M.L.; Mohnot, S.M. *Primates of South Asia — Ecology, Sociobiology and Behavior.* Cambridge, Massachusetts, Harvard University Press, 1977.

Saayman, G.S. Behavior of the adult males in a troop of free-ranging chacma baboons. *Folia Primatologica* 15:36-57, 1971.

Seyfarth, R.M. Social relationships among adult male and female baboons II: Behaviour throughout the female reproductive cycle. *Behaviour* 64:227-247, 1975.

Smuts, B.B. Dynamics of 'special relationships' between adult male and female olive baboons, pp. 112-116 in *Primate Social Relationships.* R.A. Hinde ed. Oxford, Blackwell, 1983.

Sugiyama, Y. Life history of male Japanese monkeys, pp. 225-283 in *Advances in the Study of Behaviour.* Vol. 7, J.S. Rosenblatt; R.A. Hinde; E. Shaw; C. Beer. eds. New York, Academic Press, 1976.

Takahata, Y. Social relations between adult males and females of Japanese monkeys in the Arishyama B troop. *Primates* 23:1-23, 1982.

Teas, J.; Taylor, H.G.; Richie, T.L. Seasonal influences on aggression in *Macaca mulatta,* pp. 573-576 in *Recent Advances in Primatology.* D.J. Chivers; J. Herbert. eds. New York, Academic Press, 1978.

Vandenbergh, J.G. Environmental influences on breeding in rhesus monkeys. pp. 1-19 in *Symposia of the IVth International Congress of Primatology.* Vol. 2: Primate Reproductive Behavior, Basel, Switzerland, Karger, 1973.

Wilson, A.P.; Boelkins, R.C. Evidence for seasonal variation in aggressive behavior by *Macaca mulatta. Animal Behaviour* 18:719-724, 1970.

Zuckerman, S. *The Social Life of Monkeys and Apes.* New York, Harcourt, Brace, 1932.

CHAPTER EIGHT

Why Do Adult Male and Female Rhesus Monkeys Affiliate During the Birth Season?

BERNARD CHAPAIS

INTRODUCTION

Eisenberg et al. (1972) have suggested that multimale and multifemale primate groups originated from single or age-graded male groups. While Clutton-Brock and Harvey (1977) considered the costs and benefits of the resident male of either accepting or rejecting newcomers (i.e., effects on his breeding access and inclusive fitness, and on the dynamics of intra and intergroup resource competition), Eisenberg et al. (1972) proposed that "in an evolutionary sense, the number of males in a given troop will depend on what advantages the
A reasonable assumption underlying the suggestion that the multi-
males are to the reproducing females." (p. 871). Considering the problem in this perspective, Wrangham (1980) suggested that in "female-bonded" primate groups (i.e., groups including a core of resident female relatives), females ultimately regulated male membership in order to increase the competitive ability of their group.
male type of group is an outgrowth of the one-male type is that membership in a heterosexual group is functionally advantageous for a male (compared to a solitary life or to membership in an all-male group) in terms of breeding access, predator avoidance, and access to food (Eisenberg et al. 1972). However, although the last two factors might explain why males remain attached to a heterosexual group even when no females are receptive (during the birth season), they cannot account *per se* for the existence of affiliative relationships between males and anoestrous females.

A growing body of evidence indicates that adult males and females do affiliate outside consortships in *Macaca fuscata* (Baxter and Fedigan, 1979; Grewal, 1980; Takahata, 1982), *Macaca mulatta* (Loy, 1971; Lindburg, 1971; Hill; Chapter 7 this volume), *Papio anubis* (Smuts, 1983a; 1983b; Strum, 1983) and *Papio ursinus* (Seyfarth, 1978). The principle underlying the explanations which have been offered for this phenomenon is that social relationships between distant and non-kin

Illustration 8.1 An adult male relaxes during the nonbreeding season.

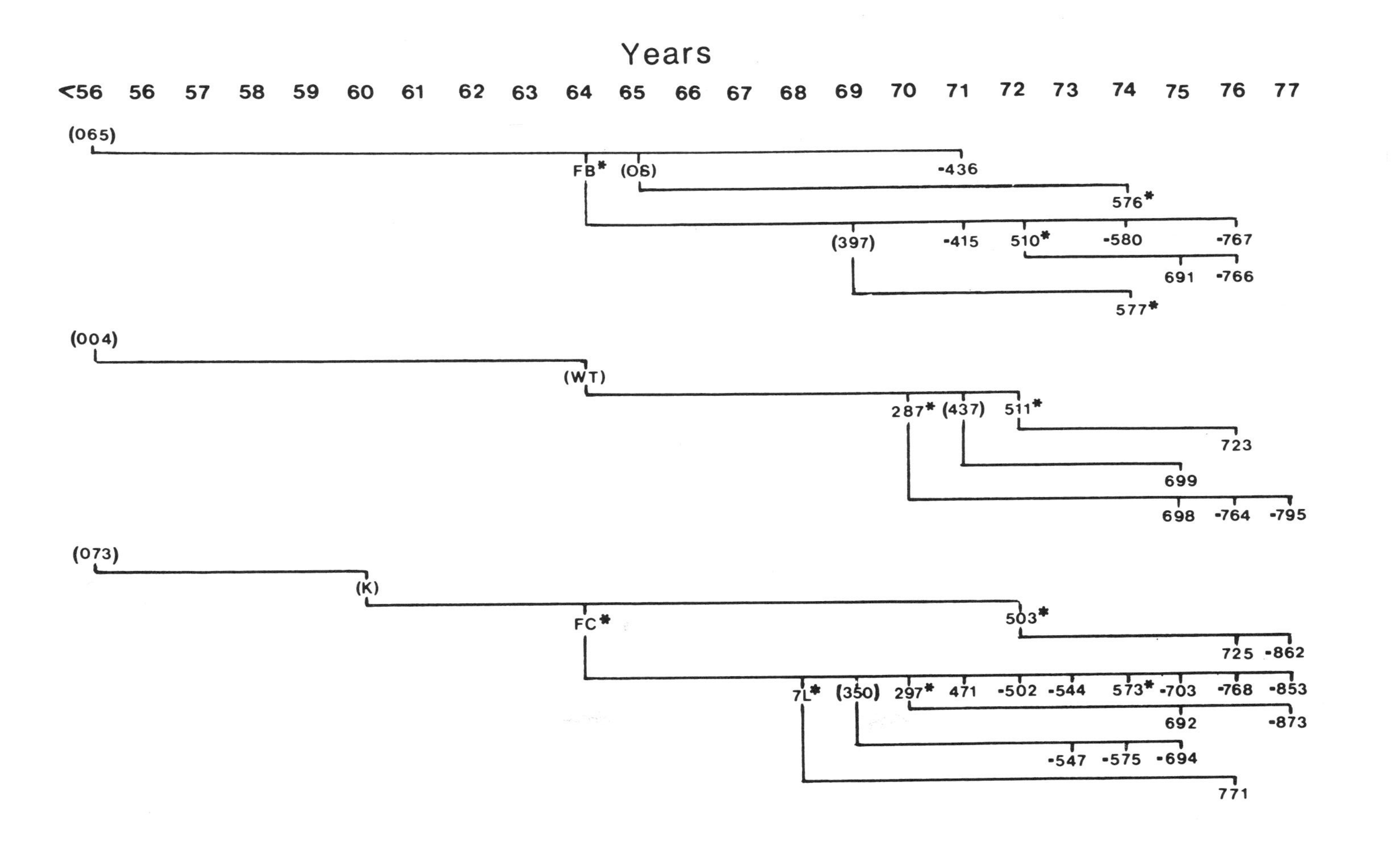
Years
<56 56 57 58 59 60 61 62 63 64 65 66 67 68 69 70 71 72 73 74 75 76 77
(065)
FB*
(OS)
-436
576*
(397)
-415
510*
-580
-767
691
-766
577*
(004)
(WT)
287*
(437)
511*
723
699
698
-764
-795
(073)
(K)
FC*
503*
725
-862
7L*
(350)
297*
471
-502
-544
573*
-703
-768
-853
692
-873
-547
-575
-694
771

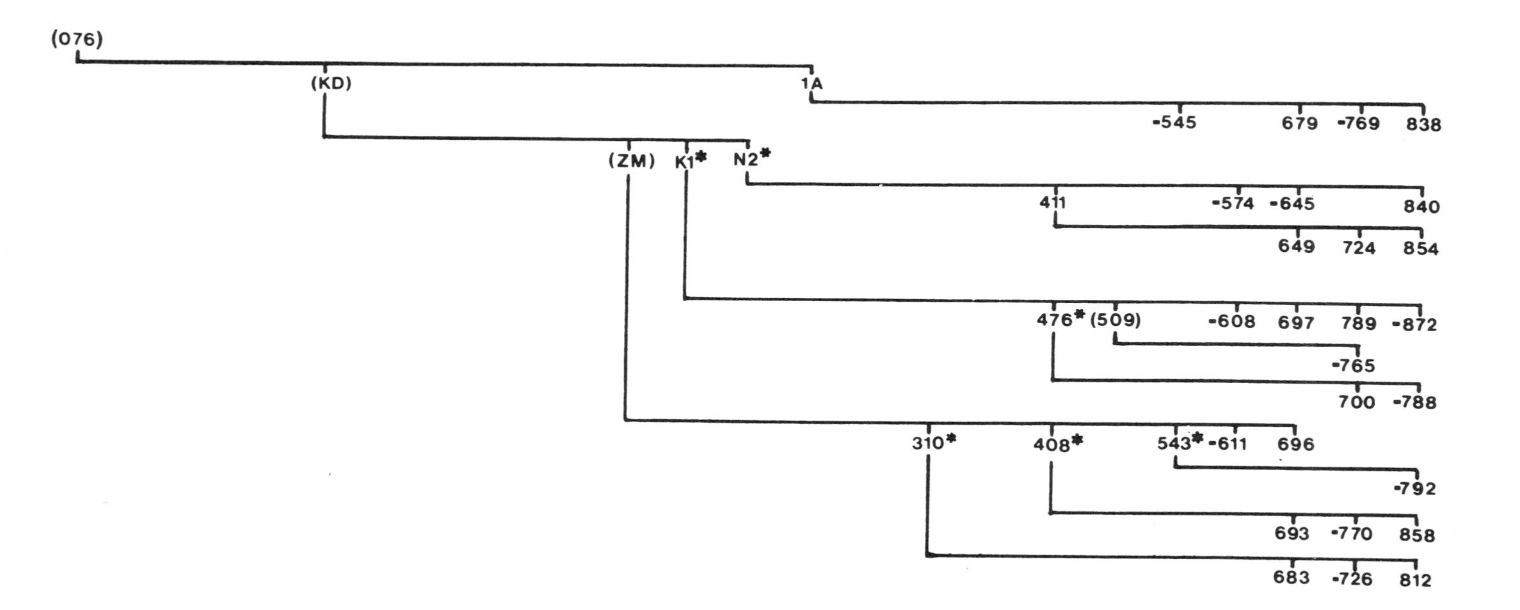

Figure 1. Date of birth and genealogical relations through the maternal line of the natal members of Group F in February 1978. The individuals who died before this date are not included except for connector females (in brackets). A dash denotes a male; an asterisk, a female who gave birth to a live offspring during the 1978 birth season.

animals (as well as between close kin) involve a reciprocal or complementary exchange of benefits, with time lags of various durations between the acts of giving and receiving. With respect to the present topic, it has been suggested that males participating in nonsexual relationships with females could benefit through: (1) an increase in their chance of mating with the females they affiliated with; (2) their use of the female as an agonistic buffer; and (3) the protection of the females' offspring, if they are likely to be the father of the latter. For their part, the females might benefit through the aid which they, or their offspring, received from the male in the context of aggressive encounters with other group members (Seyfarth, 1978; Strum, 1982, 1983; Smuts, 1983b).

The present paper describes the interactions between the adult males and adult females of Group F on Cayo Santiago, recorded during the 1978 birth season, and examines the nature of the benefits accruing to the individuals who maintained affiliative relationships.

MATERIALS AND METHODS

Subjects

Mating at Cayo Santiago is restricted to the period between July and December. Births occur from January to June with a peak in March and April. The present data were collected between February 20 and July 18, 1978; that is, during the greatest part of the birth season. In February 1978, the study Group F was comprised of four matrilines totalling 72 individuals (Figure 8.1), 12 adult non-natal males, and five non-natal peripheral males. Of the 20 adult females (age 4 or older) 18 gave birth to live offspring in 1978, one (411) delivered a stillborn, and another (471) is believed to have aborted following a severe wound. One female (1A) died on March 12, one day after giving birth, and was excluded from the present analysis.

Of the 11 natal males aged four or older (Figure 8.1) eight ranked among themselves relative to their mother's rank and could not be assigned an individual rank in relation to non-natal males (Kaufmann, 1967). Two others (the alpha male 415 and 436) were members of the central male hierarchy, along with the 12 non-natal, non-peripheral males. The remaining natal male (580) became part of the central hierarchy in May, following his rise to the second highest rank, below his brother 415 (see paragraph 1, Results). Both males were the sons of the highest-ranking female of the group (FB).

Method of Observation

The present analysis is limited to data on the interactions between the 15 males forming the central hierarchy and the 19 adult females. Observations were made between 0715 and 1145 hours on weekdays and between 0800 and 1300 hours on Saturdays and/or Sundays. Data were entered on checksheets. The 19 adult females were the object of 5-minute long focal samples (Altmann, 1974) two or three times a day, for a total of 12 times a week, or one hour per female. They were sampled following a random order renewed daily. Male-female interactions taking place during focal sampling, but not involving the focal female, were also recorded. Such concurrent sampling required a constant scanning of the group and yielded more than four times as much data as focal sampling. An attempt was made to estimate the relative observability of non-focal individuals during focal periods. On the fourth minute of each focal period, all the monkeys age three or older that were visible were recorded according to their distance from the focal female: (1) within 1.5 meters; (2) between 1, 5, and 8 m; or (3) further than 8 m. Between 10 and 45 seconds were needed to enter all monkeys. The 4,350 instantaneous scan samples recorded during the birth season yielded measures of group membership and proximity in addition to observability estimates. Finally, male-female interactions were recorded *ad libitum* outside focal samples. Since more than one-half of the total time of visual contact with group F was spent circulating through the group looking for the next focal female, this sampling source produced much data on rare or infrequent interactions.

General analytical procedure

Two analytical procedures can be used to abstract explanatory principles characterizing social structure from a given amount of social interactions of the type *A* does X to *B*, or *A* does X to *B* in relation to *C*. First, one can analyze separately the data pertaining to each dyad in an attempt to *A* does X to *B*, or *A* does X to *B* in relation to *C*. First, one can analyze separately the data pertaining to each dyad in an attempt to understand the internal dynamics of each relationship (e.g., Hinde and Stevenson-Hinde, 1976). This ideal procedure necessitates a large amount of data per dyad, which often means a small sample of subjects. Another approach, the one used in the present case, is to construct a number of unidimentional (i.e., behavior-specific) matrixes, extract from each matrix a few rules summarizing the distribution of the behavior and then attempt to relate causally these behavior-specific rules (Seyfarth, 1976; Seyfarth et al., 1978). This procedure has the advantage of not requiring large amounts of data per dyad and of being adequate for large numbers of dyads (19 x 15). It may also be possible to analyze simultaneously two or more matrixes (Pearl and Schulman, 1983). As discussed elsewhere (Chapais, 1981; 1983a), the classes of actors and recipients defined on the basis of characteristics such as sex, age, dominance, relatedness, reproductive state, etc. must be *functionally homogeneous* for the explanatory principles extracted from each behavior-specific matrix to be meaningful. For example, in the case of fight interferences, lumping the targets independently of their sex or lumping the beneficiaries independently of their degree of relatedness, etc. could be misleading.

The rhesus ethogram has been described by Altmann (1962), Hinde and Rowell (1962), and Rowell and Hinde (1962). The behaviors analyzed in the present study belong to one of the four following categories: the maintenance of mutual proximity, grooming interactions, agonism, and fight interventions.

RESULTS

Intrasexual dominance relations

Data on the age and dominance relations of males forming the central hierarchy during the birth season are presented in Table 8.1. Rank relations were linear and well defined. Recall that male 580 (the 4-year-old son of the alpha female FB and the brother of the alpha male 415) became part of the central hierarchy in May and occupied at that time the second highest rank. Male 580 is indicated as ranking second in subsequent tables.

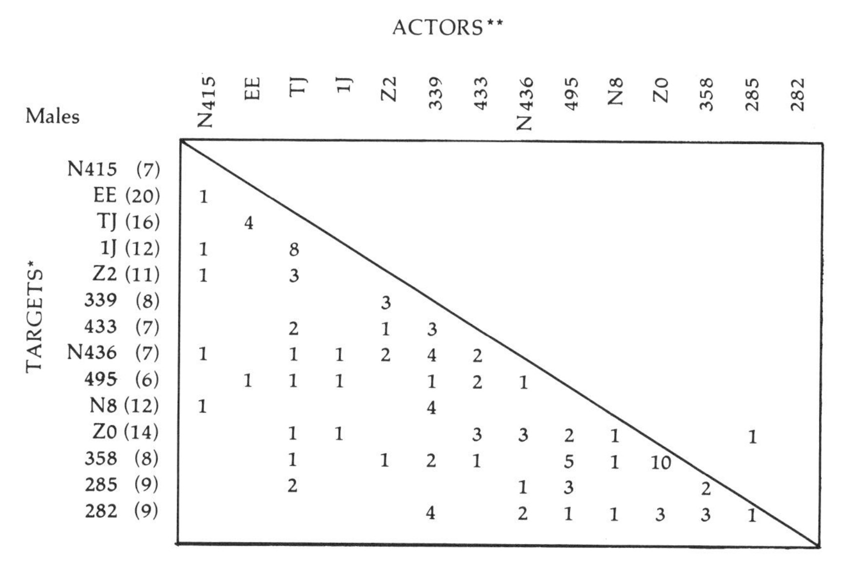

ACTORS**

TARGETS* / Males	N415	EE	TJ	1J	Z2	339	433	N436	495	N8	Z0	358	285	282
N415 (7)														
EE (20)	1													
TJ (16)		4												
1J (12)	1		8											
Z2 (11)	1		3											
339 (8)					3									
433 (7)			2		1	3								
N436 (7)	1		1	1	2	4	2							
495 (6)		1	1	1		1	2	1						
N8 (12)	1					4								
Z0 (14)			1	1			3	3	2	1			1	
358 (8)			1		1	2	1		5	1	10			
285 (9)			2					1	3			2		
282 (9)						4		2	1	1	3	3	1	

Table 8.1 — The distribution of submissive behaviors among central males, all sampling sources combined. The rank order presented is the one that produced the smallest frequency of behaviors above the diagonal.

N: Male born in group F
In brackets: The age of males in January 1978
* Approached, threatened or attacked the targets
** Fear-grinned, cowered, fled, or were supplanted.

The rank relations of female rhesus monkeys follow a highly predictable pattern which has been described by Sade (1967; 1972) and Missakian (1972). Females rank below their mother and all females ranking above her and they rank above all females ranking below their mother, except for their younger sister(s) when the latter reach puberty. Well defined, linear rank relations were extracted from data on the distribution of submissive behaviors among the 19 adult females (data and discussion in Chapais, 1981). The rank order (given in Table 8.3) presented two exceptions to the above rules, both occurring in the same matriline (FC ranking below her daughter 7L; 297 ranking above her younger sister 471).

Structure and dynamics of spatial relations

A birth season proximity score was obtained for each male-female dyad by counting the number of times each male had been recorded within 1.5 m in the instantaneous scan samples of each focal female and by dividing this frequency by the total number of scan samples (or focal periods) for this female. For example, male 415 was present within 1.5 m of FB in one out of the 230 scan samples performed when sampling FB. The resulting matrix of comparable proximity scores (Table 8.2) reveals that males differed considerably with respect to the number of females they were spatially close to. The five highest ranking males,415, 580, EE, TJ (to a lesser extent), and IJ, were close to a greater number of females and had higher proximity scores compared to lower-ranking males. This division of the adult males in two categories is also revealed by data on the distance of males to the 19 females as a group (see Table 8.2): the figures for the five highest-ranking males are consistently higher than those for lower-ranking males. The Maximum Spanning Tree or MST (see Morgan et al. 1975) derived from the matrix of Table 8.2 is shown in Figure 8.2 to further illustrate this result. Its main features are the following: First, three males are clearly central; the alpha male 415 (connected to six females), 3rd-ranking male EE (11 females), and 5th-ranking male IJ (four females). Second, the alpha female FB, connected to four males ranking 2nd to 5th, is also highly central. Third, females tend to gather around males according to their degrees fo relatedness to each other; sisters 543, 408, and 310 connected to EE; female N2, her daughter 411 and her niece 476 linked to 415; FC and her daughters 297 and 573 joined to EE.

The responsibility of individuals in the maintenance of mutual proximity was assessed on the basis of the frequencies of approaches (the distance decreasing from more to less than 1.5 m, *N*=544) and leavings (*N*=148) for each of the 30 dyads of the MST. *A* was recorded as approaching *B* only if *A*'s behavior was clearly directed at *B*. That is, only if *A* looked at *B* or sat facing *B* or directed a friendly gesture to *B*.

Table 8.2 — The proximity scores of each male-female dyad. See text for explanations.

N: males born in Group F
In brackets: the age of females in January 1978
Males and females are arranged in descending rank order (415: alpha male; FB alpha female)
* The percentage of the 4350 instantaneous scan samples (one scan per focal period) in which the male was recorded as being within 1.5 m. or between 1.5 and 8 m. or further than 8 m. from the focal female.

	Females																					
Males	FB (14)	510 (6)	577 (4)	576 (4)	511 (6)	287 (8)	503 (6)	7L (10)	FC (14)	573 (4)	297 (8)	471 (7)	N2 (12)	411 (7)	K1 (13)	476 (7)	543 (5)	408 (7)	310 (9)	x ≤1.5 m.*	>1.5 x 8m.*	x >8m.*
415	0.4	1.3	2.6	0.9	8.2	0.9	1.7	10.0	3.5	1.3	1.3	0.4	4.7	9.6	3.5	4.9	0.4			2.94	8.67	11.61
580	4.8	3.5	1.3	1.3	2.2		0.9	0.4		1.3						0.4				0.85	5.72	6.57
EE	3.0	1.3	2.2	6.1	2.6	8.3	2.6	3.9	9.6	9.7	6.1		1.3	4.8	4.0	0.9	4.4	5.7	3.5	4.21	11.93	16.14
TJ	5.7			0.4								0.9							0.9	0.41	7.68	8.09
1J	7.0	4.4	0.4	0.9	0.4	0.4	3.1			0.9		7.7						0.4	0.9	1.40	9.38	10.78
Z2		0.4																		0.02	3.06	3.08
339							0.4					0.9								0.07	1.03	1.10
433																				0.00	0.34	0.34
436				2.2		0.4				0.4			3.0		0.9					0.37	2.11	2.48
495	0.4																0.4			0.05	3.52	3.56
N8		0.4			1.3								2.6	3.0						0.39	4.21	4.60
Z0										0.4	0.4	0.9	0.4		0.4		0.4			0.16	4.09	4.25
358			0.4							0.4	0.4				0.4					0.07	2.23	2.30
285											0.4								1.3	0.09	1.08	1.17
282																0.4				0.02	1.38	1.40

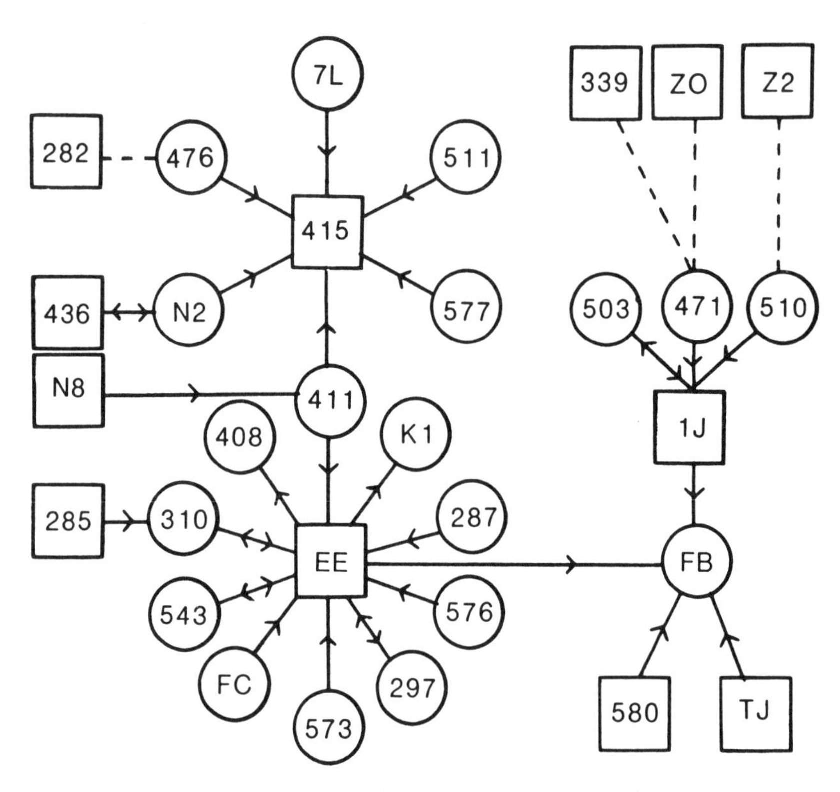

Figure 8.2 Maximum Spanning Tree constructed on the basis of a matrix of proximity for the 19 adult females and 15 central males (Table 8.2). Dash lines correspond to weak and probably nonsignificant links (see Table 8.2). The arrows originate from the individuals who played a greater role in the maintenance of mutual proximity (see text). Squares: males; circles: females. The characteristics of MSTs are as follows: (1) all individuals are included in the MST — that is, visited by at least one line; (2) no closed loops occur; and (3) the sum of the similarities composing the tree is the maximum possible sum. The spatial distribution of individuals is arbitrary. In the present case, each female is connected to a male but three males are not connected to any female: 433 was never recorded within 1.5 m. of focal females, while 495 and 358 had tied and, in any case, had very low proximity scores (see Table 8.2)

Because leavings could not be recorded exhaustively during concurrent and *ad libitum* sampling, the frequencies of leavings were very low for most dyads and it was not possible to use Hinde and Atkinson's (1970) proximity index (the percentage of approaches due to *A* — the percentage of leaving due to *A*) in a straightforward manner. Accordingly, a less direct approach was used. It was assumed, under the null hypothesis of *A* and *B* being number of approaches (or the number of leavings) executed by the members of a pair would not differ significantly. *A* was then defined as more responsible than *B* if the number of approaches due to *A* was significantly greater than the number of approaches due to *B* and if at the same time the frequency of leavings due to *A* was *not* significantly greater than the frequency of leavings due to *B*. Of the 26 dyads of the MST of Figure 8.2 for which data were available, 11 fulfilled these criteria. For another dyad (EE-287), the frequencies of approaches did not differ significantly, but the numbers of leavings did and the individual who left less often was considered more responsible for the maintenance of proximity. In one other case (EE-543), both individuals were observed to approach and leave each other equally often. In the remaining 13 cases responsibility was estimated on the basis of the direction of differences between the frequencies of each behavior.

An examination of the direction of arrows in Figure 8.2 reveals that each of the four high-ranking males connected to the alpha female FB was more responsible than she was for maintaining proximity. Excluding these four dyads and considering only the 19 dyads involving the highest-ranking males (415, EE, IJ), it is found that for four dyads the responsibilities are equal; for 13, the female is more responsible than the male and for the remaining two dyads, the male is more responsible than the female.

In summary, the five highest ranking males (415, 580, EE, TJ, and IJ) were highly central (i.e., close to the core of adult females) compared to lower ranking males, and this resulted from two factors: the attractiveness of the alpha female FB and the attractiveness of high-ranking males 415, EE, and (to a lesser extent) IJ.

The patterning of grooming relations

A grooming bout was defined as an instance where *A* groomed *B* for more than 15 seconds. If *A* stopped grooming *B* and resumed later, only one bout was counted unless: (1) the interruption lasted more than one minute, or (2). *A* interacted with a third individual; or (3) *A* and *B* were involved in an agonistic interaction. It was assumed that these criteria minimized the amount of determinancy between any two consecutive grooming bouts and justified the use of goodness-of-fit tests (see Dunbar and Dunbar, 1975). On the basis of data on the frequency and temporal patterning of grooming bouts (N=1368) for each male-female dyad (frequency of grooming directed by the male to the female and vice-versa; number of days on which grooming was recorded for this dyad; average number of days between any two successive days on which grooming was recorded), six types of dyadic grooming relations (*A* to *F*) were defined (Table 8.3).

Grooming was observed in 26.0 percent of all male-female dyads (74 out of 285), and nonoccasional grooming (i.e., type F excluded) in 49 dyads (17.2 percent). Although all males (except lowest-ranking 282) groomed and/or were groomed by females, only the five highest-ranking males were involved in long-term, high-frequency grooming relations (type *A*) (see right side of Table 8.3). Males 415 and EE in particular groomed and were groomed by the majority of females (respectively, 15 and 17). Most of the 10 lower-ranking males had either type *D* or type *F* grooming relations. For their part, all females, 577 and 297 excepted, maintained long-term grooming relations (types *A*, *B* or *C*) with males, the alpha female FB being involved in four type *A* relations with four of the five highest-ranking males.

Examining the direction of grooming within dyads (raw data not presented) it is found that males groomed the alpha female FB more often than they were groomed by her (two-tailed Wilcoxon test: N=7, T=0, $p < .02$). Thus, high-ranking males were attracted to FB, a finding which corroborates results on the responsibility in the maintenance of mutual proximity. FB excluded (in view of her special status which will be further evidenced below), it is found that females groomed first-ranking male 415 more often than he groomed them (N=14, T=9.5, $p < .01$); third-ranking males EE also received from females more grooming than he gave (N=15, T=21, $p < .05$); similarly, second-ranking male 580 tended to direct less grooming than he received, but his difference was not significant (N=7, T=3, $p > .05$) Thus, the three highest-ranking males appeared to be highly attractive for adult females. Only one female (471) did not maintain a grooming relation with either of these three males. However, 471 was involved in long-term grooming relations with the 4th and 5th-ranking males, grooming both more than vice-versa.

The three lowest-ranking males groomed females more often than they were groomed in all cases (9 dyads, 10 if FB included) while middle-ranking males directed more grooming than they received in 10 cases, and less in 7 cases (11 and 7 if FB included).

In summary, adult females were attracted to the three highest-ranking males who were themselved attracted to the alpha female along with the 4th and 5th-ranking males. Lower-ranking males affiliated comparatively little with adult females but appeared attracted to them. These results corroborate those on spatial relations.

The patterning of agonistic interactions

Dominance relations. All adult females (except the alpha female FB) were individually subordinate to all central males. FB's status was unique among females. She was the only female who did not receive aggression from any of the nine highest-ranking males, but she was displaced and threatened by *lower*-ranking males. FB was dominant over her son 580 (2nd-ranking), male EE (3rd-ranking), and IJ (5th-ranking). Although her rank relationships with some other high or middle-ranking males (e.g., TJ and Z2) were never overtly expressed,

Table 8.3 — The distribution of grooming relations among central males and adult females.

Males	FB	510	577	576	511	287	503	7L	FC	573	297	471	N2	411	K1	476	543	408	310	A	B	C	D	E	F	Total	N bouts
	Females																			Grooming relations							
415	D	F	D	F	A	F	B	A	C	C	F		C	A	D	C				3	1	4	3	0	4	15	270
580	A	B	E	B*	A*	F	D			C										2	2	1	1	1	1	8	165
EE	A		D	A*	D	A	B	C	C	A	F		C	A	C	F	B	A	C	6	2	5	2	0	2	17	471
TJ	A											B								1	1	0	0	0	0	2	122
1J	A	A*					F				F	A							D	3	0	0	1	0	2	6	190
Z2	F																			0	0	0	0	0	1	1	2
339							D					D								0	0	0	2	0	0	2	21
433						E		D												0	0	0	1	1	0	2	18
436	F		B			F							D		F			F		0	1	0	1	0	4	6	38
495														F						0	0	0	0	0	1	1	1
N8					B*								F	F	F	F				0	1	0	0	0	4	5	29
Z0																			D	0	0	0	1	0	0	1	11
358	F		F			F				D	F	F								0	0	0	1	0	5	6	18
285									F										D	0	0	0	1	0	1	2	12
282																				0	0	0	0	0	0	0	0
A	4	1	0	1	2	1	0	1	0	1	0	1	0	2	0	0	0	1	0								
B	0	1	0	2	1	0	2	0	0	0	0	1	0	0	0	0	1	0	0								
C	0	0	0	0	0	0	0	1	2	2	0	0	2	0	1	1	0	0	1								
D	1	0	2	0	1	0	2	1	0	1	0	1	1	0	1	0	0	0	3								
E	0	0	1	0	0	1	0	0	0	0	0	0	0	0	0	0	0	0	0								
F	3	1	1	1	0	4	1	0	1	0	4	1	1	2	2	2	0	1	0								
Total	8	3	4	4	4	6	5	3	3	4	4	4	4	4	4	3	1	2	4								

A: long-term, high-frequency grooming relations extending throughout the birth season, or over half of it (*)

B: long-term, low-frequency grooming relations (as for A but grooming occuring at longer intervals)

C: long-term, low-frequency grooming relations interrupted over the first weeks of lactation

D: medium-term, low-frequency grooming relations (between two weeks and two months)

E: short-term grooming relations (one week)

F: very occasional grooming bouts.

bouts: the total frequency of grooming directed by the male to the females and by the females to the male.

Males and females are arranged in descending rank order (415: alpha male; FB alpha female) various interactions suggested that these males were dominant to FB but highly attracted to her.

For example, on March 3, FB was grooming her daughter 510. Fourth-ranking male TJ approached the pair and chased 510 away. FB climbed a tree and vigorously shook branches while looking in the direction in which TJ and 510 had disappeared. A short while later, TJ was back. FB approached TJ and groomed him for a few seconds before moving off. TJ followed FB and groomed her. FB reciprocated grooming.

FB used to defend any of her relatives promptly. That she did not do so with TJ and showed eagerness to affiliate with him after the interaction constitutes an indirect proof of her subordinance to him. However, this rank relation was never expressed through aggressive or submissive behaviors, but rather, a persistent, high-frequency affiliative bond characterized the FB-TJ relationship (Table 8.3).

Male Z2's relationship with FB further points to the value of FB as a social partner from the point of view of central males. For months Z2 (6th-ranking) behaved in a way which suggested that he was attracted to FB but inhibited from interacting with her. Z2 regularly followed FB from a distance and sat alone facing the core of individuals surrounding her.

On April 20, Z2 was observed grooming FB for the first time. Both individuals were isolated from higher-ranking males and Z2 frequently glanced around when grooming. A little while later, 5th-ranking male IJ was spotted walking in their direction. Z2 stopped grooming FB at once and fear-grimaced to IJ who kept approaching. IJ lip-smacked at FB and began grooming her. At this point, Z2 jumped away from the pair and FB groomed IJ. A similar interaction was witnessed three days later.

In light of such interactions, the fact that high and middle-ranking males avoided being aggressive with FB is better interpreted as a consequence of her attractiveness and value as a social partner rather than as a manifestation of her higher dominance status.

Aggressive interactions. Many aggressive interactions took place when the animals were near the feeders inside one of the three corrals. These interactions were not included in the present analysis in order to minimize the effects of this artificial situation on the results.

Adult females were only rarely observed to initiate aggressive interactions with adult males. However they were involved in such interactions when supporting relatives (male or female) or non-relatives (male or female). On the other hand, males frequently initiated aggressive interactions with adult females. For about 60 percent of the 370 recorded aggressive acts which appeared to have been initiated by one of the 15 central males, it was not possible to determine the context in which the interaction had begun (this being due in part to the short duration of focal periods). Another 25 percent of these aggressive acts occurred in the following contexts: affiliation interferences (the male aggressed a female who was close to, or grooming, another male or a female); redirected aggression (the male aggressed a female after being attacked by a higher-ranking male; and fight interferences (the female was attacked by a male while defending an individual attacked by the same male). Finally, the remaining 15 percent of male aggressive acts appeared to be spontaneous; i.e., although the observability was good, no obvious short-term sequence of events or specific contexts could be identified.

Data on the frequency of canine slashes and punctures (muscles injured) suffered by adult females during the birth season indicate that 15 of the 19 females (the exceptions being FB, 297, N2, and 543) were severely wounded by adult males (average number of slashes and punctures per female = 1.6; range = 1 to 6). The identity of aggressors could not be determined in most cases, but a significant proportion of wounds probably occurred in the course of intergroup fights since new wounds were observed after such encounters. Thus, male aggression on females represented a significant danger.

The distribution of male aggressive acts among females is more spread out (135 dyads involved out of 285 possible dyads: 47 percent) than the patterning of grooming relations (74 dyads: 26 percent). However, not all males were aggressive or equally aggressive toward females and an examination of the distribution of aggressive acts by each male is particularly instructive.

The highest-ranking males, 415 and EE, the males most attractive to females, were among the least aggressive, ranking 11th and 12th respectively in the frequency of all aggressive acts corrected for differential observability, and 9th and 12.5th in the frequency of bites; they were also little feared (Table 8.4).

The most aggressive male, 580, systematically attacked females (his mother FB excepted) from the time he had risen to the second rank in May. 580's behavior was unique and spectacular: the great majority of his attacks on females were spontaneous (no specific, short-term context triggering them) and deliberate (e.g., preceded by his stalking the female). They were frequently severe (highest frequency of bites, Table 8.4), and they spared few females (580 bit 11 females versus a maximum of 2 for any other male). Any male who chased vigorously, wrestled with or bit a female was invariably chased by a coalition of individuals which often included males and females unrelated to the target female. On the other hand, because 580 benefitted from the support, actual or potential, from members of the first matriline, most individuals seemed inhibited from retaliating against him. Despite this high level of aggressiveness, 580 was highly sought after during the following season by the females he had bitten. Therefore, in retrospect 580's systematic aggression appears to have been a strategy of status assertion which had deferred, positive effects on his attractiveness. More specifically, females might have learned that not

Table 8.4 — Frequencies of aggressive acts directed by each male to the 19 adult females and the frequencies with which females were passively displaced by males, all sampling sources combined
* Frequency of appearances of a male in the scan samples (see Methods) of the 19 adult females (total number of scan samples = 4350).
** Corrected frequencies: raw frequencies ÷ observability

		All aggressive acts			*Bites only*			*Passive displacements*		
Males	Observability*	N female targets	Raw freq.	Corr.** freq.	N female targets	Raw freq.	Corr.** freq.	N female displaced	Raw freq.	Corr.** freq.
415	1110	8	10	0.90	1	1	0.09	4	1	0.00
580	786	16	75	9.54	11	14	1.78	15	100	0.13
EE	1534	7	11	0.72	0	0	0.00	5	5	0.00
TJ	1141	18	79	6.92	2	2	0.18	18	86	0.08
1J	1192	18	67	5.62	0	0	0.00	14	22	0.02
Z2	673	12	20	2.97	1	1	0.15	10	15	0.02
339	297	7	7	2.36	2	2	0.67	4	6	0.02
433	226	3	3	1.33	1	1	0.44	2	2	0.01
436	454	1	1	0.22	0	0	0.00	0	0	0.00
495	656	14	36	5.49	1	1	0.15	17	53	0.08
N8	527	3	3	0.57	2	2	0.38	1	1	0.00
Z0	827	15	36	4.35	0	0	0.00	13	18	0.02
358	490	10	16	3.27	2	2	0.41	9	10	0.02
285	294	4	4	1.36	0	0	0.00	3	3	0.01
282	373	0	2	0.54	0	0	0.00	2	2	0.01

paying attention to his moves was potentially harmful and that this danger could be avoided by affiliating with him.

Males TJ and IJ, the lowest-ranking of the five most central males, were often aggressive with females (ranking 2nd and 3rd in the corrected frequency of all aggressive acts, Table 8.4) but their aggressive acts were comparatively less severe (IJ was not observed to bite females and TJ ranked 6th in the corrected frequency of bites.

The 10 lower-ranking males can be divided into two categories (the nonaggressive: 433, 436, N8, 285, and 282; and the aggressive: Z2, 339, 495, Z0, and 358). This subdivision could not be clearly related to the males' rank, age, observability, or frequency of grooming with females. Most of these 10 males affiliated very little with females (Table 8.3).

In summary, male aggression on females represented a constant threat. Aggressive males affiliated relatively little with females whereas the most attractive males (415 and EE) were rarely aggressive with females (580 being an exception, as discussed above). Membership in the first matriline conferred special status: FB was highly attractive to high-ranking males; 580 could assert his rank with relative impunity; and 415 was the alpha male.

The patterning of interventions.

Male interference. Central males rarely interfered in fights opposing adult females. Thus females could not use males in this context. However, central males often interfered in fights opposing adult males and females, intervening more often in the more severe attacks. Out of 63 such interventions recorded during the birth season, males supported other males against females on seven occasions only (11.1 percent). In the remaining 56 cases, one male (two males in two instances) defended one or more females who were being attacked by a male, or supported one or more females against a male (67 male interferer-female beneficiary dyads).

All females received aid from central males but, except for one instance of aid from male 433, only the five highest ranking males aided females (Table 8.5), and they did so as dominant to the target (Table 8.6). The fact that males entered fights only against lower-ranking males had two consequences. First, male intervention terminated the agonistic interaction in most cases. Second, the lower a male ranked, the smaller the number of males he defended females against and, as a result, the less valuable an ally he was for females. This patterning is supported by data collected during the *breeding* season. In order to make these data comparable, only the instances where males aided *non*-receptive females were included. The five highest-ranking males in a hierarchy of 18 performed 80 percent of the 88 instances where adult males assisted adult females against lower-ranking adult males (128 male interferer-female beneficiary

Table 8.5 — Frequencies with which central males assisted adult females in their agonistic interactions with adult males, all sampling sources combined (56 instances corresponding to 67 male interferer-female beneficiary dyads).

Males	FB	510	577	576	511	287	503	7L	FC	573	297	471	N2	411	K1	476	543	408	310
	Females																		
415														3	1				
580	5		2			1	1		1	1		1	1						
EE				2	2		2	3	2	4	2	1			3	1	2	2	2
TJ	1			1	1		2		1		1	1	1		2			1	1
1J	2	1		1				1		1		2							
Z2																			
339																			
433					1														
436																			
495																			
N8																			
Z0																			
358																			
285																			
282																			

Table 8.6 — Frequencies with which central males threatened or attacked other males while assisting adult females, all sampling sources combined.

* TJ and IJ assisted females against 580 as dominant to him (i.e. before Mid-May)
**EE defended females against 580 twice as subordinate to him and once as dominant

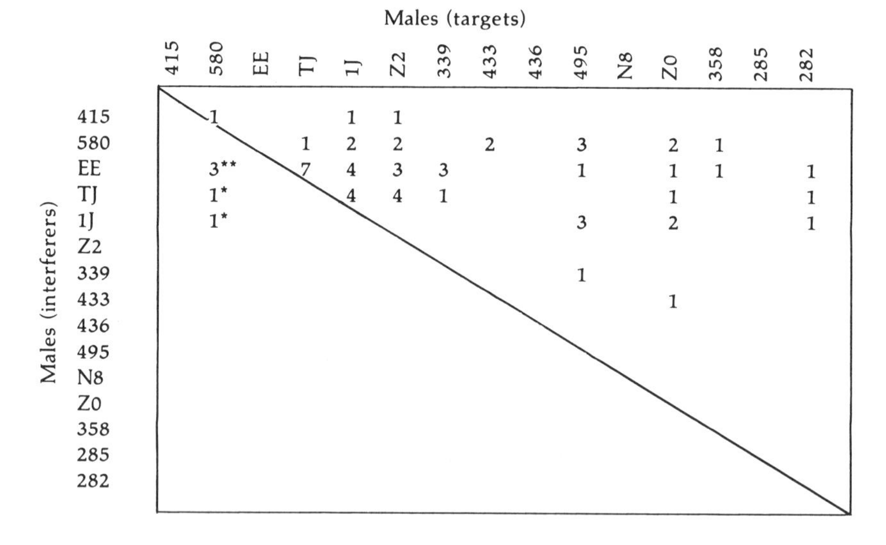

Males (interferers)	415	580	EE	TJ	1J	Z2	339	433	436	495	N8	Z0	358	285	282
	Males (targets)														
415		1			1	1									
580				1	2	2		2		3		2	1		
EE		3**		7	4	3	3			1		1	1		1
TJ		1*			4	4	1					1			1
1J		1*								3		2			1
Z2															
339										1					
433												1			
436															
495															
N8															
Z0															
358															
285															
282															

dyads). The seven highest-ranking males performed 91 percent of the 88 interventions.

Female interferences. Adult females were almost (the exception being 471) never observed to support an unrelated central male against an adult female. On the contrary, they often supported other adult females against unrelated central males. Coalitions of relatives were common and coalitions including unrelated females were not infrequent. The analysis of 61 coalitions observed during both the birth and the breeding season reveals that females of each of the four matrilines aided females belonging to all other matrilines.

Adult females were also observed to interfere in fights opposing central males. Although such fights were relatively infrequent they always elicited much excitement. In the majority of cases, females barked and hoarse-grunted at one of the two male combatants, but they sometimes intervened more directly by chasing one of the males. Combining the data of the birth and breeding seasons (in order to increase sample size), it is found that in 27 of the 29 recorded female interventions the dominant male was supported, and that 15 of the 19 adult females supported 12 adult males. Thus for a male, challenging a higher-ranking male could mean facing a coalition including adult females. Therefore, females may have contributed to stabilize male-male rank relations, and this factor might account for the low frequency of intermale fights taking place in sight of adult females.

In summary, it appears that the five highest-ranking males were useful to adult females in their agonistic relations with lower-ranking males, and that adult females, in turn, might have been useful to the highest-ranking males in their dominance interactions with lower-ranking males.

Discussion

Group F adult males and females engaged in social relationships during the birth season and this cannot be explained in terms of the continuation of sexual cycling (see Loy, 1971). First, the failure of females to conceive is not a necessary condition for the existence of male-female affiliative bonds: 17 of the 19 adult females were pregnant or lactating and regularly affiliated with one or more adult males. Although the two other females did not give birth to live offspring, they maintained long-term, high-frequency grooming relations with adult males. Second, neither the content of male-female relations (e.g., the attractiveness of high-ranking males, the distribution of aiding, etc.) nor their temporal patterning (long-term, continuous versus short-term, cyclical) are compatible with oestrous-like relationships.

Results will now be examined in an attempt to see how males and females might have mutually benefitted from social relationships maintained during the birth season.

Benefits accruing to females

Adult females frequently received aggression from males and some of these aggressive acts could result in serious wounds (e.g., the loss of an eye by female 471). Furthermore, male aggression appeared relatively unpredictable in view of the fact that a significant proportion of attacks seemed spontaneous and deliberate and that the female aggressee had hardly enough time to react in many cases. For these reasons, it is likely that aggression by males represented a constant threat so that females feared most adult males on a regular basis, as evidenced by their monitoring the activities of nearby males and directing frequent submissive gestures to them. On the other hand, males were never observed to attack a female when the latter was close to a higher-ranking male. Therefore, it is highly probable that proximity to the highest-ranking, nonaggressive males (415 and EE) had the effect of reducing the frequency of wounds that might otherwise have been inflicted to females by males. Furthermore, females were assisted by the five highest-ranking males in their agonistic interactions with lower-ranking males. Although 580, TJ, and IJ were aggressive with most females (but not with the alpha female FB) they nevertheless protected them against lower-ranking males. Thus, adult females could use the highest-ranking males, whether they associated with them or not, against lower-ranking males.

Second, affiliation with high-ranking males appeared to raise the competitive ability of adult females in certain circumstances. For example, adult females and the highest-ranking males (415 and EE in particular) regularly ate before lower-ranking males at the artificial feeders. Unfortunately, no quantitative data are available on this topic. Thus, adult females seemed to enjoy priority of access to food over middle and low-ranking males as a result of the latter avoiding high-ranking males and these tolerating adult females. In support of the present hypothesis, Watanabe (1979) suggested that alliances between the leader male and adult females in one troop of Japanese macaques might have forced adult males to stay at the periphery of the feeding ground, and Packer and Pusey (1979) reported that high-ranking male Japanese macaques often supported adult females against lower-ranking males and that only the five highest-ranking males did so.

In conclusion, adult females appeared to benefit from their relationships with high-ranking males mainly in the context of agonistic relations with lower-ranking males.

Benefits accruing to central males.

The mate selection hypothesis. It is conceivable that male-female relationships taking place during the birth season aimed ultimately at influencing later choice of mating partners. The rationale is as follows: If an oestrous female shows a marked preference for one of two male suitors (for *A* over *B*), the cost of consorting the female is higher for male *B* (Bachmann and Kummer, 1980; Strum, 1982). Moreover, if the female is reluctant to mate with male *B*, the benefits of consorting are lower for *B* than for *A*. Now, it is known that females do play an active role in mate selection in multimale primate groups (Lindburg, 1971; Stephenson, 1974; Hausfater, 1975; Packer, 1979; Tutin, 1979; Taub, 1980; Strum, 1982; Chapais, 1983d). Thus, female choice may provide an asymmetry in payoff whereby a male gains access to an oestrous female before a rebuffed higher-ranking suitor (Chapais, 1983d). On the basis of this reasoning, it is conceivable that males might attempt to create such payoff asymmetries long before the onset of consorting activity. For example, a male could groom or defend a female in order that she seeks proximity to him. If such is indeed the case, there should exist a concordance between the identity of the individuals who affiliated in nonsexual contexts and the identity of those who formed consortships. While Seyfarth (1978), Packer (1979), and Smuts (1983b) found some evidence in support of this hypothesis for baboons, Baxter and Fedigan (1979) and Takahata (1982) reported no concordance between the birth and the breeding season partners in Japanese macaques.

Data on the sexual interactions of the 19 adult females of Group F were recorded during the breeding season following the birth season under study (Chapais, 1983d). Table 8.7 compares the birth season grooming dyads (data from Table 8.3) and the breeding season sexual partners.

Male 415, who had been in close contact with six adult females (N2, 411, K1, 476, 511, 7L) throughout the birth season, consorted intensely with female 297, with whom he had not affiliated previously, and ignored his favorite birth season partner (411) for the greatest part of her oestrous. 415 mated only once with his mother (FB) whom he ignored completely during her oestrous periods.

Male 580 consorted and mated with 10 females whom he had not affiliated with during the birth season. Furthermore, 580 consorted only briefly and mated infrequently with 511, one of the two females

Table 8.7 — A comparison of the birth season grooming dyads and the breeding season sexual partners.

	Birth season		Breeding season	
Males	Grooming relation* of type A with females:	Grooming relation* of type B,C,D or E with females:	Mated** at least once with females:	Might have fertilized*** females:
415	511, 7L, 411	FB, 577, 503, 576, FC 573, N2, K1, 476, 510	FB, 297, 411	
580	FB, 511	577, 576, 503, 510 573	511, 287, 503, 7L, FC, 573, 310, 297, 471, N2, 411, 476, 543	287, 503, 7L, N2, 471, 411, 476, 543
EE	FB, 576, 287, 573, 411, 408	577, 511, 7L, 503, FC, N2, K1, 543, 310	FB, 576, 287, FC, 297, N2, 411, K1	FB, N2
TJ	FB	471	FB, 576, 287, FC, N2, 411, K1, 310	K1
1J	FB, 510, 471		FB, 576, 511, 471, 411	510, 287
Z2			FB, 577, 576, 511, 287, 503, 573, 297, 471, 411, 310	577, 287
339		503, 471	FB, 510, 577, 511, 287, 503, FC, 573, 310	510, 577, 511
433		287, 7L	510, 577, 576, 511, FC, K1, 310	510, 310
481			511, 287, 573	
285			FB, 510, 577, 576, 511, 573, N2, 411	
491			FB, 576, 503, 476, 543, 310	476
557			FB, 577, 576, N2, 543, 476	
282			FB, 576, 287, 573, 297, 476, 310	
284			503, 573	
440				
4T			FB, 287, N2, K1, 476	

Continued —

* see Table 3
** data in Chapais, 1983d
*** as measured by the proportion of observation time spent by the male consorting the female during her presumed period of conception (see legend of Table 8). A male was considered a potential father only if the proportion of time he spent consorting the female was greater than one-half the proportion achieved by the male who had the highest proportion for this female.
Males are arranged in descending rank order.

Table 8.7 Continued —

	Birth season		Breeding season	
Males	Grooming relation* of type A with females:	Grooming relation* of type B,C,D or E with females:	Mated** at least once with females:	Might have fertilized*** females:
279				
N8		511	N2, 411, K1	
358		573		
484				
436		N2, 576	411	
495				
Z0		310		

* see Table 3
** data in Chapais 1938d
*** as measured by the proportion of observation time spent by the male consorting the female during her presumed period of conception (see legend on Table 8). A male was considered a potential father only if the proportion of time he spent consorting the female was greater than one-half the proportion achieved by the male who had the highest proportion for this female.
Males are arranged in descending rank order.

he had been in close contact with during the birth season (the other being his mother whom he did not consort). 580 also avoided mating with the other females of his matriline (510, 577, and 576).

Male EE's sexual performance was far below what one might have expected on the basis of his attractiveness during the birth season. EE consorted assiduously with one of his birth season partners (alpha female FB) over her three oestrous periods and was not often active sexually with most other females with whom he had affiliated closely over the preceding months.

Males TJ and lJ had each been in close contact with females FB and 471 (and, for male lJ, with 510). If both males did mate with FB, they did not have access to her when she was likely to be ovulating nor did they have significant access to female 471 when she was receptive. However, male IJ accumulated consort time with 510 when she was likely to be ovulating. On the other hand, males TJ and lJ mated with females with whom they had not affiliated previously.

Finally, although the 18 males ranking lower than lJ had not maintained any long-term high-frequency affiliative bonds with adult females, a majority were active sexually, and four of them may even have fertilized some females. The four males who groomed females during the birth season did not mate with the same females.

In conclusion, the above analysis reveals little concordance between the patterning of the birth season male-female relationships and the patterning of sexual activity. As discussed at length elsewhere (Chapais, 1983d), the major part of the variance in male reproductive activity could be understood in terms of male-male competitive interactions (dominance) acting concurrently with the capacity of males to influence female choice (e.g., through interferences in the consortships of lower-ranking males).

If central males did not seem to benefit from their affiliative relationships with adult females in terms of sexual access, they may have taken advantage of these bonds in some other ways. For example the three top-ranking males received much grooming from adult females. However, various lines of evidence suggest that high-ranking males gained other, more important benefits from affiliating with females, namely remaining dominant over males who might outrank them on a dyadic basis.

First, as described elsewhere (Chapais, 1983b), natal males 580 and 415 owed their rank to support received from female relatives belonging to the highest-ranking matrilines.

Second, dominance relations between non-natal males were clearly of a coalitionary nature in some cases, for which relevant data are available. For example, male EE (3rd-ranking but 1st-ranking among non-natal males) was the oldest male of Group F (20), the one with the longest tenure (9½ years) and certainly not the strongest of them all. Many younger males were much bigger, more vigorous, and more agile. In view of this, it is highly probable that male EE owed his rank to the existence of a permanent coalition between himself and adult females. Lower-ranking males did not dare to challenge EE and were consequently unable to assess his real strength. Whenever EE attacked a lower-ranking male who was involved in a fight with a female or an immature monkey, many other males and females would hoarse-grunt and bark at him in support of EE, and the male would flee without even attempting to fight back. Another illustration of the polyadic nature of male-male dominance relations is provided by the behavior of a non-natal male (418) who visited Group F during the following breeding season. Male 418 defeated the most dominant males of Group F in dyadic encounters, but was overpowered by coalitions of males and females. He left the group after a few days.

Third, consistency in the direction of female interventions in fights opposing adult males lends further support to the hypothesis that adult females exerted a significant influence on the rank relations of males. While it is unlikely that females could influence the process of rank *acquisition* by non-natal males (versus natal males), they might play a determinant role in the *stability* of their rank relations.

Fourth, the alpha female's special status in relation to central males seemed to result from her high value as an ally for these males. That this phenomenon is not idiosyncratic is supported by evidence reviewed by Gouzoules (1980) on the capacity of females to influence the rank relations of adult males, and on the special status of alpha females who were observed to maintain intense affiliative relationships with high-ranking males in groups of rhesus, Japanese, and pig-tailed macaques. FB was the focus of interest of adult males and adult females as well. She received more grooming from more females than any other adult female (Chapais, 1983c). It is possible that FB's high social value stemmed from her two sons occupying the first and second highest ranks, a fact which would point to FB as the single most powerful ally. However, FB enjoyed the same status four months before her son 580 became part of the adult male hierarchy and she affiliated relatively little with her other son, the alpha male 415. Moreover, the argument could be taken the other way around: the status of FB's two sons could result from their mother's special one. Whatever the exact determinants of FB's high social value, the fact that the highest-ranking males all agreed on a strong attraction to her lends indirect support to the idea that the alpha female may play a significant role in male-male dominance relations.

Table 8.8 — The reproductive success of the central males estimated on the basis of the time they spent consorting the 18 (to-be-pregnant) females during their presumed conception period.*

Males	Reproductive success**
415	0.14
580	7.01
EE	1.44
TJ	1.72
1J	0.87
Z2	1.57
339	2.39
433	0.02
436	0.00
495	0.00
N8	0.00
Z0	0.00
358	0.00
285	0.00
282	0.00

* The fertile cycle of each female who conceived was identified by backdating 168 days from the known date of parturition; the period of conception was defined as covering the four days preceding the "attractiveness breakdown," that is, the moment presumed to coincide with the onset of the luteal phase of the cycle, and whose diagnostic criterion was the termination of consorting activity (Chapais, 1983d).

** Estimated number of offspring fathered by each male. Three males that were peripheral during the nonbreeding season had nonzero reproductive success (male 481: 0.07; male 491: 0.36; male 557: 0.19). The difference between the summation of male-specific reproductive success values (16) and the number of offspring produced (18) is due to the incompleteness of observational records during the females' conception periods.

In summary, it is suggested that the adult males who were the most central in relation to the core of adult females were in a position whereby they could postpone the moment when they would be outranked by less central and individually stronger males. The evidence available suggests that an important proximate mechanism underlying this phenomenon could be that adult females support the males most familiar to them and that, as a result, lower-ranking males are inhibited from challenging such coalitions.

The most central and highest-ranking males enjoyed a high breeding success during the following breeding season as can be seen in Table 8.8 (male 415's low reproductive success was due to the fact that he became sexually inactive after being severely wounded — he lost one testicle — by the end of the first third of the breeding season). The observation that high-ranking, central males breed more than other males, combined with the possibility that male centrality is in part a matter of long-term coalitions with females, suggests a possible answer to the question of why Group F males remained attached on a long-term basis to a group of females that were not receptive. Male-male competition for breeding access could take the form of a long-term process of competition for centrality in a group, prior to the breeding season. Thus, the observed positive correlation between estimates of rank-specific reproductive success and male rank (Chapais, 1983d) would not result simply from short-term male-male interactions, but also from longer-term male-female relationships (see Bernstein, 1976). These relationships would not determine specific mating partners since there was little concordance between the identity of the birth and breeding season partners. More generally, they would increase the likelihood of a male becoming central, enjoying a longer tenure as a high-ranking male, being active sexually, and enjoying priority of access to physical resources.

REFERENCES

Altmann, J. Observational study of behaviour: sampling methods. *Behaviour* 49:227-266, 1974.

Altmann, S.A. A field study of the sociobiology of rhesus monkeys, *Macaca mulatta*, *Ann. N.Y. Acad. Sci.* 102:338-435, 1962.

Bachmann, C. and H. Kummer. Male assessment of female choice in Hamadryas baboons. *Behav. Ecol. Sociobiol.* 6:315-321, 1980.

Baxter, M.J.; Fedigan, L.M. Grooming and consort partner selection in a troop of Japanese monkeys *(Macaca fuscata,)*, *Arch. Sex. Behav.* 8:445-458, 1979.

Bernstein, I.S. Dominance, aggression and reproduction in primate societies, *J. Theor. Biol.* 60:459-472, 1976.

Chapais, B. The adaptiveness of social relationships among adult rhesus macaques, Ph.D. thesis, University of Cambridge, England, 1981.

Chapais, B. Adaptive aspects of social relationships among adult rhesus monkeys. In *Primate Social Relationships: An Integrated Approach*. R.A. Hinde, (ed.) Oxford: Blackwell Scientific Publications. pp. 286-289, 1983a.

Chapais, B. Matriline membership and male rhesus reaching high ranks in their natal troop. In *Primate Social Relationships: An Integrated Approach*. R.A. Hinde, (ed.) Oxford: Blackwell Scientific Publications. pp. 171-175, 1983b.

Chapais, B. Dominance, relatedness and the structure of female relationships in rhesus monkeys. In *Primate Social Relationships: An Integrated Approach*. R.A. Hinde, (ed.) Oxford: Blackwell Scientific Publications. pp. 209-219, 1983c.

Chapais, B. Reproductive activity in relation to male dominance and the likelihood of ovulation. *Behav. Ecol. Sociobiol.* 12:215-228, 1983d.

Clutton-Brock, T.H. and P.H. Harvey. Primate ecology and social organization. *J. Zool. Lond.* 183:1-39, 1977.

Dunbar, R.I.M. and P.E. Dunbar. Social dynamics of Gelada baboons. *Contrib. Primatol.* Vol. 6, Basel: S. Karger, 1975.

Eisenberg, J.F., N.A. Muckenhirn and R. Rudran. The relation between ecology and social structure in primates. *Science* 176:863-874, 1972.

Gouzoules, H. The alpha female: observations on captive pigtail monkeys, Folia Primatol. 33:46-56, 1980.

Grewal, B.S. Social relationships between adult central males and kinship groups of Japanese monkeys at Arashiyama with some aspects of troop organization. *Primates* 21:161-180, 1980.

Hausfater, G. Dominance and reproduction in baboons (*Papio cynocephalus*) *Contrib. Primatol.* 7:1-150, 1975.

Hinde, R.A. and S. Atkinson. Assessing the roles of social partners in maintaining mutual proximity as exemplified by mother-infant relations in rhesus monkeys. *Anim. Behav.* 18:169-176, 1970.

Hinde, R.A. and T.E. Rowell. Communication by postures and facial expressions in the rhesus monkey *(Macaca mulatta)). Proc. Zool. Soc. Lond.* 138:1-21, 1962.

Hinde, R.A. and J. Stevenson-Hinde. Towards understanding relationships: dynamic stability. In *Growing Points in Ethology*. P.P.G. Bateson and R.A. Hinde, (eds.) Cambridge: Cambridge University, 1976.

Kaufmann, J.H. Social relations of adult males in a free-ranging band of rhesus monkeys. In *Social Communication Among Primates*. S.A. Altmann, (ed.) Chicago: University of Chicago Press, 1967.

Lindburg, D.G. The rhesus monkey in North India: an ecological and behavioral study. In *Primate Behavior*. Vol. 2 L.A. Rosenblum, (ed.) New York: Academic Press. pp. 1-106, 1971.

Loy, J.D. Estrous behavior of free-ranging rhesus monkeys *(Macaca Mulatta). Primates* 12:1-31, 1971.

Missakian, E.A. Genealogical and cross-genealogical dominance relations in a group of free-ranging rhesus monkeys *(Macaca mulatta). Primates* 13:169-180, 1972.

Morgan, B.J.T., M.J.A. Simpson, P. Hanby and J. Hall-Craggs. Visualizing interactions and sequential data in animal behaviour: theory and application of cluster-analysis methods. *Behaviour* 56:1-43, 1975.

Packer, C. Male dominance and reproductive activity in *Papio anubis Anim. Behav.* 27:37-45, 1979.

Packer, C. and A.E. Pusey. Female aggression and male membership in troops of Japanese macaques and Olive baboons. *Folia Primatol.* 31:212-218, 1979.

Pearl, M. and S.R. Schulman. Techniques for the analysis of social structure in animal societies. *Adv. Study Behav.* 13:107-146, 1983.

Rowell, T.E. and R.A. Hinde. Vocal communication by the rhesus monkey *(Macaca mulatta) Proc. Zool. Soc. Lond.* 138:279-294, 1962.

Sade, D.S. Determinants of dominance in a group of free-ranging rhesus monkeys. In *Social Communication Among Primates.* S.A. Altmann, (ed.) Chicago: University of Chicago Press. pp. 99-114, 1967.

Sade, D.S. A longitudinal study of social behavior of rhesus monkeys. In: *Functional and Evolutionary Biology of Primates.* R. Tuttle, (ed.) Chicago: Aldine, 1972.

Seyfarth, R.M. Social relationships among adult female baboons. *Anim. Behav.* 24:917-938, 1976.

Seyfarth, R.M. Social relationships among adult male and female baboons, II, Behaviour throughout the female reproductive cycle. *Behaviour* 64:227-247, 1978.

Seyfarth, R.M., D.L. Cheney and R.A. Hinde. Some principles relating social interactions and social structure among primates. In *Recent Advances In Primatology.* Vol. 1 D.J. Chivers and J. Herbert, (eds.) London: Academic Press. pp. 39-52, 1978.

Smuts, B. Dynamics of 'special' relationships between adult male and female olive baboons. In: *Primate Social Relationships: An Integrated Approach.* R.A. Hinde, (ed.) Oxford: Blackwell Scientific Publications. pp. 112-116, 1983a.

Smuts, B. Special relationships between adult male and female olive baboons: selective advantages. In: *Primate Social Relationships: An Integrated Approach.* R.A. Hinde, (ed.) Oxford: Blackwell Scientific Publications. pp. 262-266, 1983b.

Stephenson, G.R. Social structure of mating activity in Japanese macaques. In *Proc. Symp. 5th Cong. Int. Primatol. Soc.* S. Kondo, M. Kawai, A. Ehara and K. Kawamura, (eds.) Japanese Science Press, Tokyo. pp. 63-115, 1974.

Strum, S.C. Agonistic dominance in male baboons: an alternative view. *Int. J. Primatol.* 3:175-202, 1982.

Strum, S.C. Use of females by male olive baboons. *Am. J. Primatol.* 5:93-109, Takahata, Y. Social relations between adult males and females of Japanese monkeys in the Arashiyama B troop. *Primates* 23:1-23, 1982.

Taub, D.M. Female choice and mating strategies among wild Barbary macaques (Macaca sylvanus L.) *In The Macaques: Studies in Ecology, Behavior, and Evolution.* D.C. Lindburg, (ed.) Van Nostrand, Reinhold, N.Y. 1980. of wild chimpanzees *(Pan troglodytes schweinfurthii). Behav. Ecol. Sociobiol.* 6:29-38, 1979.

Tutin, C.E.G. Mating patterns and reproductive strategies in the community of wild chimpanzees *(Pan troglodytes schweinfurthu). Behav. Ecol. Sociobiol.*

Watanabe, K. Alliance formation in a free-ranging troop of Japanese monkeys in the Arashiyama B troop *Primates* 20:459-474, 1979.

Wrangham, R.W. An ecological model of female-bonded primate groups. *Behaviour* 75:262-300, 1980.

CHAPTER NINE

Lineage-Specific Mating: Does It Exist?

CAROL A. McMILLAN

INTRODUCTION

In nearly all species of group-living primates and other mammals whose behaviors have been studied in the wild, one sex generally leaves the natal group at puberty (Baker, 1978). This sexual migration acts ultimately to reduce the probability of inbreeding in a group. It is possible that patterns of emigration also affect the genetics of a population in other ways. This chapter gives evidence for intergroup group transfer by related males and for a lineage-specific mating system which may act to increase the genetic distances between matrilineages of rhesus monkeys within a group.

In most species, it is the males who transfer groups (Clutton-Brock and Harvey, 1976). Rhesus monkeys are not an exception to this general rule. Male transfer has been reported in free-ranging colonies (Boelkins and Wilson, 1972; Drickamer and Vessey, 1973; Hausfater, 1972; Sade, 1968). There is increasing evidence that male rhesus monkeys do not transfer groups at random (Sade, 1968; Meikle and Vessey, 1981), but tend to join the same social group as their maternally-related older brothers. Drickamer and Vessey (1973) found that, on la Parguera, four out of nine pairs of brothers, separated in age by less than fifteen months, changed groups together. Of the remaining five pairs, in four cases the older brother shifted groups at between three and four years of age, before the younger brother reached sexual maturity.

On Cayo Santiago, when males leave their natal troop they tend to enter all-male subgroups where they are usually tolerated by the other males (Boelkins and Wilson, 1972). They are accepted by forming a relationship with a male already established in the central group. Wilson (1968) found some evidence that the males who give support are usually relatives, even brothers. Once males are established in the group, Boelkins and Wilson (1972) found that in three cases where known maternal half-sibs were in the same non-natal group, "the males were very close associates" (p. 135). Kaufmann (1967) and Miller, Kling, and Dicks (1973) also found more positive social interactions between relatives. There is evidence that

Illustration 9.1 A consorting male and female at the end of a series mount.

brothers disrupt each other's interactions with estrous females less frequently than expected (Meikle and Vessey, 1981). They also tend to remain in the group longer than do males without brothers in the group.

While group transfer may be seen as an evolutionary mechanism for reducing inbreeding within a population by reducing the chance of parent-offspring matings, the question arises as to whether group transfer by related individuals affects mating patterns in any other way. Since 1956, data have been collected on the Cayo Santiago rhesus colony which have permitted determination of the matrilineal relationships for nearly every animal alive in 1980. In a previous study done on blood group genetic data from the colony McMillan (1979; McMillan and Duggleby, 1981) found that genetic distances (Nei, 1972) between the matrilineages within one troop were no smaller than genetic distances between matrilineages across troop boundaries. If a male and his brothers entering a troop mated at random across matrilines, it would be expected that the genetic distances between those matrilineages would decrease due to the shared paternal genes of the offspring. Matrilineages within a troop should, therefore, become genetically more similar to each other than to lineages in other troops. Since this was not found to be the case, a lineage-specific mating hypothesis was proposed as an explanation for the patterning of genetic distances which was found (McMillan, 1979; McMillan and Duggleby, 1981). If a male (and his brothers) tended to mate within one matrilineage, the matrilineages within one group would remain relatively distant from one another genetically.

Theoretically, lineage-specific mating among matrilines could occur in one of four different ways. First, a male might do most of his mating with females of one lineage. Second, a female could tend to mate with males from a single matriline. Third, related males might mate within one matriline, and, fourth, conversely, related females might tend to mate with males of a single matriline.

MATERIALS AND METHODS

During the 1980 mating season, behavioral data were collected to help confirm or deny this hypothesis. The data were collected during 2,555 hours of observation by four observers on social Group I, one of six groups of the island at that time. Focal animal samples were collected on the 48 females showing multiple signs of estrous, based on behavior, body coloration and perineal swelling. Periods of most probable fertility were later calculated for each female based on a combination of two methods reported by Gordon (1981) and by Riopelle and Hale (1975). A detailed explanation of the use of these methods in this study has been described previously (see McMillan, 1982a). In Group I, 41 males were observed to copulate during the 1980 mating season. A total of 1,796 mating bouts were observed with

816 ejaculations. Census data were taken weekly on the troop.

RESULTS

In order for all but one of the lineage-specific mating hypotheses to be supported it would be necessary for groups of related males to be present in the troop. Since most natal males transfer out of their troop at puberty, males joining another troop would have to be related to one another. To some extent this was found to be the case for Group I in 1980. During the 1980 mating season, the males in Group I were not a random sample from all the lineages on Cayo Santiago. There were 21 lineages on the island having two or more potential breeding males (four years or older) in them. Only 10 of these lineages were represented in Group I in 1980, three of them being the natal lineages. This lack of representation of all lineages was not due to the other eleven lineages being small; 47 percent of the non-natal males on the island belonged to one of the 11 lineages which had no members in Group I. For the lineages which were represented by copulating males in Group I, it was necessary to determine whether their numbers were proportional to the size of the group in relation to the total male population of the island. A chi square test was done for the five non-natal lineages which were large enough for the statistic to be meaningfully applied (Cochran, 1954). Table 9.1 shows the distribution of males (at least four years of age) within these lineages. The null hypothesis that the lineages were proportionally represented in the group was rejected at the .01 level (x^2 = 16.3 with 4 *df*. *C* = .40).

Having found that only certain lineages were represented in Group I and that these were represented disproportionately, it was also found that the relationships of many of the males within each lineage were quite close. Hereafter, a group of related males may be referred to as a *male lineage*. It must be remembered, however, that these are lineages of males related through their female ancestors, the male ancestors being unknown. Of the 26 non-natal males, 17 (65.4 percent) had at least one brother in the group. There were five sets of two brothers, one set of three brothers, and one set of four brothers. Two

Table 9.1 The distribution of males over three years of age who were natal to lineages not in Group I

Lineage	Number of Males in Group I	Number of Males not in Group I	Total
022	8	3	11
092	5	19	24
AC	6	12	18
076	4	9	13
073	2	20	22

individuals from the set of four brothers were not included in Table 9.1 because they were too young to mate. Fifteen males (51.7 percent) had at least one nephew in the group, and there were 12 first cousins: one set of two cousins, one set of three cousins, and one set of seven cousins. Each non-natal male in Group I had an average of over half his brothers (56.0 percent) in the group, one-third of his male first cousins (33.3 percent) and nearly one-fourth of his uncles (23.9 percent) Because there were five other groups on the island at the time, and because Group I represented only 20 percent of the population, these figures were all higher than would have been expected with a random distribution of males around the island.

Since groups of related males were found to be in Group I during the 1980 mating season, the questions of all four types of lineage-specific mating could be addressed. Three tests were done on the overall mating patterns to determine whether mating was random with regard to the lineages of the partners. First, the numbers of completed copulations observed during focal animal sampling were cross-tabulated between the male and female lineages and a chi square test was performed (Table 9.2). The results showed that the number of copulations were not randomly distributed with respect to lineage (x^2 = 54.1 with 8 *df.* $p < .001, C = .32$). Similar results were obtained when the same test was done on a cross-tabulation of ejaculations observed during the periods of most probable female fertility (Table 9.3). Copulations completed during fertile periods were found to be nonrandom with respect to the lineages of the partners (x^2 = 24.9 with 8 *df*, $p < .001, C = .32$). Finally, the number of individuals within the three female and five male lineages who were observed to have mated during the 1980 mating season were cross-tabulated by the lineages of their respective partners (Table 9.4). A chi square test on these data showed that the pairing of individuals among lineages was not random (x^2 = 16.4 with 8 *df.* $p = .04, C = .21$).

After establishing the matings were not randomly distributed among lineages with regard to the total number of ejaculations, the number of ejaculations during a female's period of probable fertility, nor the numbers of mating partners in each lineage, it was necessary to look more closely to determine whether these results were due to more specific patterns which might be in harmony with a lineage-specific mating hypothesis.

A simple lineage-specific mating system would involve each male or female tending to mate with partners who are related to each other. In order to test these hypotheses, a one-way chi square test was done for each of 15 males and 15 females who had a sufficient number of mating bouts for the test to be valid (Cochran, 1954). The null hypothesis that there was no difference between the expected and observed number of matings by each individual with partners in each lineage was rejected at the .001 level for each test, and the contingency

Table 9.2 The distribution of completed copulations observed during focal sampling: female lineages with male lineages

Female Lineages		Male Lineages 091	022	076	AC	092	Total
091	Observed	64	93	14	42	16	229
	Expected	49.4	91.8	27.6	37.0	23.2	
DM	Observed	25	18	18	16	13	90
	Expected	19.4	36.1	10.9	14.5	9.1	
116	Observed	11	75	24	17	18	145
	Expected	31.3	58.1	17.5	23.4	14.7	
		100	186	56	75	47	464

x^2 =54.1 with 8 *df*, $p < .001$, $C = .32$.

Table 9.3 Distribution of copulations during periods of probable female fertility across male and female lineages

Female Lineages		Male Lineages 091	022	076	AC	092	Total
091	Observed	47	109	25	15	7	203
	Expected	31.8	104.7	35.3	16.5	14.7	
DM	Observed	6	21	20	4	10	61
	Expected	9.5	31.5	10.6	5.0	4.4	
116	Observed	1	48	15	9	8	81
	Expected	12.7	41.8	14.1	6.6	5.9	
		54	178	60	28	25	345

x^2 = 24.9 with 8 *df*, $p < .001$, $C = .32$

Table 9.4 Distribution of mating pairs across male and female lineages

Male Lineages		Female Lineage 091	DM	116	Total
091	Observed	58	17	14	89
	Expected	44.7	18.0	26.2	
AC	Observed	29	13	14	56
	Expected	28.2	11.4	16.5	
022	Observed	56	22	43	121
	Expected	60.8	24.5	35.6	
076	Observed	21	12	19	52
	Expected	26.1	10.5	15.3	
092	Observed	17	9	16	42
	Expected	21.1	8.5	12.4	
		181	73	106	360

x^2 = 16.4 with 8 *df*, $p = .04$, $C = .21$

coefficients were all greater than .5. Thus, males and females were both found to mate nonrandomly with respect to their partner's lineage.

A possible confounding factor to these analyses would be individual partner preference; if one male and one female did most of their mating with each other, then mating would not appear to be random among the lineages for either partner. For instance, although a cross-tabulation table might show that most mating was done by a female within a certain male lineage, that effect might be due to the fact that all her mating had been done with just one male in that lineage. It was thus necessary to determine not just the lineage in which the most mating was done, but whether that lineage contained several partners with whom an individual mated more than he or she did with partners outside that lineage. To determine this, a separate Mann-Whitney U test was done for each female with all the males belonging to the five lineages. For each test, the males were ranked according to the number of times they mated with the female. A separate Mann-Whitney U test was then done on each female comparing these ranks for males in the lineage in which the female did her most mating with the ranks for males outside that lineage. Only seven of the 42 females who mated with more than three males (this included all the females observed to mate 10 or more times) showed a significant ($p < .05$) trend toward mating within one male lineage.

Similar tests were done for each male across the female lineages. The females were ranked separately for each male according to their rates of mating with that male. Separate Mann-Whitney U tests were then done comparing the ranks of the females in the lineage in which the male mated most often with ranks of females outside that lineage. The results for males were stronger than the results for the females; close to half of the males (12 of 29) who mated with more than three females (this included all males observed to have mated 10 or more times) were found to mate most often with significantly more females from one lineage (not necessarily the same lineage for each male). For all 12 males, the probability that the distribution of their matings among the female lineages occurred by chance was less than .05, for nine of these males it was less than .01. Thus, 16.8 percent of the females and 41.4 percent of the males tested showed a nonrandom distribution of their matings among the lineages of the opposite sex.

The next question could then be asked: are these tendencies shared among lineage mates? Figure 9.1 is a sociogram illustrating the answer to this question. It shows the direction and significance of these tendencies among the lineages. Each arrow indicates one individual's

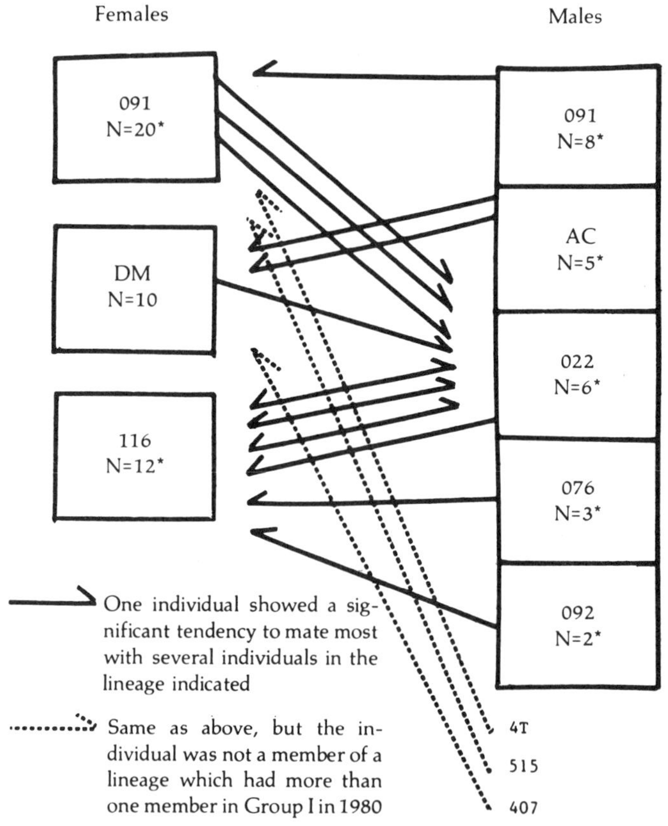

Figure 9.1 Sociogram of the tendencies of individuals to mate across male and female lineages based on Mann-Whitney U Tests.

'perference' based on the results of the Mann-Whitney U tests. Not only was the 'preferred' male lineage the same among females from the same lineage, it was the same for *all* females. The lineage containing males with whom the females engaged in significantly more matings was the same for all seven females: lineage 022. These females represented all three of the female lineages: three were from 091, one was from DM, and three were from 116.

For the males the picture was more complex. Four of the 12 males showing preferences were from lineage 022. This constituted two-thirds of the males in that lineage who were seen to mate with more than three females. All four of these males showed a significant

---

[1]The term 'preference' is used in this section for ease of discussion; it does not reflect any measure of possible emotion or the degree of goal direction in observed behavior; it only reflects a higher than expected frequency of matings between individuals or groups of individuals.

tendency to mate most with females of lineage 116. Two out of the five males in lineage AC showed preferences; both were for females from lineage DM. Only one of eight males in natal lineage 091 showed a statistically significant preference, and this was for the females of his own lineage. Each of the remaining males was the only one in his lineage exhibiting a preference and/or the only male of his lineage in the group. Two preferred the females from lineage 091, two preferred 116 females, and one preferred DM females. Thus, there were two female and one male lineages with members who exhibited significant tendencies to mate within the same lineage as their lineagemate or mates. For one female and one male lineage (116 and 022), the preference was reciprocal. In no case did two lineagemates choose different lineages.

These data tend to support some aspects of lineage-specific mating. The question of whether a group of related individuals tended to mate within one lineage received some support. There were no results contradictory to this hypothesis. Three out of the eight lineages tested could be shown to follow such a pattern to some degree. For the females, the question was complicated by the fact that all females for whom the test results were significant showed a tendency to mate most within the same male lineage — lineage 022.

A word must be added here concerning the fact that all the females who showed a tendency to mate within a group of related males *chose* the same group. This raises the question of what characteristic a lineage might have which would affect the amount and patterning of the matings in which its members engaged. To answer this question the lineages were analyzed for characteristics which might be associated with disproportionate mating frequencies.

Dominance. Dominance rank was only found to be a possible factor among the females. Within a troop, female lineages may be ranked by social dominance. Lineage 091, the lineage of the highest ranking females, was preferred only by three males: one was a young male member of that lineage, and two were old males with no other relatives in the group. The middle-ranking lineage was preferred by males from only one lineage and by one older male who had no relatives in the group. The lowest ranking lineage, 116, was preferred by males from three lineages.

Although the female lineages could be ordered by rank, the male lineages could not. The number of males in each lineage was too small for a significant test of male lineage and dominance rank, but it appears from Figure 9.2 that there was no association between the two variables. It also seems that no association existed between the rank of males in a lineage and the number of females mating within that lineage. The preferred lineage, 022, had a mixture of males of all ranks.

Figure 9.2 Number of males of each rank in each lineage.

Male Dominance Rank	Male Lineage 091	022	076	AC	092
High	XX	XXX	XX	XXX	
Medium	XX	XXX	X	X	XX
Low	XXXX	XX		X	X

Natality. The natality of a male lineage was a factor which could only affect members of lineage 091, the only natal members. No female showed a significant preference for mating with lineage 091 males. In fact, it can be shown that the amount of mating done with lineage 091 males was less than expected for all female lineages. Table 9.5 shows the results of one-way chi square tests for each female lineage on the number of completed copulations involving each of the five male lineages. The expected number of matings did not equal the observed number for any lineage. The expected number of matings between lineage 091 males and lineage DM females (based on numbers of individuals in each lineage) was 18.5 percent higher than the observed number. For 091 females and males of their own lineage the expected number was three times higher than the observed number, and the expected number for lineage 116 females with 091 males was seven times higher than the observed numbers. No other male lineage had numbers of matings lower than expected for all three female lineages. It appeared that females from all three lineages avoided mating with the natal males.

Table 9.5 One-way chi square tests of the distribution of fertile ejaculations in female lineages across male lineages

Female Lineage	X^2	df	*Probability Associated with the Occurrence of H_0	N	C
091	31.8	4	<.001	122	.45
DM	18.6	4	<.001	44	.55
116	32.0	4	<.001	51	.62

Expected frequencies are adjusted for the number of males in each lineage:

$$\frac{\text{\# males in lineage}}{\text{total \# of males}} \times \text{Total \# of matings in the female lineage}$$

*H_{0_j} = The females in each lineage mated the expected amount with the males in each male lineage.

Figure 9.3 Female age with lineage.

Age (years)	091	DM	116
21			X
20			
19			
18			X
17			
16	X		X
15			
14			
13			
12			
11	X	X	X
10	X		X
9	XX	X	XX
8	XX	X	X
7	XXX		
6	XXXX		
5	XXX	X	X
4	XXX	XX	X
3	XX	XX	XXX
2	XXX	X	X
	M = 6.2	*M* = 5.4	*M* = 8.6

Female Lineages

Age. Age did not seem to be a factor in male choice of females of any particular lineage. It can be seen from Figure 9.3 that age distributions were very similar within each of the female lineages, making it impossible for the males to choose a lineage on the basis of female age. Although the mean ages of the females in each lineage differed (6.2 years for lineage 091, 5.4 years for lineage DM, and 8.6 years for lineage 116), a median test of age showed no significant difference between the female lineages (x^2 = 2.11 with 2 *df*. p = .555, *n.s.*).

It appears that age may play a more important part among male lineages than among female lineages. Male age was found to be the significant factor in the female preference for lineage 022 males. The

Figure 9.4 Male age with lineage.

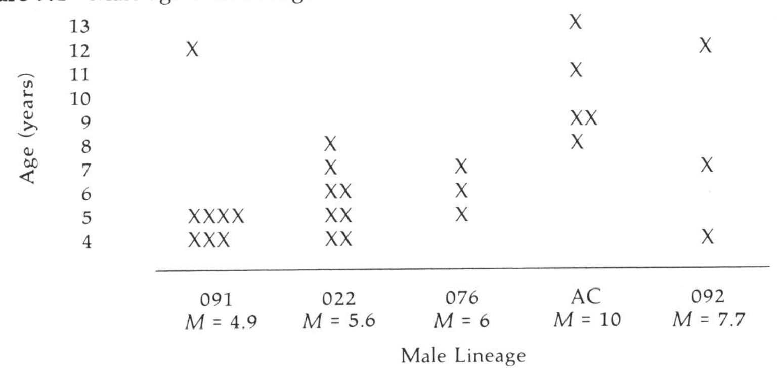

number of males in each lineage was too small for statistical tests to be valid, but the graph in Figure 9.4 shows that the age distribution of males varied greatly among the five lineages. Lineage 022, the lineage in which the seven females did proportionally the greatest amount of mating, was the lineage with the largest number of six, seven, and eight-year-old males. It was shown in a previous paper (McMillan, 1982b) that males of these ages are the most active sexually. This raises a pertinent question: did the females mate more with lineage 022 males because of the ages of the males within that lineage, or were the six, seven, and eight-year-old males found to mate most because they were nearly all in lineage 022, and females preferred males of that lineage for some other reason?

In order to distinguish between these two possibilities the Mann-Whitney U tests were repeated on each female after the scores of the six, seven, and eight-year-old males were removed from the data. If females still were found to mate most with the remaining four males in lineage 022, it would be expected that some factor other than age was causing the increased number of matings in that lineage. However, after the data on young adult males were removed from the analyses, no significant preference was found of any female for males of any lineage. It may thus be assumed that the factor associated with the disproportionate number of males in lineage 022 doing large amounts of mating was due to their age and not to some other factor unique to that lineage. Male age may also be the important factor in the low frequency of matings observed between all three female lineages and the one natal male lineage, 091. All but one of the males in lineage 091 was subadult, and it has been shown in the previous paper (McMillan, 1982b) that subadult males mate less than adults. Therefore, natality may not be the pertinent factor, but, once again, it is probable that male age was causing the disproportionately low number of matings observed with lineage 091 males.

DISCUSSION

The data gave some support to a simple lineage-specific mating hypothesis. Forty percent of the males who could be tested showed a significant tendency to do most of their mating with females of one lineage (the lineages differed among the males). The same trend was exhibited by 16.8 percent of the females tested, but all the females preferred the same male lineage, and this was the lineage which contained nearly all of the six, seven, and eight-year-old males. Male age was found to be the significant variable, and the lineage of the partner was discounted as not having a significant effect on mating for females. However, a partner's lineage was considered to be of moderate importance in male mating patterns.

The proximate causes of a tendency toward lineage-specific mating may be relatively simple. Males may choose to enter the same troop as their close male relatives simply because these are the only individuals outside the natal troop with whom a male is familiar. Since male relatives appear to form affiliative relationships with each other in a new troop, the chance of a male and one or more of his relatives mating with the same female is increased by the proximity of related males to each other. On the other hand, since female members of a matrilineage generally remain in close proximity to one another (McMillan, 1982a), a male who mates with one female may be likely to mate with her relatives simply because they are nearby. If he forms a close association with one female he will be more likely to be near her relatives than to be near other females who come into estrous.

Many factors may act together as ultimate causes of lineage-specific mating; lineage-specific mating fits neatly into almost everyone's theory. While its overall impact on the gene pool and, therefore, its evolutionary consequences have yet to be modeled, several speculations are possible.

Parental Investment Theory. The results of this study are consistent with parts of Trivers' (1972) parental investment theory. A female can bear only one offspring after a single mating season, but a male might produce several. If a female mated with males who were closely related to each other, their relationship would not be reflected in her offspring, since she would still only have one per year. Thus it would not particularly benefit one female to mate with related males. In this study, females did not show a strong tendency to mate with related males. Such was not the case for males, however. A male may produce several offspring in one season; some advantages may exist if these offspring are borne by females who are related to one another. Since groups of related females tend to stay in close proximity to one another, a male who had more than one offspring in one year could assist in the protection of those offspring more easily by mating primarily with females who were related to one another. Thus, any assistance he gave the lineage would benefit him more than if he had only one offspring in the group, since that assistance would increase the likelihood of several of his offspring surviving, not of just one. Selection would, therefore, favor males who, for whatever proximate reason, assisted in groups containing more than one of their offspring. Since individual paternities have yet to be ascertained on Cayo Santiago, it has not been demonstrated that males remain in a group and assist their probable offspring. However, Hamilton and Busse (1980) have observed that chacma baboon males having infant offspring in a troop emigrate significantly less than expected if paternity were not a factor.

If probable paternity does increase the likelihood of a rhesus male remaining in a group and giving aid after the mating season has ended, then related females who mate with one male might benefit from the possibility that his increased investment would cause him to remain nearby to assist and protect his larger number of possible offspring. Selection would favor females who acted in a way which encouraged a male to remain with the lineage and to aid her with her offspring. Evidence for a mechanism which increases the likelihood of males mating with females who are related to each other was reported earlier for the Cayo Santiago population (McMillan, 1981). It was found that the mating runs of related females within sublineages tended to be synchronized, making it easy for a male to mate with several related females in a short time period.

Thus, from the point of view of parental investment theory, selection should favor a system where one male mates with related females. This was the strongest of the lineage-specific mating patterns which were found in this study.

Kin Selection Theory. The data from this study also lent some support to the hypothesis that significant preferences would be shared among lineagemates. Kin selection theory (Maynard-Smith, 1964) may offer some explanation of this trend for a male to mate within the same lineage chosen by his brother or other close relatives. In such a case, if a male assisted in the protection and/or rearing of offspring in the lineage, even if he were not supporting his own offspring, the chances would be good that he would be supporting infants who carried some of his genes which had been passed on to them by his relatives.

Stirring the Gene Pool. The evidence found for a lineage-specific mating system has already been discussed in terms of parental investment and kin selection theory. An argument can also be made that such a system, in the long run, would be beneficial to the group as a whole. Macaques appear to have had a history as a colonizing species, entering many new habitats. In such species it is advantageous to keep large genetic variability in the population. Inbreeding in a large gene pool may cause a possibly beneficial new gene to be lost, but a lineage-specific mating system might act to localize new genes while avoiding the problems of inbreeding.

Although the pattern found in this study was only moderate and only involved certain lineages, if it were to continue over time, it would result in the lineages within each troop being relatively genetically isolated from one another. If a gene were expressed in heterozygotes, the lineages might act as genetic test groups where a new gene would have a chance of expression even though inbreeding were being avoided. A beneficial gene would act to increase the number of individuals in the lineage and, thereby, spread to a new lineage when males with the gene left the group to mate. Deleterious genes, on the other hand, would be most likely to decrease the size of only one or two lineages, so their effect, to some degree, would tend to be contained in a small subpopulation of the group. Such a mating system would act to keep many diverse genes localized within the population. In a species such as the rhesus, which has colonized many diverse environments, the evolution of such a mating system would have been advantageous. Although arguments such as this for species and group selection are difficult to model (see Boorman and Levitt, 1972; and Gilpin, 1975 for group selection models), the ultimate advantages which observed patterns may give to the species as a whole should not be overlooked. A more thorough modeling of the effects of such a mating system would be a valuable topic for future research.

CONCLUSIONS

The simplest form of lineage-specific mating was supported by data from the males, but not by the female data. Over 40 percent of the males tested each chose one female lineage in which they did the majority of their mating (these lineages were not the same for every male). However, the data showing that 16.8 percent of the females showed a similar preference for a male lineage were probably reflecting an association between male age and mating success and not any actual preference for a particular lineage. Male age appeared to be more significant than lineage membership in determining female preferences. For male preference, however, female lineage membership appeared to play a large part.

Proximity and familiarity of individuals may be the proximate causes of lineage-specific mating. In order to determine which ultimate causes may be affecting lineage-specific mating, continuing studies should be done to determine whether males who mated often within a lineage tended to remain during the following years and protect and/or assist with the offspring. The data in this study are cross-sectional, and verification can only occur with a longitudinal study of the same population. Genetic studies should also be done to determine whether related males tended to father offspring in one lineage, and, over time, how such mating patterns would affect the distribution of genes in the lineages.

ACKNOWLEDGEMENTS

I would like to thank the Caribbean Primate Research Center of the University of Puerto Rico and my three assistants, Cheryl Alongi, Phil Jackson, and Peter Killoran for making this study possible. This work was supported in part by NIH contract NO1-RR-7-2115 to the Caribbean Primate Research Center of the University of Puerto Rico.

REFERENCES

Baker, R.R. *The Evolutionary Ecology of Animal Migration,* Hodder and Stroughton, London, 1978.

Boelkins, R.C., and Wilson, A.P. Intergroup social dynamics of the Cayo Santiago rhesus *(Macaca mulatta)* with special reference to changes in group membership by males. *Primates* 13:125-140, 1972.

Boorman, S.A., and Levitt, P.R. Group selection on the boundary of a stable population. *Proc. Nat. Acad. Sci.* 69:2711-2713, 1972.

Clutton-Brock, T.H., and Harvey, P.H. Evolutionary rules and primate societies. In Bateson, P.P., and Hinde, R.A. (eds.) *Growing Points in Ethology,* Cambridge University Press, Cambridge, pp. 195-238, 1976.

Cochran, W.G. Some methods for strengthening the common chi square tests. *Biometrics* 10:145-158, 1954.

Drickamer, L.C., and Vessey, S.H. Group changing in free-ranging male rhesus monkeys. *Primates* 14:359-368, 1973.

Gilprin, M.E. *Group Selection in Predator-prey Communities,* Princeton University Press, Princeton, 1975.

Gordon, T.P. Reproductive behavior in the rhesus monkey: social and endocrine variables. *Amer. Zool.* 21:185-195, 1981.

Hamilton, W.J. III, and Busse, C. Male transfer and offspring protection in chacma baboons. Paper presented at the VIIth International Congress of Primatology, Florence, 1980.

Hausfater, G. Intergroup behavior of free-ranging rhesus monkeys *(Macaca mulatta). Folia Primatol.* 18:78-107, 1972.

Kaufmann, J.H. Social relations of adult males in a free-ranging band of rhesus monkeys. In Altmann, S.A. (ed.) *Social Communication Among Primates,* University of Chicago Press, Chicago, pp. 73-98, 1967.

Lindburg, D.G. Rhesus monkeys: mating season mobility of adult males. *Science* 166:1176-1178, 1969.

Maynard-Smith, J. Group selection and kin selection. *Nature* 201:1145-1147, 1964.

McMillan, C.A. Genetic differentiation of female lineages in rhesus macaques on Cayo Santiago. *Am. J. Phys. Anthropol.* 50:461-462, 1979.

McMillan, C.A. (1979). Synchrony of estrus in macaque matrilines at Cayo Santiago. *Am. J. Phys. Anthropol.* 54:251, 1981.

McMillan, C.A. Factors affecting mating success among rhesus macaque males on Cayo Santiago. Ph.D. dissertation, State University of New York at Buffalo, UMI #DA8303226, 1982a.

McMillan, C.A. Male age and mating success among rhesus macaques. Paper presented, at the IXth Congress of the International Primatological Society, Atlanta, GA., 1982b.

McMillan, C.A., and Duggleby, C.R. Interlineage genetic differentiation among rhesus macaques on Cayo Santiago. *Am. J. Phys. Anthropol.* 56:305-312, 1981.

Meikle, D.B., and Vessey, S.H. Nepotism among rhesus monkey brothers, *Nature* 294-160, 1981.

Miller, M.H., Kling, A., and Dicks, D. Familial interactions of male rhesus monkeys in a semi-free-ranging troop. *Am. J. Phys. Anthropol.* 38:605-612, 1973.

Nei, M. Genetic distance between populations. *Am. Nat.* 106-283-292, 1972.
Neville, M.K. A free-ranging rhesus monkey troop lacking adult males. *J. Mammalol.* 49:771-773, 1968.
Riopelle, A.J., and Hale, P.A. Nutritional and environmental factors affecting gestation length in rhesus monkeys. *Am. J. Clin. Nutr.* 28:1170-1176, 1975.
Sade, D.S. Inhibition of son-mother mating among free-ranging rhesus monkeys. *Sci. Psychoanal.* 12:18-38, 1968.
Trivers, R.L. Parental investment and sexual selection. In Campbell, B. (ed.), *Sexual Selection and the Descent of Msn 1871-1971,* Aldine, Chicago, pp. 136-179, 1972.
Wilson, A.P. Social behavior of free-ranging rhesus monkeys with an emphasis on aggression. Ph.D. Thesis, University of California at Berkeley, 1968. Cited in Crook, J.H. *Anim. Behav.* 18:197-209, 1970.

Illustration 10.1 A young female forages vegetation.

CHAPTER TEN

Hair Mineral Content as an Indicator of Mineral Intake in Rhesus Monkeys (Macaca mulatta)

BERNADETTE M. MARRIOTT, J. CECIL SMITH, JR., RICHARD M. JACOBS, ANN O. LEE JONES, RICHARD G. RAWLINS, MATT J. KESSLER

INTRODUCTION

The annual capture of rhesus monkeys on Cayo Santiago provides scientists with the opportunity to obtain anthropometric and selected physiological data with which to correlate behavioral observations. In our studies of the behavior and food intake patterns of rhesus monkeys in different environments, the Cayo Santiago groups serve as an ideal intermediate in our three populations: a free-feeding monkey troop in Nepal, free-ranging provisional animals on Cayo Santiago, and group-housed monkeys who only have access to a commercial diet in Maryland. As part of this comparative study, we have collected hair and serum samples to provide physiological parameters with which to attempt correlation with food and nutrient intake.

Hair is a tissue which incorporates trace minerals into its structure during the growing process. In humans, scalp hair grows at a rate of approximately 0.4 mm/day and the developing hair is exposed to the metabolic environment before the outer layers form a seal containing the trace metals accumulated during formation. Trace minerals are found in relatively high levels in hair. Unlike blood values, hair samples are not affected by short-term changes in diet but reflect longer-term overall intake. Recognition of these advantages has lead to the use of hair samples as a tool in forensic medicine (Perkons and Jervis, 1962) for the detection of prolonged administration of hazardous elements such as arsenic.

Additional advantages such as ease of collection and storage have lead nutritionists to remain interested in hair as a potential noninvasive biopsy material to measure mineral nutritional status in humans. The assumption is that hair mineral values will reflect excessive or deficient intakes of minerals. In humans, the correlation of hair mineral concentration with blood values is confounded by hair treatment. Contamination from metals applied to the hair surface through use of permanents, dyes, and shampoos cannot be readily assessed (McKenzie, 1978). Natural hair color and texture (Schroeder and Nason, 1969), as well as sampling location along the hair shaft or head (Hambidge, 1973; Swift and Brown, 1972) may also lead to variability in mineral concentration. These human subject variables and the difficulty with which one can separate endogenous from exogenous sources of minerals have recently lead Hambidge (1982) and others (cf. DeAntonio et al., 1982; Laker, 1982; Hilderbrand and White, 1974) to question the validity of the use of hair samples for measuring mineral nutritive status in humans. It is well known that samples from individual human subjects are not reliable measures of the nutrient status for many elemental nutrients. However, when carefully sampled, prepared, and analyzed human hair can be useful as an indicator of mineral nutrition status in population or longitudinal studies (Gibson, 1980; Hambidge, 1982).

Analysis of hair, fur, and feathers has been commonly used as an indicator of exposure of animals to essential and nonessential elements (Combs et al., 1982). As in humans, element concentrations in animal pelage and plumage are generally 10 times higher than that found in plasma or serum (Maugh, 1978). Hair has been viewed as a biopsy material which has great potential for assessing long-term mineral status or mineral contamination in livestock and wildlife. O'Mary et al., (1970) reported that concentration of Cu in hair of Holstein and Hereford cattle was affected by dietary Cu levels. Anke (1966) found that hair Mg in cattle receiving Mg dietary supplements was higher than during nonsupplementation. Miller (1970) found that dietary Zn levels of cattle and goats were reflected in hair concentrations. Most such reports, however, stress the specificity of their findings and the necessity for careful sample collection and preparation. Even with nonhumans, who do not treat their pelage with exogenous meterials, the utilization of hair to determine mineral status remains controversial due to insufficient baseline information and controlled experimentation.

The usefulness of hair analysis as an indicator of exposure to environmental contaminants has had variable value dependent upon the animal species and metals involved. Kelsall et al. (1975a; 1975b) and Pannekoek et al. (1974) have found that the feather mineral content may be useful in determining the origins of wild lesser snow geese. The hair of experimental rabbits reflected the amount of radioactive Pb injected subcutaneously (Jaworowski et al, 1966). However, Dorn et al. (1974) found that high hair Pb concentration in cows grazing near a Pb smelter was due to exogenous contamination and did not correlate with blood Pb levels. Similarly, Huckabee et al. (1973) reported a higher concentration of Hg in the hair of coyotes compared to the hair of the rodents upon which they preyed.

In captivity, the rhesus monkey is the most widely used nonhuman primate model of human nutrition and disease processes. Rhesus monkeys have a digestive system which is similar to humans. They have also adapted to a wide geographic range in Asia and their omnivorous diet includes a diverse sampling of temperate and tropical plant and insect species. Unlike human hair, monkey hair is not subject to many of the confounding treatment variables mentioned for humans. It may thus prove a suitable tool to study the relationship between hair minerals and mineral nutrition status in a primate model (Pories and Strain, 1966).

There are few studies which have analyzed a broad spectrum of elements or had a sample size which was large enough to examine age and sex variability. The work of Clark and Huckabee (1977) with *Macaca fuscata* represents the only large scale sampling of hair minerals in a nonhuman primate.

The purpose of this chapter is to present an overview of hair mineral content in our initial sample of monkeys from Cayo Santiago. Rhesus monkey hair concentrations of Se (Loew et al., 1975) and Zn (Sandstead et al., 1978; Swenerton and Hurley, 1980) have been previously described for small numbers of animals housed individually in stainless steel cages. Besides providing comparative levels of hair Zn for large groups of freely mobile animals, this study provides hair levels of Cu, Ca, P, Mg, and Mn for comparison with hair levels on other animal species and other physiological measurements on the rhesus monkey.

Materials and Methods

The sample population was composed of 208 free-ranging rhesus monkeys on Cayo Santiago. There were approximately 700 animals on the island at the time of sampling in January 1981, divided into six distinct troops. Genealogy and exact age were known for all individuals. A detailed description of the sample population and habitat have been previously reported (Kessler and Rawlins, 1983). Animals are supplied *ad libitum* with standard commercial high (24-26 percent) protein monkey diets and they also consume plant parts and soil (Sultana and Marriott, 1982). The monkeys were captured, chemically restrained with a 10 mg/kg dose of ketamine HCl, and weighed prior to hair sample collection. All animals selected for study were healthy, vigorous individuals at the time of sampling, which was conducted between the breeding and birth season. Female reproductive state was determined by rectal palpation at the time of sample collection and pregnancy confirmed by subsequent outcome during the birth season. Day of gestation at sample collection was calculated by subtraction on the basis of 168±5 days gestation for the rhesus monkey (van Wagenen, 1954).

Hair samples were taken, using vinyl gloves and stainless steel scissors, by clipping hair as close to the skin as possible from the left shoulder of each animal. Samples were stored in sealed plastic bags or paper envelopes until they were processed for analysis. Prior to analysis, the samples were individually washed using a non-ionic alcohol detergent[1] and deionized water solution (1:10 vol. x vol.) in a mechanical shaker for 30 minutes, rinsed 10 times in deionized water, and oven dried at 70° C in plastic Petri dishes to a stable weight. Hair was then totally digested to a clear, colorless solution in a 1:2 mixture of 70 percent perchloric acid and 30 percent hydrogen peroxide. Care was taken to eliminate sources of metal contamination from collection and processing and to use metal free materials throughout. Hair Ca, Cu, Fe, Mg, Mn, Na, P, and Zn concentrations were quantified by inductively coupled argon plasma atomic emission spectrometry (ICP-AES). Blank samples of the digestion mixture, pooled samples of hair and reference standards were interspersed with samples to monitor precision and accuracy of the analyses. Samples were analyzed in a random order and coded numerically so that the sample identity was unknown to the analyst. Data were analyzed using a two-way analysis of variance (ANOVA), simple linear regression (Nie et al., 1975) and a Student's t-test (Snedecor and Cochran, 1967). The minimum level of statistical significance accepted was $p < 0.05$.

RESULTS

Table 10.1 presents the mean body weights in kilograms for sample animals. Accurate body weights for our entire sample were not available. For example, weight for only one of the 12 nonpregnant adult females is reported.

Table 10.1 Mean Weights of Rhesus Monkeys from Cayo Santiago on Date of Hair Sampling

Age/Sex Class	n/total[1]	Body weights Mean + ISD (Kg)
Juvenile Male (< 3 years)	50/51	2.6 ± 0.7
Adult Male (> 3 years)	32/43	11.0 ± 2.3
Juvenile Female (< 3 years)	76/76	2.1 ± 0.6
Adult Female (> 3 years)		
Pregnant	16/26	10.7 ± 2.4
Not Pregnant	1/12	9.6 ± 0
Total	175/208	

[1]n/totals = number of animals who were weighed at the time of hair sample collection/ total number of animals sampled.

[1]. Acationex, Scientific Products, McGaw Park, IL. 60085.

In the following tables, the sample size ranged from 198 to 208 individuals. Of the essential elements, eight were quantifiable in rhesus monkey hair using the ICP-AES. Of these eight elements the sodium concentration was extremely variable, probably reflecting the high degree of solubility (Clark and Huckabee, 1977) of this element in hair. Therefore, the results for sodium are not reported.

Table 10.2 illustrates the mean ± SEM for the hair concentration of Ca, Cu, Fe, Mg, Mn, P, and Zn.

Table 10.2 Hair Mineral Concentration ($\bar{x}$ ± SEM) in Rhesus Monkeys from Cayo Santiago.

Mineral	n[1]	Mean
		µg/g ± SEM
Calcium	208	415 ± 13.4
Phosphorous	204	193 ± 3.0
Magnesium	208	138 ± 6.7
Manganese	194	2.84 ± 0.150
Iron	208	62 ± 4.2
Copper	208	8.59 ± 0.130
Zinc	208	159 ± 2.6

[1]n = number of samples. A sample was excluded only if sample size was too small for accurate determination for that element.

Table 10.3 Sex Differences in Hair Mineral Concentration ($\bar{x}$ µg/g ± SEM) of Rhesus Monkeys from Cayo Santiago

Mineral	Male	Female
Calcium***,[1]	453 ± 24.1	384 ± 13.7
Phosphorous**,[1]	185 ± 4.4	200 ± 3.8
Magnesium***	174 ± 11.17	109 ± 6.8
Manganese**,[1]	3.29 ± 0.280	2.46 ± 0.129
Iron	65 ± 7.6	60 ± 4.4
Copper	8.32 ± 0.199	8.82 + 0.168
Zinc	161 ± 4.6	158 + 2.9

*** $p \leq .001$
** $p \leq .01$
[1] ANOVA age interaction also statistically significant ($p < .05$)

The large variation of mean values was associated with significant sex or age effects as shown in Tables 10.3 and 10.4 Male hair concentration of Ca, Mg and Mn was significantly higher than female hair concentration ($p \leq .001$) whereas hair phosphorus was found in significantly higher concentrations in female hair samples ($p \leq .01$). The significant main female hair concentration($p \leq .001$), whereas hair phosphorus was found in significantly higher concentrations in female hair samples ($p \leq$

tively). For hair Ca and Mn, subsequent statistical analyses indicated that mean adult hair levels were comparable for both sexes but the mineral concentrations in hair for male juveniles were significantly greater than for females ($p < .001$; $p < .001$). For hair P there were significant differences in the juvenile and adult levels across males and females ($p < .005$; $p < .001$, respectively).

Table 10.4 Age Differences in Hair Mineral Concentration of Rhesus Monkeys from Cayo Santiago

Mineral	Age			
	Juvenile (0-3 years)		Adult (>3 years)	
	$\bar{x}$ µg/g ± SEM			
Calcium***[1]	469	± 19.0	330	+ 12.4
Phosphorous	194	± 3.8	191	± 4.7
Magnesium	128	± 7.1	154	± 12.8
Manganese**[1]	3.24	± 0.234	2.28	± 0.124
Iron**[1]	71	± 6.4	45	± 3.4
Copper***	9.54	± 0.14	7.11	± 0.127
Zinc***	168	± 3.79	145	± 2.4

*** $p \leq .001$
** $p \leq .01$
[1] ANOVA sex interaction also statistically significant ($p < .05$)

As indicated in Table 10.4, overall hair concentrations for Ca ($p < .001$), Cu ($p < .001$), Fe ($p < .01$), Fe ($p < .01$), Mn ($p < .01$) and Zn ($p < .001$) were statistically significantly lower for adults when compared with juveniles. Significant linear regression of hair mineral values on chronological age was subsequently shown for Cu (p .001) and Zn ($p < .001$). Inspection of significant age versus sex interactions for hair Ca ($p < .01$), Fe ($p < .05$), and Mn ($p < .05$) indicated that hair mineral concentrations of these elements declined more rapidly with age in males than in females. In all cases linear regression of hair mineral on chronological age was statistically significant for males (Ca: $p < .001$; Fe: $p < .005$; Mn: $p < .005$) and for females for Ca ($p < .005$).

Table 10.5 Pregnancy Status and Hair Mineral Concentration of Adult Female Rhesus Monkeys from Cayo Santiago.

Mineral	Adult Females: Not Pregnant n[1] = 12	Adult Females: Pregnant n = 26
	$\bar{x}$ µg/g ± SEM	
Calcium	326 ± 10.2	327 ± 10.6
Phosphorous	232 ± 6.2	203 ± 7.0
Magnesium	86 ± 6.0	94 ± 6.2
Manganese	1.97 ± 0.871	2.30 ± 0.998
Iron***	66 ± 14.5	49 ± 4.1
Copper	7.39 ± 0.391	7.41 ± 0.249
Zinc	152.4 ± 8.54	138.2 ± 4.54

[1]n = number of samples
***$p < .001$

Table 10.5 presents the results of analysis of hair concentrations in pregnant and nonpregnant adult females. Hair Fe concentrations were significantly lower in pregnant versus nonpregnant adult females using a Student's *t*-test ($p < .001$). Differences in hair levels for Cu and Zn across these two groups were not significantly different, however, concentration of Zn and Cu declined as gestation advanced.

Monkey adult hair Zn concentrations from this study and comparison with data available in the literature are presented in Table 10.6. A close correspondence of hair Zn levels among adult free-ranging rhesus macaques in Puerto Rico, free-ranging Japanese macaques in

Table 10.6 Adult Hair Zinc Concentration

Species Description	n[1]	Zinc $\bar{x}$ µg/g	Source
Macaca mulatta Free Ranging	81	145	Present Study
Macaca mulatta[2] Stainless Steel Caged	4	155	Sanstead *et al.* (1978)
Macaca radiata Stainless Steel Caged	4	312	Swenerton & Hurley (1980)
Macaca fuscata Free Ranging	80	125	Clark & Huckabee (1977)
Homo sapiens Free Living	——	150	Shapcott (1982)

[1]n = number of samples
[2]all pregnant females

Texas, and normal rhesus monkeys housed individually in stainless steel cages in Montana (Sandstead et al., 1978) is evidenced in this table. A hair Zn concentration of 145 µg/g may thus be fairly representative of a normal value for adult rhesus or Japanese macaques. Shapcott (1982) recently reviewed the literature on human hair zinc levels and concluded that a representative mean value for adult humans was 150 µg/g which is comparable to the macaque concentrations.

In Table 10.7 the hair mineral concentrations reported by Clark and Huckabee (1977) for *Macaca fuscata* are listed with the present data for *Macaca mulatta*. The differences in values between species represent potential variability in species, diet, hair preparation, and, in particular, analysis technique (Shapcott, 1982). For adult animals where sample size is comparable, hair concentrations of Fe and Zn are very similar.

DISCUSSION

This study provides baseline data for eight essential elements in hair in a large population of free-ranging monkeys with a well defined background. The sample population was comprised of animals from one troop on the island — Group L. As such, the sample is more representative of familial rather than cross-lineage patterns. These animals were healthy and well exercised with free access to both a commercial diet and forage. Therefore, these values could be considered as an estimate of normal hair concentrations of these selected minerals. However, before the normal ranges for these elements could be defined, other biological parameters that are known to indicate the nutritional status with regard to these elements must be examined. Additionally, experimental nutritional studies must be carried out to evaluate the ability of hair to reflect changes in dietary intake and nutritional status. Possible confounding effects from other nutrients must also be examined. Also, these data must be considered in light of the sample distribution and the biological as well as statistical significance of the numerical values.

The sample of 208 individuals was distributed fairly evenly across age classes from nine months to 26 years. Although the definition of an aged rhesus monkey remains unclear, in assessing the Cayo Santiago population Kessler and Rawlins (1983) established ≥ 10 years as an "aged animal." Rhesus monkeys on Cayo Santiago have a considerably shorter lifespan than those of captive caged animals, but this observation appears to be typical for many mammalian species. It appears that the high prevalence and cumulative risk of tetanus on Cayo Santiago (Rawlins and Kessler, 1982) may be a factor in the shorter lifespans on Cayo Santiago. There is little information on the lifespan of rhesus monkeys in their natural habitats. So while this sample

Table 10.7 A Comparison of Hair Concentration[1] ($\bar{x}$ µg/g ± SEM) for Free-ranging *Macaca fuscata* and *Macaca mulatta*.

Species/age	Analysis Method	Ca (n)[2]	Mn (n)	Fe (n)	Cu (n)	Zn (n)	Source
Macaca fuscata							
Juveniles	NAA[3]	677±121 (5)	4.0±0.9 (8)	91 ± 15 (5)	13 (2)	125±5 (8)	Clark & Huckabee (1977)
Adults	NAA	466±26 (69)	1.9±0.12 (79)	50±4 (48)	13±2.7 (34)	125±2 (80)	
Macaca mulatta							
Juveniles	ICP-AES[4]	469±19.0 (127)	3.11±0.234 (79)	71±6.4 (127)	9.54±0.143 (127)	168±3.8 (123)	Present study
Adults	ICP-AES	331±12.4 (81)	2.19±0.124 (39)	46±3.4 (81)	7.11±0.127 (81)	145±2.4 (81)	

1 — Samples were taken from the same location on the body in both studies.
2 — number of samples
3NAA = neutron activation analysis
4ICP-AES = inductively-coupled argon plasma atomic emission spectrometry.

population may be proportionally representative of the age distribution of animals on the island of Cayo Santiago, it may not reflect the usual age distribution of animals in captive caged colonies. Due to the constraints of trapping and the above mentioned population parameters, the sample population was not specifically age/sex balanced. The sample population was divided for statistical purposes into pre- and postpubertal animals to elucidate any major age-related trends imitating procedures in the published literature for humans (Laker, 1982; Petering et al., 1971; Klevay, 1970; Kelsall et al., 1975a). In addition, the sample was comprised of animals from one troop on the island — Group L. As such the sample is more representative of familial rather than cross-lineage patterns.

Within these sampling constraints this study is the first large scale analysis of hair minerals in rhesus monkeys. Because the only previous multi-element analysis of nonhuman primate hair was from *M. fuscata*, a species which is not used as extensively in biomedical research as *M. mulatta*, hair mineral values for the rhesus monkey aid in the development of the biological characteristics of this species.

From the present survey the mineral concentrations of adult rhesus monkey hair for Ca, P, Mg, Mn, Fe, Cu, and Zn often closely paralleled values reported for adult humans (Iyengar et al., 1978). Significant changes in hair Cu with age and hair Zn with pregnancy parallel some reports for human subjects (cf. Hambidge et al., 1983), but not others (Petering et al., 1971; Vir et al., 1981). The mean values for most elements, with the exception of Zn, were lower but within the range of the values reported by Clark and Huckabee (1977) for free-ranging *M. fuscata*. Even though differences in hair sample preparation and analysis can substantially alter results (Hambidge, 1982; Shapcott, 1982), the values for these two macaque populations, and particularly the adults, are remarkably similar. The differences in hair Zn between these two studies may reflect different dietary intake of Zn by these two populations. Miller (1970) found that hair Zn concentrations more consistently represented dietary Zn levels than Zn concentration in other tissues for goats and cattle.

The interpretation and use of hair analysis for assessment of an individual's nutritional status is controversial (cf. Hambidge, 1982; Laker, 1982 Combs et al., 1982).Much of this controversy focuses on the variability introduced by hair treatment by subjects, hair sample processing, and the inconsistencies between specific hair mineral concentrations and other known indicators of nutritional status (e.g. blood concentrations). As previously mentioned, data from this study was not confounded by methodological factors or other biological factors that usually affect the interpretation of human hair analysis. Published reports of research on livestock have found fairly consistent correlations between mineral supplementation experiments and increased hair levels for trace minerals and heavy metals (cf. Combs et al., 1982). Hair analysis may have practical value for farmers attempting to monitor mineral status in livestock. The similarity between the biology of the rhesus macaque and that of humans suggests that the rhesus monkey could be used effectively in determining the value of hair analysis as a nutritional assessment indicator.

CONCLUSION

These data provide an initial multielement survey of hair mineral concentration on a well documented free-ranging group of rhesus monkeys. Because the rhesus monkey is widely used in biomedical research as a surrogate model for the study of nutrition and disease in man, and because many of the confounding variables common to human hair analysis are not a problem in the rhesus monkey, they may also prove to be a useful model for evaluating the potential efficacy of hair analysis used as an indicator of individual nutritional status in humans. In addition, this study illustrates the potential value offered by the Cayo Santiago colony for combined animal model and behavioral studies by utilizing the yearly trapping as point source for biological samples on a population with a known genetic history.

REFERENCES

Anke, M. Major and trace elements in cattle hair as an indicator of Ca, Mg, P, K, Na, Fe, Zn, Mn, Cu, Mo and CO .3. Effect of additional supplements on mineral composition of cattle hair. *Arch. Tierzucht.* 16:57, 1966.

Bate, L.C. Absorption and elution of trace elements in human hair. *J. Appl. Radiation Isotopes* 17:417, 1966.

Clark, T. W.; Huckabee, J.W. Elemental hair analysis of Japanese macaques transplanted to the United States. *Primates* 18(2):299-303, 1977.

Combs, D.K.; Goodrich, R.D.; Meiske, J.C. Mineral concentrations in hair as indicators of mineral status: a review. *J. Anim. Sci.* 54(2): 391-398, 1982.

DeAntonio, S.M.; Katz, S.A.; Scheiner, D.M.; Wood, J.D. Anatomically-related variations in trace-metal concentrations in hair. *Clin. Chem.* 28:2411-2413, 1982.

Dorn, R.C.; Phillips, J.O.; Pierce, I.I.; Chase, G.R. Cadmium, copper, lead and zinc in bovine hair in the new lead belt of Missouri. *3Bull. Environ. Contam. Toxicol.* 12:626-632, 1974.

Gibson, R.S. Hair as a biopsy material for the assessment of trace element status in infancy. A review. *J. Human Nutr.* 34:405-416, 1980.

Hambidge, K.M. Increase in hair copper concentration with increasing distance from scalp. *Am. J. Clin. Nutr.* 26:1212, 1973.

Hambidge, K.M. Hair analyses: worthless for vitamins, limited for minerals. *Am.J. Clin. Nutr.* 36:943-949, 1982.

Hambidge, K.M.; Krebs, N.F.; Jacobs, M.A.; Favier, A.; Guyette, L.; Ikle, D.N. Zinc nutritional status during pregnancy: a longitudinal study. *Am. J. Clin. Nutr.* 37:429-442, 1983.

Hilderbrand, D.C.; White, D.H. Trace elements analysis in hair. An evaluation. *Clin. Chem.* 20:148-151, 1974.

Huckabee, J.W.; Cartan, F.O.; Kennington, G.S. Environmental influence on trace elements in hair of 15 species of mammals. ORNL-TM 3747, Oak Ridge National Laboratory, Oak Ridge, Tennessee, 1977.

Huckabee, J.W.; Cartan, F.O.; Kennington, G.S.; Camenzind, F.J. Mercury concentration in the hair of coyotes and rodents in Jackson Hole, Wyoming. *Bull. Environ. Contam. Toxicol.* 9:37-43, 1973.

Iyengar, G.V.; Kollman, W.E.; Bower, W.J.M. *The Elemental Composition of Human Tissues and Body Fluids.* Verlag Chemie, New York, New York, 1978.

Jaworowski, J.; Bilkiewicz, J.B.; Kostanecki, W. The uptake of $_{210}$Pb by resting and growing hair. *Int. J. Radiat. Biol.* 11:563-566, 1966.

Kelsall, J.P.; Pennekoek, W.J.; Burton, R. Chemical variability in plumage of wild lesser snow geese. *Canada J. Zool.* 52(9):1369-1375, 1975a.

Kelsall, J.P.; Pannekoek, W.J.; Burton, R. Variability in the chemical content of waterfowl plumage. *Canada J. Zool.* 53(10):1379-1386, 1975b.

Kessler, M.J.; Rawlins, R.G. The hemogram, serum biochemistry, and electrolyte profile of the free-ranging Cayo Santiago rhesus macaques *(Macaca mulatta.) Am. J. Primatol.* 4:107-116, 1983.

Klevay, L.M. Hair as a biopsy material, II. Assessment of copper nutriture. *Am. J. Clin. Nutr.* 23:1194-1202, 1970.

Laker, M. On determining trace element levels in man: the use of blood and hair. *Lancet* 8292:260-262, 1982.

Loew, F.M.; Olfert, E.D.; Schiefer, B. Chronic selenium toxicosis in Cynomolgus monkeys. *Lab. Prim. Newsl.* 14(4):7, 1975.

McKenzie, J.M. Alteration of the zinc and copper concentration of hair. *Am. J. Clin. Nutr.* 31:470-476, 1978.

Maugh, T.H. Hair: A diagnostic tool to complement blood serum and urine. *Science* 202:1271-1273, 1978.

Miller, W.J. Zinc nutrition of cattle: A review. *J. Dairy Sci.* 53:1123-1135, 1970.

Nie, N.N.; Hull, C.H.; Jenkins, J.G.; Steinbrenner, K.; Bert, D.H. *Statistical Package for the Social Sciences.* 2nd ed., McGraw Hill, New York, N.Y. 1975.

O'Mary, C.C.; Bell, M.C.; Snead, N.N.; Butts, W.T., Jr., Influence of ration copper on minerals in the hair of Hereford and Holstein calves. *J. Anim. Sci.* 31:626, 1970.

Pannekoek, W.J.; Kelsall, J.P.; Burton, R. Methods for analyzing feathers for elemental content. *Canad. Fisheries and Marine Service Technical Report No.* 498, 1974.

Perkons, A.K.; Jervis, R.E. Applications of radioactivation analysis in forensic investigations. *J. Forensic Sci.* 7:449-464, 1962.

Petering, H.G.; Yeager, D.W.; Witherup, S.D. Trace metal content of hair. I. Zinc and copper content of human hair in relation to age and sex. *Arch. Environ. Health* 23:202-207, 1971.

Pories, W.J.; Strain, W.H. Zinc levels of hair as tools in zinc metabolism. p.363-382 in *Zinc Metabolism.* A.S. Prasad, ed. Springfield, IL, C.C. Thomas Press, 1966.

Rawlins, R.G.; Kessler, M.J. A five-year study of tetanus in the Cayo Santiago rhesus monkey colony: Behavioral description and epizootiology. *Am. J. Primatol.* 3:23-39, 1982.

Sandstead, H.H.; Strobel, D.A.; Logan, G.M.; Marks, E.O.; Jacob, R.A. Zinc deficiency in pregnant rhesus monkeys: effects on behavior of infants. *Am. J. Clin. Nutr.* 31:844-849, 1978.

Schroeder, H.A.; Nason, A.P. Trace minerals in human hair. *J. Invest. Dermatol.* 52:71-78, 1969.

Shapcott, D. Hair and plasma in the diagnosis of zinc deficiency. p. 121-129 in *Clinical Application of Recent Advances in Zinc Metabolism.* A.S. Prasad, I.E. Dreosti, B.S. Hetzel, eds. New York, Alan R. Liss Inc., 1982.

Snedecor, G.W.; Cochran, W.G. *Statistical Methods,* 6th ed., Iowa State University Press, Ames, IA, 1967.

Sultana, C.J.; Marriott, B.M. Geophagia and related behavior of rhesus monkeys *(Macaca mulatta)* on Cayo Santiago Island, Puerto Rico. *Internat. J. Primatol.* 3:338, 1982.

Swenerton, H.; Hurley, L.S. Zinc deficiency in rhesus and bonnet monkeys, including effects on reproduction. *J. Nutr.* 110:575-583, 1980.

Swift, J.A.; Brown, A.C. The critical determination of fine changes in the surface architecture of human hair due to cosmetic treatment. *J. Soc. Cosmetol. Chem.* 23:695-702, 1972.

Van Wagenen, G. Body weight and length of the newborn laboratory rhesus monkey *(Macaca mulatta) Fed. Proc.* 13:157, 1954.

Vir, S.C.; Love, A.H.G.; Thompson, W. Zinc concentration in hair and serum of pregnant women in Belfast. *Am. J. Clin. Nutr.* 34:2800-2807, 1981.

Illustration 11.1 A female stares at detritus groomed from her infant.

CHAPTER ELEVEN

Age-Dependent Impairments of the Rhesus Monkey Visual and Musculoskeletal Systems and Apparent Behavioral Consequences

C. JEAN DEROUSSEAU, LASZLO Z. BITO, PAUL L. KAUFMAN

INTRODUCTION

Studies of aging in nonhuman primates suggest that many of the age-related disabilities that occur in human beings may also characterize other primate species (Schultz, 1969; Bowden, 1979). Thus, adult biological systems once thought to be relatively invariant within species may instead show significant age-related variation. Furthermore, by analogy to the human condition, normal or pathological age-dependent changes can be expected to significantly affect behavior and to be related to the overall adaptive strategy of a species. However, the difficulties of measuring age-related variation on a population scale and of determining age in wild-born animals, have slowed the development of aging studies in nonhuman primates. Only recently have suitable noninvasive techniques been developed and large populations of captive animals with documented histories become available for such studies.

Since 1976, more than 600 rhesus monkeys at two of these facilities, the Caribbean Primate Research Center (CPRC) and the Wisconsin Regional Primate Research Center and University of Wisconsin Primate Laboratory (WI), have been examined by multidisciplinary teams in order to characterize aging in two biological systems that are critical to the primate adaptation—the visual and musculoskeletal systems. This chapter reviews some of the findings from this research, and explores their significance for the understanding of primate behavior and social organization.

AGE-RELATED DECLINE IN THE MUSCULOSKELETAL SYSTEM

Among the musculoskeletal disorders associated with human aging, osteoarthritis and osteoporosis are perhaps the most apparent and debilitating. Osteoarthritis, or degenerative joint disease, is a noninflammatory disorder of movable joints characterized by stiffness and pain, and by gradual degeneration of articular cartilage, sclerosis of juxta-articular bone, and growth of osteophytes (Sokoloff, 1969). Osteoporosis is generally diagnosed when fractures, especially of the hip and spine, result from excessive loss of bone; it is most commonly observed in postmenopausal women (Heaney, 1983). Although each of these conditions may be related to underlying normal aging processes, they are generally recognized as disorders only when a significant loss of function occurs (see Trueta, 1968; Aegerter and Kirkpatrick, 1975; Garn, 1975; Tonna, 1977).

Osteoarthritis and osteoporosis, by limiting the activities of the older person, frequently result in absenteeism and a considerable decrease in productivity. Ironically, these pathologies are often most pronounced in parts of the body stressed by a particular type of labor or set of habitual activities (Mintz and Fraga, 1973). Thus, they may limit, most of all, the capacity to perform the activities at which an individual is most skilled. Also notable among the problem sites in human beings are areas that are weight bearing and involved in bipedal locomotion, such as the spine, hip, and knee (Huskisson and Hart, 1973; Avioli, 1983).

Because of advances in medical technology, severe loss of locomotor function in contemporary societies can be restored to varying degrees through surgical procedures such as the replacement of osteoarthritic joints or fusion of vertebral bodies. However, even with the availability of such procedures, age-related musculoskeletal disorders handicap a significant proportion of the elderly. If these conditions exist in nonhuman primates, they must certainly limit the normal behavioral repertoire of older animals. This implies that some age-related changes in primate behavior patterns could have a biological rather than a social origin, and that an understanding of biological aging must be regarded as a prerequisite for the understanding of behavioral aspects of aging.

Osteoarthritis has been observed in the skeletons of dinosaurs (Moodie, 1923), and in a wide variety of mammals (Fox, 1939) including primates (Schultz, 1969). Bone loss, if not osteoporosis, can also be demonstrated in laboratory animals (Andrew, 1971; Young, Niklowitz, and Steele, 1983). Although this broad distribution suggests that older animals might routinely demonstrate loss of function with the development of age-related disorders, it is generally accepted that natural selection will eliminate individuals, including the aged, who demonstrate any true handicap. Yet, because of the difficulty of determining age in wild populations, very few population studies have actually documented the patterning, incidence, and significance of these age-related disorders in nonhuman species.

In 1972, a program was established to maintain a skeletal collection at the Cayo Santiago (CS) rhesus monkey colony of the CPRC. Since that time, animals from the colony have been systematically collected at death and their skeletons added to the CS collection, which now numbers over 600 individuals. Because the collection is drawn from the well-censused CS colony (Sade et al., 1977; Rawlins, 1979), most skeletons are of known sex and age-at-death, and are extremely valuable for studies of skeletal aging.

During the mid- and late-1970's, the skeletons of 154 adults from this collection were examined to determine the incidence of osteoarthritis, age-related bone loss and possible osteoporosis in the CS population (DeRousseau, 1978; 1985b). These initial studies showed that degenerative joint disease was present in the CS population, that it appeared to increase in frequency with age at all the major appendicular joints, and that pronounced degeneration consistent with a loss of function was present in almost one-third of the individuals between 10 and 15 years of age. At all ages, females tended to exhibit higher frequencies of degeneration and a more even distribution through the body than males did (DeRousseau, 1978).

Females also appeared to be at greater risk of developing osteoporosis than males, since an apparent loss of cortical bone with age was observed in females over nine years old, but not in males of comparable ages (DeRousseau, 1985b). However, previous research on this skeletal collection (Buikstra, 1975) had shown no indication that females exhibit a higher frequency of healed fractures in the appendicular skeleton than males do. Thus, the relationship between bone loss and fractures seen in cases of human osteoporosis was not apparent. It should be noted, however, that in human females bone loss does not typically become pathological until after menopause. Although rhesus monkeys do apparently go through menopause (Hodgen, et al., 1977; Robinson, et al., 1982), in the CS colony few monkeys achieve the advanced ages at which indications of menopause have been observed in this species. Thus, bone loss observed in these rhesus females is consistent with the bone loss that precedes osteoporosis in human beings.

In summary, studies of the CS skeletal collection demonstrate that bone loss in females and degenerative joint disease in both sexes are similar to the conditions observed in human beings. Although functional limitations due to osteoporosis were not apparent in the study sample, frequencies of degenerative joint disease did indicate that a significant proportion of CS monkeys might have reduced locomotor functions by the middle of the second decade of life. Thus, in nonhuman primates age-related disorders would appear to present behavioral limits not only for the rare aged animal, but also for at least a proportion of relatively young adults.

Although functional impairment does not always accompany morphological degeneration (e.g., Gresham and Rathey, 1975), concurrent research on the behavior of the living CS population (Rawlins, 1976; 1978) did reveal variability in positional behavior consistent with a loss of function with age. Locomotor activities, especially on arboreal substrates, decreased markedly with age in both sexes, although females were more active than males at all ages (see also DeRousseau, Rawlins, and Denlinger, 1983; Rawlins, n.d.). However, reductions in activity were most apparent in juveniles. Because these findings were based on a sample of living animals, the relationship between skeletal and behavioral observations was difficult to establish.

In order to clarify this relationship, an initial survey of anesthetized animals from the CS and La Parguera populations of the CPRC was conducted in 1978 in conjunction with the ocular survey described below. Since loss of passive mobility with age has been observed in several human populations (e.g., Beighton, Solomon, and Siskolne, 1973), and linked in some groups to the development of osteoarthritis (Allander, 1974), passive joint excursions were measured to evaluate potential ranges of mobility at the major appendicular joints and their relationships to anatomical constraints, such as palpable osteophytic development.

In the CS animals, loss of passive joint mobility at most appendicular joints appeared to begin well before the development of osteoarthritis and generally corresponded to age-related changes in active locomotion (DeRousseau, Rawlins, and Delinger, 1983; see also Chapter 12, this volume). In addition, females generally showed greater passive mobility at every age than did males, a pattern consistent with behavioral observations, but apparently inconsistent with the observed incidence of degenerative joint disease. This suggests that joint degeneration, at least during its early phases, may not be the primary limitation to joint mobility and function, although results do not rule out the possibility that later adult loss in passive mobility may be related to joint degeneration. At most joints, the decrease in passive mobility with age appeared to continue linearly throughout the lifespan, resulting in a loss of one to two degrees annually (Figure 11.1). This central trend may represent a response to one set of aging processes, while variation around the average may indicate the interaction of additional variables such as degenerative joint disease.

From these considerations, we can conclude that loss of function associated with musculoskeletal aging in rhesus monkeys is probably a complex interaction of several aging processes. A gradual decrease in passive joint mobility appears to begin very early in ontogeny, perhaps

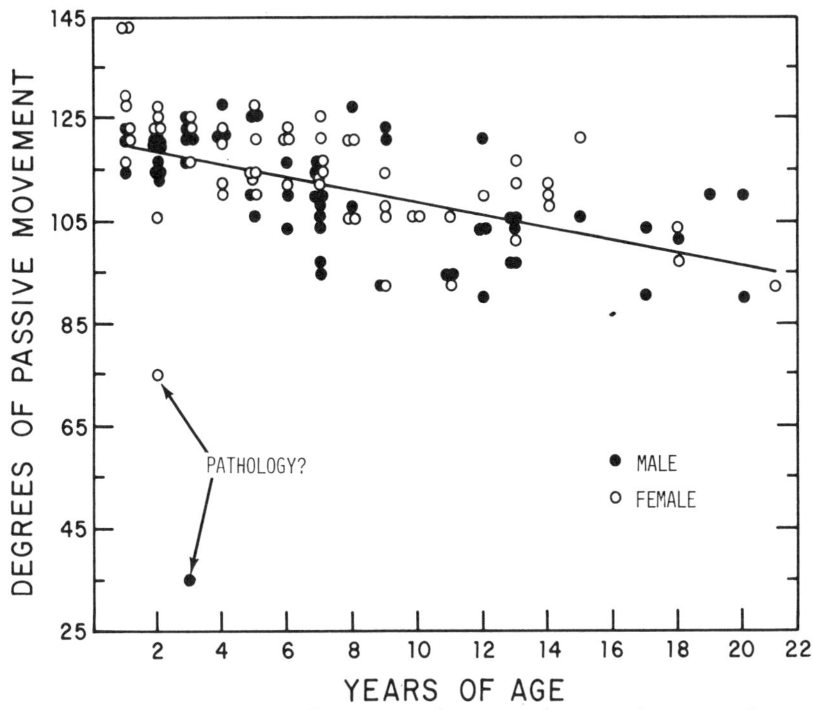

Figure 11.1 Maximum wrist flexion as a function of age in rhesus monkeys.

at birth, and is probably related to the growth process. As structures mature and remodel, perhaps in response to increasing body size and weight, loss in passive mobility continues. In older animals, age-related disorders, such as osteoarthritis, may be superimposed upon this underlying age change to further restrict mobility. Loss of function resulting from the biomechanical insufficiencies of osteoporotic bone may be an additional locomotor handicap at more advanced ages than those usually attained by CS monkeys.

In order to examine animals in the last trimester of life and to investigate the role of activity levels in both the development of human osteoarthritis and osteoporosis, further survey work was conducted at the WI facility. Animals from this colony are housed indoors in cages or gang pens, presumably subject to different activity levels, different biomechanical environments, and different locomotor requirements than are free-ranging animals at CS. In addition, the oldest animals in the WI colony are over 30 years old and can be considered to be postmenopausal.

Initial radiographs of WI animals suggested that loss of cortical bone with age does lead to osteoporosis in females over 25 years of age (Gorman, 1983), but the degree to which animals exhibit this loss is extremely variable and currently under investigation. The WI data also showed that degenerative changes in the spine, like those observed in the appendicular skeleton of CS animals, appeared to increase in frequency with age; most WI animals showed at least some evidence of vertebral disk and/or body collapse by the middle of the second decade of life (DeRousseau, 1982; 1985a; 1985b). Interestingly, the most pronounced degeneration occurred in the lower thoracic vertebral region, a pattern that may be related to the stresses of habitual quadrupedal locomotion or the habitual hunched sitting posture of the rhesus monkey.

To date, 2 surveys of CS and 4 surveys of WI animals have been DeRousseau, 1982; 1985a; 1985b), most of whom were examined for our ocular studies as well (described below). These measurements have over 500 animals (see DeRousseau, Rawlins, and Delinger, 1983; DeRousseau, 1982; 1984), most of whom were examined for our ocular studies as well (described below). These measurements have been aimed at evaluating biomechanical influences on the development of osteoarthritis and osteoporosis, at comparing rates of skeletal aging in the two study populations, and at assessing the potential impact of normal aging upon locomotor functions. Our research design includes longitudinal evaluations of specific animals, retrospective analyses of risk factors associated with disease, and the generation of predictive models of aging. Furthermore, because we are examining many parameters in each animal, we are also beginning to evaluate whether observed age changes are the result of widespread fundamental aging processes, and/or of specific processes particular to each organ system. These questions are currently under active investigation, but some preliminary findings from the most recent CS survey are relevant to the present discussion.

As in the WI animals, spinal degeneration increases with age in the CS free-ranging animals (Table 11.1), seems to be most prominent in the thoracic region, and may be relatively advanced toward the middle of the second decade of life. No differences in spinal degeneration by sex were observed, although CS males and females differ significantly in several measures of body size including weight and calf circumference, and in other functional measures such as passive mobility at the shoulder and hip (Table 11.1).

These latter measures also showed significant age-related variation in the CS sample. Both weight and calf circumference, the latter reflecting primarily muscle development, appear to increase from growing to adult animals, but then to decrease in old adults; joint mobility decreases with age in both sexes (Table 11.1). Thus, juvenile and growing animals appear to have the greatest potential for locomotor behaviors requiring flexibility and young adults the greatest potential for behaviors requiring strength. In contrast, animals over 15 years of age exhibit musculoskeletal characteristics inconsistent with the ability to perform locomotor functions

Table 11.1 Age-related changes in Cayo Santiago rhesus monkeys (mean ± s.d.) Accommodative amplitude was calculated as the maximum difference in refractive errors measured before and after application of carbachol. Measurements of both rotatory movements were taken with the animal supine and limbs flexed at 90 degrees in the parasagittal plane. The knee or elbow was also flexed at 90 degrees, and the distal segment of the limb moved medially and laterally to quantify degrees of medial and lateral rotation, respectively. Scores of spinal osteoarthritis (0 to 4) were averaged for all intervertebral symphyses between the ninth thoracic and the first sacral vertebrae.

	Males			Females		
Age (yr):	0-8	9-16	17-24	0-8	9-16	17-24
No:	(23)	(19)	(5)	(28)	(15)	(8)
Accommodative Amplitude (diopters)	30.4±6.0	23.9±4.5	7.6±5.8	32.1±7.2	22.4±6.3	13.1±10.8
Shoulder Rotation (degrees)	195±38	161±12	154±18	222±46	185±14	163±17
Hip Rotation (degrees)	164±42	125±16	111±17	185±49	146±11	127±23
Spinal Osteoarthritis	0.1±0.2	1.2±0.8	2.5±0.3	0.1±0.1	1.2±0.6	2.2±0.4
Mid-calf Circumference (cm)	14.1±3.7	18.1±0.9	15.6±0.9	13.5±2.9	16.4±1.2	15.3±1.0
Body weight (kg)	6.6±3.4	12.6±1.7	9.7±0.9	5.9±3.5	10.0±1.5	9.4±2.1

optimally. It seems very likely that tendencies to negotiate difficult substrates, to travel for long periods of time, etc., would be affected by these age-related changes.

In summary, all evidence analyzed to date confirms the hypothesis that age-related disorders of the musculoskeletal system occur in nonhuman primates with sufficient frequency and severity to result in a considerable loss of function. Furthermore, these changes can be expected to limit locomotor behavior not only in the obviously aged monkey, but also in at least some relatively young adults.

AGING IN THE VISUAL SYSTEM

Age-related disorders of the human eye include presbyopia, the development of cataracts, glaucoma, and retinal degeneration. Retinal, especially macular, degeneration and glaucoma must be regarded as the most severe clinical problems, but their frequencies are relatively low. In contrast, based on a large study (Duane, 1912) that does not necessarily measure up to current biostatistical standards (Bito, et al., 1982 ; Kaufman, Bito, and DeRousseau, 1983), presbyopia, or loss of accommodative amplitude, has been considered to be the most predictable and inevitable consequence of aging in human beings. In fact, loss of accommodative amplitude is said to begin before adolescence. However, this loss of function becomes noticeable and is commonly referred to as presbyopia, only when objects of importance such as newsprint, can no longer be brought into focus at an acceptable distance. Some degree of lenticular opacity is common in individuals over the age of 60 or 65 (Cotlier, 1975), but even in the oldest age groups, cataracts inconsistent with normal vision are not regarded to be as predictable as is the development of presbyopia in middle age.

In technologically developed societies, presbyopia is not regarded to be a major visual handicap because it can be easily corrected by the use of reading glasses. Cataracts represent a more serious problem, as removal of the cataractous lens requires surgery. However, the resultant refractive deficit can also be corrected by the use of spectacles, contact lenses, or surgically implanted intraocular lenses. If other primates share with human beings such age-dependent afflictions of the visual system in the absence of corrective measures, these disorders could seriously limit normal visual function and ultimately result in behavioral limitations. Thus, the potential impact of these age-dependent visual handicaps can be expected to be much greater in nonhuman primates than in contemporary human societies.

A survey of the CS and La Parguera rhesus populations of the CPRC conducted in January of 1978 revealed the existence of some retinal and vitreal abnormalities (El Mofty, et al., 1978; Denlinger, Eisner, and Balazs, 1980), but potentially handicapping age-related ocular disorders such as senile cataracts or ocular hypertension indicative of glaucoma were not found. In fact, juveniles were found to have significantly higher intraocular pressures than adults (Bito, Merritt, and DeRousseau, 1979). In a second survey of the CS population, tonographic measurements of intraocular pressure were combined with ultrasonic measurements of the axial dimensions of the globe, and a relationship between pressure and ocular growth was established (DeRousseau and Bito, 1981).

These ultrasonic measurements also indicated that the lens of the rhesus monkey continues to grow throughout adulthood. Since continual growth of the lens is assumed to be one, and possibly the primary, contributory factor in the development of human presbyopia, the next survey was designed to include studies aimed at revealing the existence and the natural history of presbyopia. This survey was conducted at the WI facility, which houses large populations of rhesus monkeys over the age of 20 years and some over the age of 30. While in this and in subsequent surveys the animals were subjected to a thorough ophthalmic examination and a variety of ocular parameters were studied (Bito et al., 1982; Kaufman and Bito,

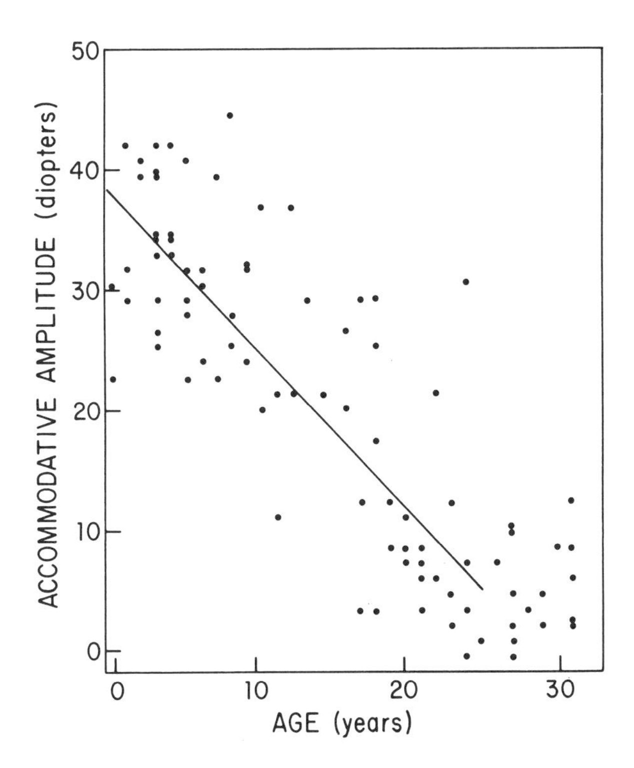

Figure 11.2 Accomodative amplitude as a function of age in rhesus monkeys.

1982), this discussion will focus on the occurrence of presbyopia and cataracts.

Accommodative range of the WI monkeys ranging in age from 0.5 months to over 30 years was estimated by measurements of refractive error before and after topical application of carbachol. The decline in this drug-induced accommodative amplitude with age appears to be a linear function, reaching zero in some individuals by the beginning of the third decade of life (Figure 11.2). A 1983 survey at CS of 100 rhesus monkeys ranging in age from 0.5 to 24 years showed a rate of loss of accommodative amplitude comparable to that shown in Figure 11.2 for the same age range of the WI population (Table 11.1).

The observation that very young animals may show a remarkable 35-40 diopters (D) of accommodation (Table 11.1; Figure 11.2), as compared to the 18-20 D of accommodation reported for children (Duke-Elder, 1958), deserves discussion. It should be noted that the supramaximal doses of carbachol used in our survey may overestimate the true accommodative range of rhesus monkeys since such doses of carbachol are expected to cause a spasm of the ciliary muscle. However, preliminary experiments indicate that, at least in the 20-25 D range, drug-induced accommodation exceeds centrally stimulated accommodation by only a few diopters. It is reasonable to assume therefore that the physiologically available accommodative amplitude in juvenile rhesus monkeys exceeds that of the human youngster.

Such large accommodative amplitudes are not unexpected if we consider the very small body size and some of the typical behavior patterns of the newborn rhesus monkey. Values of 30-40 D imply that infant monkeys can focus on an object as close to its eyes as one inch. This means that a newborn rhesus monkey can bring into sharp focus its mother's body as it holds onto her. Newborn or juvenile rhesus monkeys may also spend considerable time in the quadrupedal position, looking at objects on the ground. In this position the eyes of a young juvenile are about 12 cm from the ground: (arm length)-(occiput to cornea distance), and even small objects or normal ground vegetation may come within a few cm of the eye (Figure 11.3a). While the social or behavioral importance of the ability to focus at these distances may not always be obvious, accommodative amplitudes of this magnitude may be of great developmental importance; there is accumulating evidence that focused and fused retinal images during the first months of life are required for the normal development of the visual system (Eggers and Blakemore, 1978; Blakemore and Eggers, 1978a, 1978b; Hendrickson et al., 1982).

Because the interpupillary distance of a juvenile rhesus is only about 20-25 mm, fusion of such close images would not require a great degree of convergence. The interpupillary distances of human infants are considerably larger (about 65 mm) than those of infant monkeys. Thus, for the same degree of convergence, children require a smaller

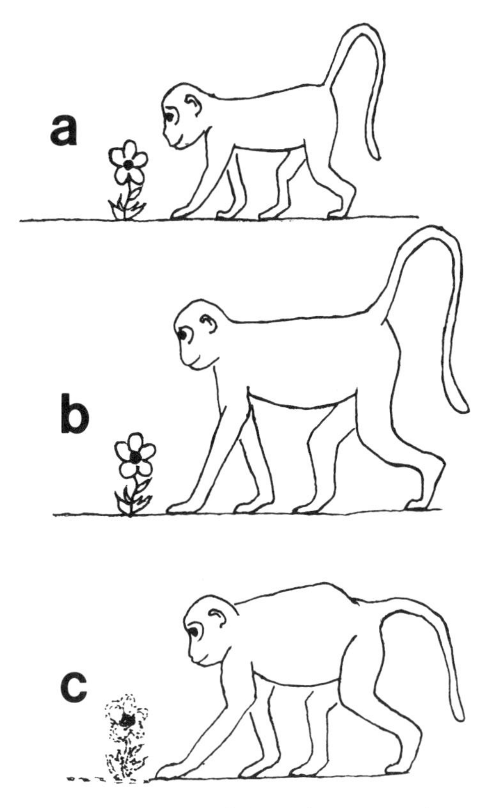

Figure 11.3 Functional consequences of loss of accommodative amplitude in rhesus monkeys.

b. An animal at the beginning of the second decade of life has lost about half of its maximal accommodative amplitude, but can still function normally.

c. An old animal with no ability to accommodate cannot bring objects at arm's length, or directly in front of its hand on the ground, into focus.

dioptric change in accommodation than rhesus monkeys to bring an object into focus.

Loss of accommodation during the first trimester of life would appear to be synchronized with an increase in body size (Figure 11.3a versus 11.3b). As an animal grows in quadrupedal standing height, its ability to focus on objects on the ground remains unimpaired, even if some accommodative amplitude is lost during the growth period. However, during the second trimester of life, loss of accommodative amplitude may, at least in some animals, begin to present a visual handicap; and by the age of 20-25 most animals have little drug-induced, and perhaps no physiologically available accommodative amplitude left. While in growing animals, increase in stature can compensate for accommodative loss, in aging animals reach and standing height may in fact be decreased due to joint degeneration and loss of joint mobility. Thus, an old animal with the ability to accommodate less than 2 or 3 D could not bring an object held at arm's length into focus, nor could it focus on an object on the ground directly in front of its hand while in a quadrupedal position (Figure 11.3c). Reduced speeds of locomotion, less time spent in arboreal activities (Rawlins, n.d.), and changes in grooming orientation (Sade, 1965) may, at least in part, be the behavioral consequences of the development of presbyopia. The decreased performance level of older monkeys on visually-dependent psychological or psychophysiological testing could also, at least in part, be due to their decreased visual capacities rather than to decreased motivation.

While the refractive error measured during drug-induced accommodation indicates an animal's ability to focus on near objects, the resting refractive error indicates whether an animal is emmetropic (i.e., can bring an image at infinity into focus on the retina), myopic (nearsighted), or hypermetropic (farsighted). Our baseline measurements indicate that most monkeys in the WI samples appear to be myopic (Bito et al., 1982). This finding is consistent with previous reports on nonhuman primates, in which such myopia has been interpreted to be a consequence of the near environments of small cages (Young and Leary, 1973). However, our measurements of the free-ranging CS animals with no history of caging showed that these animals were also predominantly myopic, although to a somewhat lesser extent than the caged animals at WI. While myopia is not an uncommon affliction in human beings, it must be noted that measurements taken on monkeys under ketamine anesthesia are not strictly comparable to those obtained on human beings who are conscious and can be instructed during examination to focus on particular targets. Under some conditions, such as in the absence of adequate visual stimulation, emmetropic human subjects may also appear to be myopic (Owens and Liebowitz, 1980; Westheimer and Blair, 1973). Thus, resting refraction in anesthetized monkeys may not represent the true refractive error of the eye (Bito et al, 1982).

Nonetheless, some animals both at CS and at the WI facilities, showed refractive errors too great to be explained by such factors as anesthesia (Bito et al., 1982), suggesting that some animals may indeed be nearsighted. Preliminary analysis of results obtained to date suggest that the extent of apparent myopia and, in fact, the frequency of true myopia (resting refractions more than two standard deviations from the mean resting refraction of the population) may indeed be greater among the caged animals in WI than among the free-ranging animals of CS. However, myopia sufficient in extent to be incompatible with normal far vision was also observed in a few CS animals, including some juveniles; thus, the frequency distribution and extent of myopia at CS may be comparable to that of some human populations. Some behaviors, such as recognition of animals or food at a distance or the perceptions of subtle social signals, may be affected by myopia and should be taken into consideration in the interpretation of behavioral peculiarities of even juvenile and young adult monkeys.

In WI, where sufficient animals over the age of 25 were available for study, the development of cataractous changes could also be followed. All rhesus monkeys that reached the age of 28 showed lenticular opacities indistinguishable from the typical opacities that characterize human senile cataracts (Kaufman and Bito, 1982). Based on clinical experience with human subjects we would not expect such cataracts to allow normal visual acuity. In fact, some advanced cataracts were comparable to opacities associated with less than 20/70 vision in human beings.

Frank senile cataracts have not yet been observed at CS. This is not entirely surprising, because none of the animals in the CS population are in the age group that showed this pathology at WI. However, precataractous changes similar to those observed in WI animals between 15 and 25 years of age were evident in this age group at CS. Thus, our findings on cataractous changes together with the above presented data on presbyopia suggest that lenticular aging is grossly similar in the CS and WI populations.

The possibility that there are subtle differences in lenticular aging between animals that live in the bright sunlight and high temperatures at CS as compared to the caged WI animals that are maintained in air conditioned rooms under artificial light is, however, a question of great importance. It has been proposed that intense light and/or high environmental temperatures accelerate lenticular aging in human beings (Miranda, 1979). Controlled populations of rhesus monkeys allow a much more accurate evaluation of such environmental influences than heterogeneous human populations do, but larger numbers of animals, both at WI and CS will have to be included in our studies before we are able to draw a definitive conclusion about such environmental factors.

We can conclude, however, on the basis of data already collected, that as is the case in human beings, the ability to accommodate is the only known nonreproductive physiological function in rhesus monkeys that is completely lost well before the maximum life span of the species, and in many cases during the reproductive life span of the individual. We can further conclude that severe age-dependent visual handicaps may restrict the behavioral repertoire of animals well before other signs of aging are apparent and that such visual handicaps may be among the factors that set limits to the lifespan of free-ranging and/or feral animals.

AGE-RELATED CHANGE IN OTHER ORGAN SYSTEMS

Many other human organ systems and tissues show functional decline with age (Finch and Hayflick, 1977), none more readily apparent than changes in the skin. Skin becomes increasingly wrinkled and stretched, in part due to repeated muscle activity and in part due to time-dependent changes in the chemical composition of cutaneous and/or subcutaneous tissues (Rossman, 1977). Preliminary results from examination of CS rhesus monkeys showed that skin elasticity, measured by the time it takes a skin fold pinched under constant pressure to disappear, decreases with age (r=0.435, $p <$ 0.001). More precise measures of skin elasticity using a controlled vacuum to raise the skin, combined with ultrasonic measurement of skin displacement are under development. Hypothetically, as skin becomes less resilient, it may become more vulnerable to trauma, perhaps as an animal moves through brush or is attacked by another animal. Skin that is loosely bound to its substrate may also allow easier spread of infection following penetrating wounds, and like other progressive age changes may increasingly become a liability. Further surveys of both the WI and CS populations are planned to detail the effects of aging on the parameters discussed in this paper and also to examine other organ systems that are likely to contribute to the decline of overall functional abilities as a consequence of aging.

EVOLUTIONARY CONSIDERATIONS

Cartmill (1972; 1974) has suggested that primate evolution may have centered on visual and motor mechanisms for catching prey, and secondarily for locating other foodstuffs in a discontinuous habitat. This perspective and others describe the primate adaptation as one that emphasizes visual acuity and binocularity, musculoskeletal features that reflect a tendency toward upright posture, grasping digits, and other characteristics. These features are generally perceived as somewhat variant between primate species, but constant within species, characterizing the econiche of each particular taxa.

In the human, we recognize a considerable amount of intra-population variability in those features that are important to the primate adaptation, for example, in visual acuity and capacity for speed and endurance in locomotion. While some of this variation reflects genetic differences, one of the most obvious sources of variability is ontogenetic change, including processes of growth and aging. In human populations, this variability is embedded in a protective social matrix. Furthermore, human societies use technology to take care of members that show immature or impaired functions. Yet, we know very little about this variation in nonhuman species, or about its relationship to social organization.

Maxim (1979) has suggested that different nonhuman primate societies tend to favor survival of the aged of one or both sexes. Thus, processes of development, aging, and longevity may be tied to social systems of support, even in nonhuman primates. Primate sociality may buffer varying degrees of functional competency (see Chapter 4, this volume), and in turn may be reinforced by the long-term survival of important social networks.

From the present findings, loss of accommodative amplitude, decreased joint mobility, and the development of osteoarthritis would seem to increasingly handicap the aging rhesus monkey. Maxim's (1979) model would predict that, in spite of such aging processes, the matrilineal social organization of the rhesus monkey should lead to enhanced longevity, especially in females.

While this paper does not address demography *per se*, we know that animals in both colonies studied survive to advanced ages, despite the functional limitations implicit in the measures discussed above. WI animals are provided with veterinary care in the case of morbidity, perhaps explaining their continued survival into old age. Animals at CS, however, are maintained under a veterinary policy of nonintervention, so that a long life is not insured by medical attention, although other human behaviors (e.g., provisioning and predator removal), may have some effect on longevity. Alternatively, it may be that, as Maxim predicts, social organization at CS does provide an environment that further enhances survival and buffers the effects of physical aging.

Yet, it would seem that whatever the influences on longevity are, their effects are ultimately limited, more so without medical intervention. At some point, accumulated age changes may present so many physical handicaps to the aging monkey that, despite social or even medical support, limitations not only to the performance of certain behaviors, but also ultimately to life span become apparent.

Undoubtedly, further work is necessary to clarify the relationship of aging processes to mortality, and the importance of social factors as mediators in (or modulators of) that relationship. In order to fully understand the rhesus monkey adaptation, functional decline with age, life span, and social systems must all be considered as part of a total adaptive strategy.

CONCLUSIONS

Over the last eight years, surveys of the Cayo Santiago skeletal collection, the Cayo Santiago and La Parguera populations, animals at the Wisconsin Regional Primate Research Center and the University of Wisconsin Primate Laboratories have revealed remarkable similarities between aging in the rhesus monkey and many of the well-known parameters of aging in human beings. Age-related disorders observed in the rhesus visual and musculoskeletal systems closely resemble the disorders and afflictions that result in behavioral limitations in human populations. These include the loss of accommodative amplitude, the development of cataracts, loss of passive joint mobility and the development of degenerative joint disease. By analogy with the human experience, we conclude that such age-related changes in rhesus monkeys must result in functional limitations, and must be taken into consideration when interpreting age-dependent changes in the behavior and social organization of this species.

ACKNOWLEDGEMENTS

We are especially grateful to the Caribbean Primate Research Center and the Wisconsin Regional Primate Research Center and University of Wisconsin Psychology Laboratories for permission to study their colonies, and for the invaluable assistance with our surveys provided by their staffs. We also gratefully acknowledge the important contributions of our staffs at Columbia University and the University of Wisconsin and of students from New York University. These studies were supported in part by NIH grant EY-04146 to Columbia University, RR-07062 to New York University, RR-00167 to the Wisconsin Regional Primate Research Center, and NIH grant no. RR-01293 and contracts RR-52110 and 223-79-1029 to the Caribbean Primate Research Center.

REFERENCES

Aegerter, E.; Kirkpatrick, J.A., Jr. *Orthopedic Diseases. Physiology, Pathology, Radiology.* Philadelphia, W.B. Saunders Co., 1975.

Allander, E.; Bjornsson, O.J.; Olafsson, O.; Sigfusson, N.; and Thorsteinsson, J. Normal range of joint movements in shoulder, hip, wrist and thumb with special reference to side: a comparison between two populations. *International Journal of Epidemiology* 3(3):253-261, 1974.

Andrew, W. *The Anatomy of Aging in Man and Animals.* New York, Grune and Stratton, Inc., 1971.

Avioli, L.V., ed. *The Osteoporotic Syndrome, Detection, Prevention, and Treatment.* New York, Grune and Stratton, Inc., 1983.

Beighton, P .; Solomon, L.; Soskolne, C.L. Articular mobility in an African population. *Annals of Rheumatic Disease* 32:413-418, 1973.

Bito, L.Z.; DeRousseau, C.J.; Kaufman, P.L.; Bito, J.W. Age-dependent loss of accommodative amplitude in rhesus monkeys: An animal model for presbyopia. *Investigative Ophthalmology & Visual Science* 23(1):23-31, 1982.

Bito, L.Z.; Merritt, S.Q.; DeRousseau, C.J. Intraocular pressure of rhesus monkeys (*Macaca mulatta*). *Investigative Ophthalmology & Visual Science* 18:785-793, 1979.

Blakemore, C.; Eggers, H.M. Animal models for human visual development, pp. 651-659 in *Frontiers in Visual Science.* Cool, S.J., Smith III, E.L., eds. Springer-Verlag, 1978a.

Blakemore, C.; Eggers, H.M. Effects of artificial anisometropia and strabismus on the kitten's visual cortex. *Arch. Ital. Biol.* 116:385-389, 1978b.

Bowden, D.M., ed. *Aging in Nonhuman Primates.* New York, Van Nostrand Reinhold Co., 1979.

Buikstra, J.E. Healed fractures in *Macaca mulatta:* Age, sex and symmetry. *Folia Primatologica* 23:140-148, 1975.

Cartmill, M. Arboreal adaptations and the origin of the order Primates, pp. 97-122 in *The Functional and Evolutionary Biology of Primates.* Tuttle, R., ed. Chicago, Aldine-Atherton, 1972.

Cartmill, M. Rethinking primate origins. *Science* 184:436-443, 1974.

Cotlier, E. The lens, pp. 275-297 in *Adler's Physiology of the Eye.* Moses, R.A., ed. St. Louis, C.V. Mosby Company, 1975.

Denlinger, J.L.; Eisner, G.; Balazs, E.A. Age-related changes in the vitreous and lens of rhesus monkeys (*Macaca mulatta*). I. Initial biomicroscopic and biochemical survey of free-ranging animals. *Experimental Eye Research* 31:67-79, 1980.

DeRousseau, C.J. *Osteoarthritis in Non-Human Primates: A Locomotor Model of Joint Degeneration.* Northwestern University, Ph.D. Dissertation, 1978.

DeRousseau, C.J. Patterns of skeletal aging in rhesus monkeys. *American Journal of Physical Anthropology* 57:180, 1982.

DeRousseau, C.J. Aging in the musculoskeletal system of rhesus monkeys: II. Degenerative joint disease. *American Journal of Physical Anthropology,* 67:177-184, 1985a.

DeRousseau, C.J. Aging in the musculoskeletal system of rhesus monkeys: III. Bone loss. *American Journal of Physical Anthropology* 68:157-167, 1985b.

DeRousseau, C.J.; Bito, L.Z. Intraocular pressure of rhesus monkeys (*Macaca mulatta*). II. Juvenile ocular hypertension and its apparent relationship to ocular growth. *Experimental Eye Research* 32:407-417, 1981.

DeRousseau, C.J.; Rawlins, R.G.; Denlinger, J.L. Aging in the musculoskeletal system of rhesus monkeys: I. Passive joint excursion. *American Journal of Physical Anthropology* 61:483-494, 1983.

Duane, A. Normal values of the accommodation at all ages. *Transactions of the American Medical Association, Section on Ophthalmology,* pp. 383-391, 1912.

Duke-Elder, S. *System of Ophthalmology. Vol. 1. The Eye in Evolution.* St. Louis, C.V. Mosby Co., 1958.

Eggers, H.M.; Blakemore, C.B. Physiological basis of anisometropic amblyopia. *Science* 201:264-267, 1978.

El-Mofty, A.A.M.; Gouras, P.; Eisner, G.; Balazs, E.A. Macular degeneration in rhesus monkeys (*Macaca mulatta*) *Experimental Eye Research* 27:499-502, 1978.

Finch, C.E.; Hayflick, L., eds. *Handbook of the Biology of Aging.* New York, Van Nostrand Reinhold Co., 1977.

Fox, H. Chronic arthritis in wild mammals. *Transactions of the American Philosophy Society* 31:73-148, 1939.

Garn, S.M. Bone-loss and aging, pp. 39-57, in *The Physiology and Pathology of Human Aging*. Goldman, R.; Rockstein, M., eds. New York, Academic Press, 1975.

Gorman, L. Bone loss and osteoporosis in rhesus monkeys. *American Journal of Physical Anthropology*, 60:200-201, 1983.

Gresham, G.E.; Rathey, U.K. Osteoarthritis in knees of aged persons. Relationship between roentgenographic and clinical manifestations. *Journal of the American Medical Association* 233(2):168-170, 1975.

Heaney, R.P. Prevention of age-related osteoporosis in women, pp. 123-144 in *The Osteoporotic Syndrome, Detection, Prevention, and Treatment*, Avioli, L.V., ed. New York, Grune and Stratton, Inc., 1983.

Hendrickson, A.; Movshon, J.A.; Eggers, H.M.; Gizzi, M.S.; Korrpes, L.; Boothe, R.G.: Anatomical and physiological effects of early unilateral blur. *Investigate Ophthalmology Supplement*. 22(3):237, 1982.

Hodgen, G.D.; Goodman, A.C.; O'Connor, A.; Johnson, D.K. Menopause in rhesus monkeys: Model for study of disorders in the human climacteric. *American Journal of Obstetrics and Gynecology* 127:581-584, 1977.

Huskisson, E.C.; Hart, F.D. *Joint Disease: All the Arthropathies*. Baltimore, Williams and Wilkins Co., 1973.

Kaufman, P.L.; Bito, L.Z. The occurrence of senile cataracts, ocular hyptertension and glaucoma in rhesus monkeys. *Experimental Eye Research* 34:287-291, 1982.

Kaufman, P.L.; Bito, L.Z.; DeRousseau, C.J. The development of presbyopia in primates. *Transactions of the Ophthalmological Societies of the United Kingdom* 101, Pt. 3:323-326, 1983.

Maxim, P.E. Social behavior, pp. 56-70, in *Aging in Nonhuman Primates*. Bowden, D.M., ed. New York, Van Nostrand Reinhold Co., 1979.

Mintz, G.; Fraga, A. Severe osteoarthritis of the elbow in foundry workers. *Archives of Environmental Health* 27:78-80, 1973.

Miranda, M.N. The geographic factor in the onset of presbyopia. *Transactions of the American Ophthalmological Society* 77:603-621, 1979.

Moodie, R.L. *Paleopathology: An Introduction to the Study of Ancient Evidences of Disease*. Urbana, University of Illinois Press, 1923.

Owens, D.A.; Liebowitz, H.W. The intermediate resting point of accommodation: recent evidence and applications. *Proceedings of the International Society for Eye Research* 1:71, 1980.

Rawlins, R.G. Locomotor ontogeny in *Macaca mulatta;* I. Behavioral strategies and tactics. *American Journal of Physical Anthropology* 44:201, 1976.

Rawlins, R.G. Locomotive and manipulative use of the hand in *Macaca mulatta*. *American Journal of Physical Anthropology* 48:427-428, 1978.

Rawlins, R.G. Forty years of rhesus research. *New Scientist* 82:108-110, 1979.

Rawlins, R.G. Locomotor ontogeny in *Macaca mulatta:* A behavioral and morphological analysis. Unpublished manuscript on file at the Caribbean Primate Research Center, n.d.

Robinson, J.A.; Chandrashekar, V.; Vernon, M.W.; Bridson, W.E. The menopause in the female rhesus. *The Gerontologist* 22(5):100, 1982

Rossman, I. Anatomic and body composition changes with aging, pp. 189-221, in *Handbook of the Biology of Aging*, Finch, C.E.; Hayflick, L., eds. New York, Van Nostrand Reinhold Co., 1977.

Sade, D.S. Some aspects of parent-offspring and sibling relations in a group of rhesus monkeys, with a discussion of grooming. *American Journal of Physical Anthropology* 23:1-17, 1965.

Sade, D.S.; Cushing, K.; Cushing, P.; Dunaif, J.; Figueroa, A.; Kaplan, J.R.; Lauer, L.; Rhodes, D; Schneider, J. Population dynamics in relation to social structure on Cayo Santiago. *Yearbook of Physical Anthropology* 20:253-262, 1977.

Schultz, A.H. *The Life of Primates.* New York, Universe Books, 1969.

Sokoloff, L. *The Biology of Degenerative Joint Disease.* Chicago, The University of Chicago Press, 1969.

Tonna, E.A. Aging of skeletal-dental systems and supporting tissues, pp. 470-495, in *Handbook of the Biology of Aging,* Finch, C.E.; Hayflick, L., eds. New York, Van Nostrand Reinhold Co., 1977.

Trueta, J. *Studies of the Development and Decay of the Human Frame.* Philadelphia, W.B. Saunders Co., 1968.

Westheimer, G.; Blair, S.M. Accommodation of the eye during sleep and anesthesia. *Vision Research* 13:1035-1040, 1973.

Young, D.R.; Niklowitz, W.J.; Steele, C.R. Tibial changes in experimental disuse osteoporosis in the monkey. *Calcified Tissue International* 35:304-308, 1983.

Young, F.A.; Leary, G.A. Visual-optical characteristics of caged and semifree-ranging monkeys. *American Journal of Physical Anthropology* 38:377-382, 1973.

Illustration 12.1 A juvenile male descends a tree.

CHAPTER TWELVE

Joint Mobility as a Function of Age in Free-Ranging Rhesus Monkeys (Macaca mulatta)

JEAN E. TURNQUIST

INTRODUCTION

The reduction in joint mobility which characterizes geriatric rhesus macaques (see Chapter 11, this volume) is not a phenomenon of old age *per se*, but a continuation of a process seen throughout the life cycle. In the Cayo Santiago colony, DeRousseau, Rawlins, and Denlinger (1983) measured the amount of passive movement (in degrees) at specific joints and found a steady decline with advancing age. This study recorded degrees of motion as the amount of motion between maximum extremes of the range (i.e., between maximum flexion and maximum extension) without reference to the actual positions of the limb segments. For most of the analysis, the population was divided in four age groups: juveniles (0-3 years), adolescence (4-6 years), young adulthood (6-9 years), and adults (≥10 years).

The evidence of decreasing joint mobility with age in rhesus monkeys closely parallels evidence in man which has "generally... assumed that in the majority of cases, limitation of joint movements is a result of normal aging or some kind of degenerative joint disease or post-traumatic state and not inflammatory joint disease." (Allander et al., 1974, page 253). Despite this assumption, however, there are few studies, even in man, which provide data on the effect of age on the normal range of joint mobility. All populations show considerable variability in joint mobility (Wood, 1971; Allander et al., 1974), yet the actual ranges of motion may vary between races or ethnic groups (Harris, 1949; Santos and Azevedo, 1981; Schweitzer, 1970; Walker, 1975). The same general trends, however, are seen in all populations in that females generally have greater joint mobility than males and joint mobility decreases with age (Beighton et al., 1973; Allander et al., 1974). Studies which have included subjects of all ages have shown that the most rapid decline in joint mobility occurs in early childhood, after which the rate gradually decreases even though joint mobility continues to decline as a person ages (Wynne-Davies, 1971; Silverman et al., 1975; Ellis and Bundick, 1956; Beighton et al., 1973).

The limited data available on joint mobility in nonhuman primates parallels the findings in man. It shows considerable variation in joint mobility within a population, differences in ranges between various populations, generally greater mobility in females than males, and a general decline in joint mobility with age (DeRousseau et al., 1983; Turnquist, 1983).

In order to further document the effects of age on joint mobility in a single population, the study reported here was undertaken. Its purpose was to determine if age affects both extremes of the range of mobility (as measured with reference to a fixed zero degree) equally and if early changes in passive joint mobility can be correlated with developmental and/or locomotor changes.

MATERIALS AND METHODS

In January 1983, 163 rhesus monkeys *(Macaca mulatta)* on Cayo Santiago were surveyed for passive joint mobility. The sample included both males and females and ages ranged from seven months to 25 years (Figure 12.1). Nine parameters of motion (five on the forelimb: shoulder flexion and extension[1], elbow extension, and wrist flexion and extension; and four on the hindlimb: hip extension, knee extension, and ankle dorsiflexion and plantarflexion) were measured on monkeys restrained with ketamine HCl. Passive joint mobility of the right extremity was measured with a goniometer using the technique described in Turnquist (1983). A zero degree reference point (see Figure 12.2) was used to document the part of the range which was most affected by age.

The sample population was divided into yearly age groups and the mean for each parameter plotted. Males and females were plotted separately because of the different maturational rates and previously noted sex differences in joint mobility (DeRousseau et al., 1983; Turnquist, 1983; Beighton et al., 1973; Allander et al., 1974). A DEC 2020 Digital Computer was used for data analysis.

---

[1]The terms shoulder flexion and shoulder extension are used here to refer to anterior and posterior movement in the parasagittal plane. These movements are the same as those referred to in comparative anatomy as shoulder protraction and shoulder retraction, respectively.

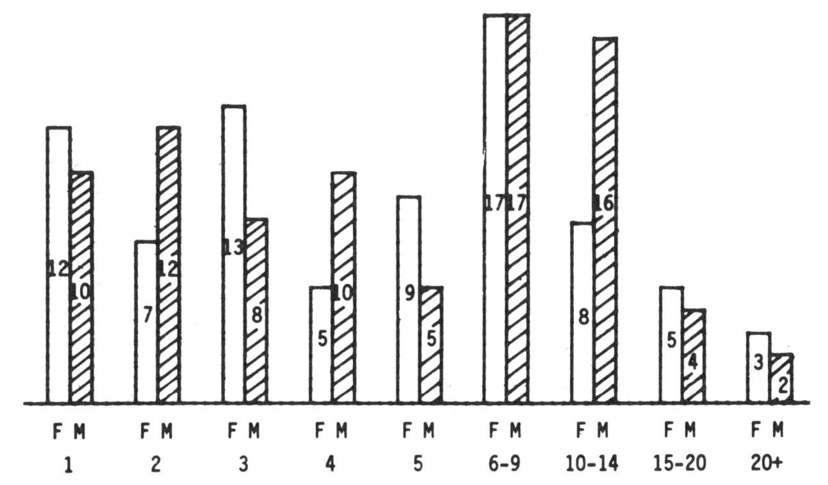

Figure 12.1 Age and sex distribution of Cayo Santiago population sampled January 1983. The number in each bar indicates the number of animals in the category. Age is in years.

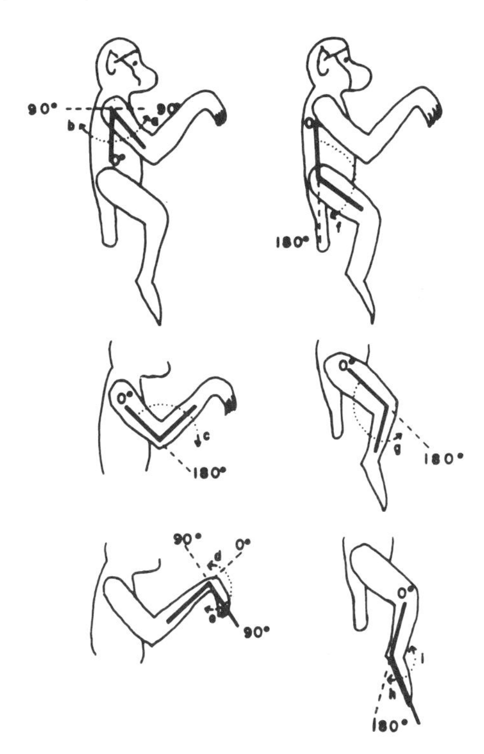

Figure 12.2 Position of the goniometer. The position of 0° and 90° or 180° are marked for each motion of the joint. Mobility was measured in the directions indicated. a) shoulder flexion, b) shoulder extension, c) elbow extension, d) wrist extension, e) wrist flexion, f) hip extension, g) knee extension, h) ankle plantarflexion, and i) ankle dorsiflexion. (Modified from Turnquist, 1983, p. 213.

RESULTS AND DISCUSSION

The results of this study are typified by Figure 12.3 which plots the degree of hip extension for each age/sex group. This plot of only one extreme of the range of joint mobility (measured from a zero degree reference point) shows that here age changes in mobility are characterized by a period of rather rapid decline early in life—in the case of hip extension, between birth and around two years of age in females and five years of age in males. This is followed by an extended period of little or no change in joint mobility during the middle part of the life cycle. At approximately 15 years of age, a second period of rapid decline in joint mobility occurs.

With advancing age the range of passive elbow extension (Figure 12.4) has the same general pattern of decline as that of hip extension; i.e., a period of rapid decline early in life followed by an extended period of general stability, and then a decline again in the teens and into the twenties. Comparisons between the pattern of changes in joint mobility early in life and Cheverud's (1981) published accounts of the fusion of epiphyseal plates in the Cayo Santiago skeletal collection show some interesting parallels. Cheverud found a marked difference in the age of fusion of various epiphyses around this joint. Specifically, the distal humeral epiphyses fused a little before the age of two years (22 months) in females and a little after the age of two (25-26 months) in males. The proximal ulnar and radial epiphyses, however, did not fuse until approximately four years (45-48 months) in females and five years (59 months) in males. Thus, he found two distinct times for the epiphyseal fusions around this joint. The first fusions in the area were within 3-4 months of each other in males and females, but the timing of the second set of fusions differed by a year between the two sexes.

Comparing these ages of epiphyseal fusions with the joint mobility data shows that the most rapid period of decline in elbow extension occurs before the age of two, which is also the age of fusion of the distal humerus in both sexes. This rapid decline before the age of two is the only statistically significant ($p < 0.05$) yearly change in this parameter at any time during the life cycle. After the age of two, the rates of epiphyseal fusion in the two sexes diverge, as does joint mobility. In the female, the final epiphyseal fusions at the elbow occur at the age of four, and by this age the passive joint mobility in the female has already stabilized. In the male, however, the proximal radial and ulnar epiphyses do not fuse until five, and this is the age at which the male range of elbow mobility stabilizes at the adult level. Thus, at the age of fusion, joint mobility is already stabilized at the adult level.

Although the parallel between the changes in joint mobility and the rate of epiphyseal fusion are apparent, this does not imply that the greater mobility in youth is a result of any type of movement at the

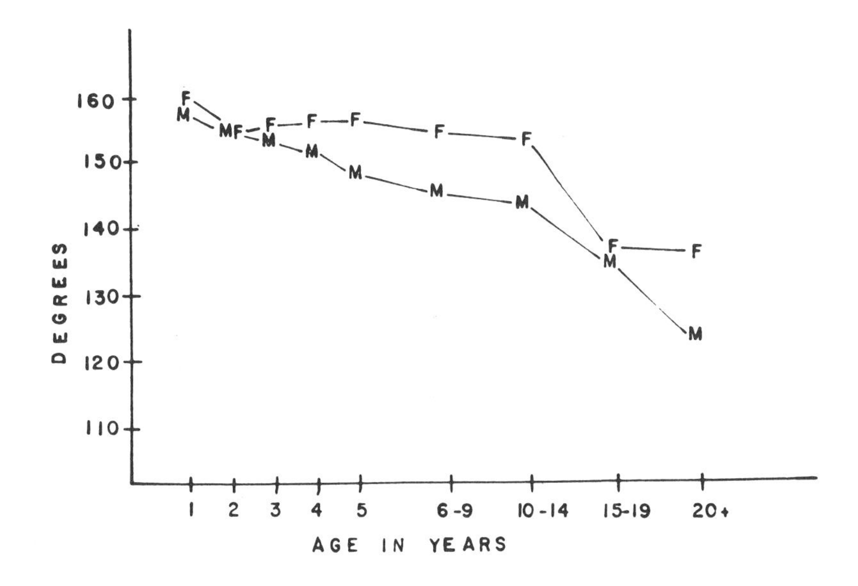

HIP EXTENSION

Figure 12.3 Hip extension. Degrees are mean value of limit of passive mobility. Age is in years. F=female, m=male.

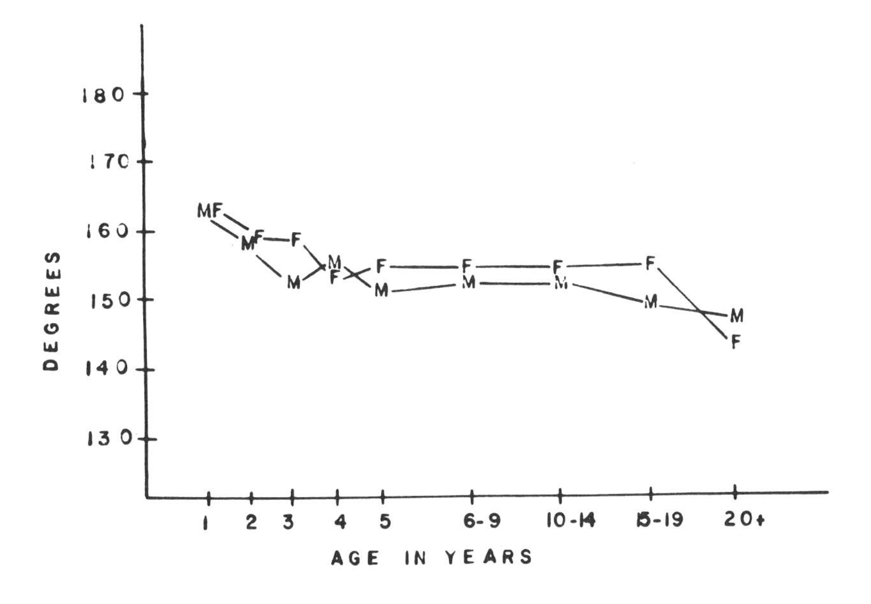

ELBOW EXTENSION

Figure 12.4 Elbow extension. Degrees are mean value in degrees of limit of passive mobility. Age is in years. F=female, M=male.

epiphyseal plate. Instead, the lack of fusion is merely an excellent index of the immaturity of the joint structure and the presence of a greater amount of cartilage than would be found in a mature joint. It is this presence of more immature soft tissue, not the lack of fusion *per se*, which probably partially accounts for the increased mobility prior to the cessation of long bone growth.

Both hip and elbow extension have marked age-related changes in passive joint mobility but not all parameters of motion are affected equally by age. Ankle plantarflexion (Figure 12.5), for example, shows the same general decline in mobility with age, yet the overall change is less than 10° from infancy to old age. This, however, does not imply that joint mobility in the ankle is relatively constant throughout life because ankle dorsiflexion (Figure 12.6), and thus the mobility of the ankle as a whole, is greatly influenced by age. By measuring the extremes of motion relative to a 0° reference point, as opposed to simply measuring the range of total motion, it is evident that the age-related changes in ankle mobility are at the dorsiflexion, not the plantarflexion, end of the range of movement.

These relationships between passive joint mobility and age, and passive joint mobility and epiphyseal fusion define the potential for movement, but not the actual use of the limbs. The frequencies of locomotor versus postural behavior and the relative frequences of

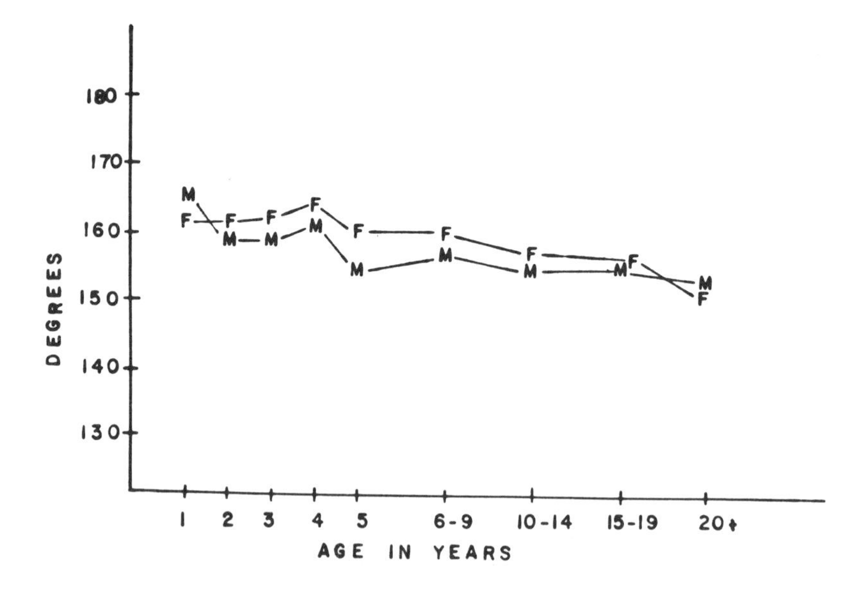

Figure 12.5 Ankle planterflexion. Degrees are mean value in degrees of limit of passive mobility. Age is in years. F=female, M=male.

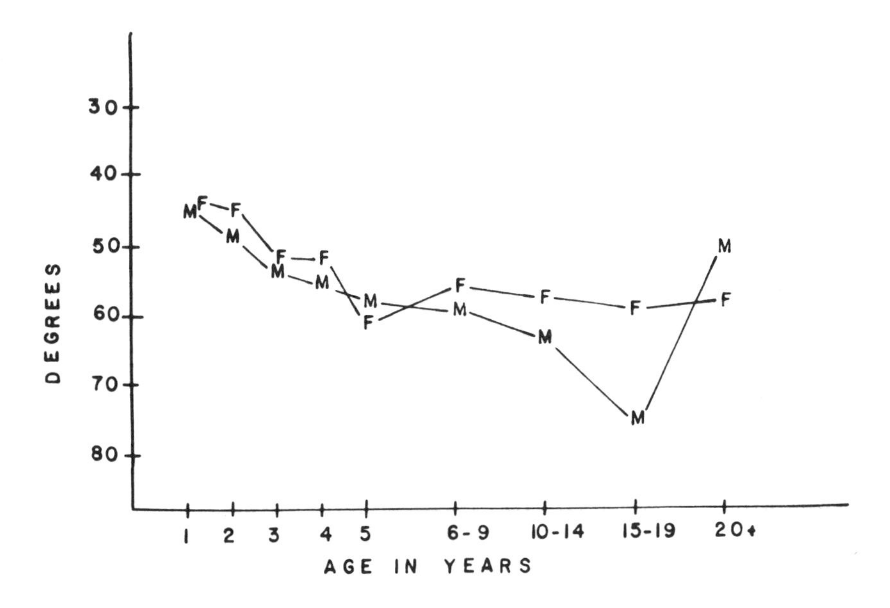

Figure 12.6 Ankle dorsiflexion. Degrees are mean value in degrees of limit of passive mobility. Age is in years. F=female, M=male.

different locomotor patterns for different aged monkeys on Cayo Santiago has been studied in detail by Rawlins (1976; 1977; n.d.). This research quantified the behavior of six age groups of animals: infants under one year, yearlings, preadolescence (two years), adolescence (three years), young to middle aged adults (four to 10 years), and old age adults (>10 years). Analysis included consideration of age, body size, weight, sex, rainfall, reproductive cycles, and social rank. Age was the major determinant in the changes in patterns of locomotor frequencies in the Cayo Santiago rhesus monkeys. The frequency of locomotor behavior increased very rapidly during infancy and continued to increase up until the age of two to three years, after which it gradually declined. Rawlins found that the changes in locomotor frequencies paralleled developmental changes and that the functional changes appeared to precede the structural changes. The current data on joint mobility supports his findings, in that the rapid decline in joint mobility, which characterizes the early years, is not completed until after the peak in locomotor activity between the ages of two and three. Rawlins clearly correlated increasing body size and growth with decreased locomotor activity. Joint mobility does not show the same clear correlation with body size *per se*, but the obvious large muscular development, particularly in mature male free-ranging rhesus does affect and limit joint mobility. An example of how muscle development and tone can effect passive joint mobility can be seen in

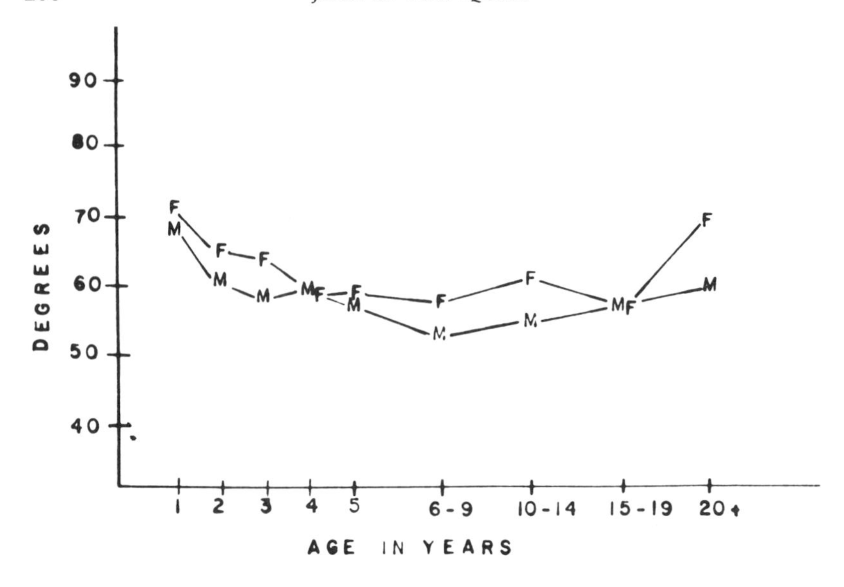

WRIST EXTENSION

Figure 12.7 Wrist extension. Degrees are mean value in degrees of limit of passive mobility. Age is in years. F=female, M=male.

AGE AND JOINT MOBILITY

IN FREE-RANGING RHESUS MONKEYS

(CAYO SANTIAGO -- FEMALES)

	shoulder		elb	wrist		hip	knee	ankle	
	fl	ex	ex	fl	ex	ex	ex	df	pf
1 vs. 2 Years	-	**	**	-	**	-	-	-	-
2 vs. 3 Years	-	-	-	-	-	-	-	**	-
3 vs. 4 Years	**	-	-	-	-	-	-	-	-
4 vs. 5 Years	-	-	-	-	-	-	-	**	-
5 vs. 6-9 Years	-	-	-	-	-	-	-	-	-
6-9 vs. 10-14 Years	-	-	-	-	-	-	-	-	-
10-14 vs. 15-19 Years	-	-	-	-	-	**	**	-	-
15-19 vs. 20+ Years	-	-	-	-	-	-	-	-	-

Significance by Student's "t" Test (two-tailed)
Significant $\leq$ .05
** Younger > Older
oo Older > Younger

Figure 12.8 Influence of age on joint mobility in females. Elb=elbow, fl=flexion, ex=extension, df=dorsiflexion, pf=plantarflexion.

wrist extension (Figure 12.7). This motion is primarily limited by muscle tone and strength of the flexor group rather than by any bony or ligamentaous obstruction. Here, there is the usual decline of mobility in the early years, but then an increase in old age as the muscles lose their strength.

The effect of age on joint mobility in females in the Cayo Santiago population is summarized in Figure 12.8 which compares the range of joint mobility between adjacent age groups. This study of the extremes of the range of mobility (as measured with reference to a fixed 0°) shows that the decline in passive joint mobility which characterizes the life cycle of these animals is not a gradual one from birth to death, but rather follows a pattern of very rapid decline in early life, a stable period during adulthood, and another period of decline at the end of the life cycle. The lack of greater statistical significance during this latter decline is in part due to decreasing sample sizes in the older animals.

CONCLUSIONS

1. In free-ranging rhesus monkeys on Cayo Santiago the decrease in passive joint mobility with age does not affect both extremes of the range of mobility equally.
2. In general, passive joint mobility as measured with reference to a fixed 0° declines rapidly early in life, is relatively stable during young adulthood and middle age, and declines more rapidly again in old age.
3. The duration of the decline in joint mobility early in life can be correlated with physical changes associated with maturation such as epiphyseal fusion and body growth.
4. The decline in joint mobility early in life can also be correlated with behavioral changes in locomotor frequencies.

ACKNOWLEDGEMENTS

I would like to thank Dr. Matt Kessler and the staff at the Caribbean Primate Research Center for their assistance in handling the animals. Cayo Santiago was supported by USPHS, NIH, contract #RR7-2115 and grant RR-01293 to the University of Puerto Rico. I would also like to thank Dr. Matt Kessler and Dr. Richard Rawlins for helpful comments and discussions during the preparation of this chapter.

REFERENCES

Allander, E.; Bjornnsson, O.J.; Olafsson, O.; Sigfusson, N.; and Thorsteinsson, J. Normal range of joint movements in shoulder, hip, wrist, and thumb with special reference to side: A comparison between two populations. *Int. J. Epidemiol.* 3:253-261, 1974.

Beighton, P.; Solomon, L.; and Soskolne, C.L. Articular mobility in an African population. *Ann. Rheum. Dis.* 32:413-418, 1973.

Cheverud, J.M. Epiphyseal union and dental eruption in *Macaca mulatta*. *Am. J. Phys. Anthropol.* 56:157-167, 1981.

DeRousseau, C.J.; Rawlins, R.G.; and Denlinger, J.L. Aging in the musculoskeletal system of rhesus monkeys. I. Passive joint excursion *Am. J. Phys. Anthropol.* 61:483-494, 1983.

Ellis, F.; and Bundick, W.R. Cutaneous elasticity and hyperelasticity. *Arch. Derm.* 74:22-32, 1956.

Harris, H.; and Joseph, J. Variation in extension of the metacarpo-phalangeal and interphalangeal joints of the thumb. *J. Bone Joint Surg.* 31B:547-559, 1949.

Rawlins, R.G. Locomotor ontogeny in *Macaca mulatta* I: Behavioral strategies and tactics. *Am. J. Phys. Anthropol.* 44:201, 1976.

Rawlins, R.G. Locomotor ontogeny in *Macaca mulatta* II: Locomotor dimensions of the shoulder girdle and forelimb. *Am. J. Phys. Anthropol.* 47:256, 1977.

Rawlins, R.G. Locomotor ontogeny in *Macaca mulatta:* A behavioral and morphological analysis. Unpublished manuscript on file at the Caribbean Primate Research Center, no date.

Santos, M.C.N.; and Azevedo, E.S. Generalized joint hypermobility and black admixture in school children of Bahia, Brazil. *Am. J. Phys. Anthropol.* 55:43-46, 1981.

Schweitzer, G. Laxity of metacarpo-phalangeal joints of fingers and interphalangeal joint of the thumb: A comparative inter-racial study. *South Afr. Med. J.* 44:246-249, 1970.

Silverman, S.; Constine, L.; Harvey, W.; and Grahame, R. Survey of joint mobility and in vivo skin elasticity in London school children. *Ann. Rheum. Dis.* 34:177-180, 1975.

Turnquist, J.E. Influence of age, sex, and caging on joint mobility in the patas monkey (*Erythrocebus patas*). *Am. J. Phys. Anthropol.* 61:211-220, 1983.

Walker, J.M. Generalized joint laxity in Igloolik Eskimos and in Island Lake Amerindians. *Hum. Biol.* 47:263-275, 1975.

Wood, P.H.N. Is hypermobility a discrete entity? *Proc. Royal Soc. Med.* 64:690-693, 1971.

Wynne-Davis, R. Familial joint laxityj. *Proc. Royal Soc. Med.* 64:689-690, 1971.

CHAPTER THIRTEEN

The Golden Rhesus Macaques of Cayo Santiago

MATT J. KESSLER, RICHARD G. RAWLINS, AND PAUL L. KAUFMAN

INTRODUCTION

Pickering and van Wagenen (1969) published the original and only previous detailed and illustrated report on the golden rhesus macaque. The condition was characterized by a light red or golden coat color with light skin pigmentation. Golden rhesus monkeys were otherwise considered normal. Studies conducted over a period of years revealed that the developmental and reproductive features of the golden macaques were identical to those of normal rhesus monkeys. An experimental, controlled breeding program established that the mode of transmission of the golden macaque trait was probably through a single autosomal recessive gene, and that all monkeys with the phenotype are purely homozygous. It was estimated that the phenotypic frequency is 1:10,000 births (0.01 percent) in feral populations, assuming a random distribution of matings.

Recently, golden macaques were reported for the first time in the Cayo Santiago rhesus monkey colony (Kessler and Rawlins, 1983; Rawlins and Kessler, 1983). The purposes of this chapter are to describe and picture the golden rhesus macaques of Cayo Santiago, to trace the inheritance of this phenotype through the matrilines in the population, and to present new information on the ocular manifestations of this rare phenotype of *Macaca mulatta*.

MATERIALS AND METHODS

Cayo Santiago has been occupied by an introduced population of free-ranging rhesus monkeys (*Macaca mulatta*) since 1938 (Carpenter, 1972; Rawlins, 1979; Windle, 1980; see also Chapters 1 and 2, this volume). Monkeys have been removed periodically from the colony, but no new stock has ever been added except through births. Between

1972 and 1984 the colony was left intact and maintained under seminatural conditions to provide for long-term studies of behavioral ontogeny and population dynamics. The monkeys were provisioned with commercial high protein monkey chow (Agway, Inc., Syracuse, NY), and had water available *ad libitum*. Abundant natural vegetation provided a routine dietary supplement and the monkeys were geophagic (Sultana and Marriott, 1982). All monkeys were of known identity, age, sex, and maternal lineage. Paternity was not known. At the end of the 1983 birth season, the colony was organized into six naturally formed social groups and a small band of peripheral males. The demographic records for the population comprised the data base for this chapter.

Each January the yearling monkeys are tattooed, earnotched, and bled for genetics studies, and an entire social group is captured for ongoing research. Thus in January 1983, the authors had an opportunity to handle some of the golden macaques. Each monkey was given an intramuscular restraint dose (10 mg per kg body weight) of ketamine hydrochloride (Vetalar®, Parke-Davis, Detroit, MI), and subjected to physical and ophthalmological examinations. The ophthalmological examinations were conducted by an American Board of Ophthalmology certified ophthalmologist (PLK) experienced in the study of both the human and rhesus monkey eye. The examination consisted of Goldman applanation tonometry, anterior and posterior segment slitlamp biomicroscopy, and indirect ophthalmoscopy through dilated pupils using 10 percent phenylephrine hydrochloride and 1 percent tropicamide topically.

RESULTS

Nine golden rhesus macaques (5 male, 4 female) were born on Cayo Santiago from phenotypically normal parents between 1972 and 1984. At birth, the golden macaque infants typically had reddish or blond hair and very pale skin. They were easily distinguishable from normal rhesus monkey infants which are dusky brown or grey. At approximately one to two months of age, the hair began to develop a richer gold or blond color. As the golden monkeys age, the hair coat becomes two-toned in appearance, with the hair on the lower back, rump, and thighs a darker shade of gold or blond than the remainder of the body. This color distribution has developed in most of the Cayo Santiago golden macaques and was characteristic of another golden rhesus monkey (116) born in India and resident in the La Parguera, Puerto Rico colony for many years (see color plate).

Eight of the nine golden monkeys on Cayo Santiago were born in the 1981 through 1984 birth seasons. The overall frequency for the phenotype, based on a total of 1722 births in the 13 years (1972-84), was 0.52 percent. The mean (± 1*SD*) frequency of the trait in the 1981

A. 554 ♀ with 1983 infant H10 ♂ (35 days). Note blond hair and pale skin.

B. 262 ♀ with D92 ♀ (9 months). Note development of two-toned hair coloration.

C. 262 ♀ with her two blond offspring, D58 ♂ (23 months) and D92 ♀ (9 months). (Photo by B. Marriott.)

D. C57 ♀ (18 months). Note blond coat color.

E. D07 ♀ (17 months), illustrating golden coat color and two-toned appearance.

F. 15-year-old golden ♂ 116 from La Parguera colony. Note distinctive two-toned color distribution. (Photo by M. Evans.)

through 1984 birthcrops was 1.07 percent (± 0.57 percent) compared with 0.13 percent (± 0.38 percent) for the preceding nine years. Five of the nine births were confined to a single social group (Group I), and four of the five births in Group I were in the DM matriline. Female 262 had two consecutive golden infants in 1981 and 1982, and 262's daughter B60 had a golden infant (I31) in 1984. The remaining four golden monkeys were born in Group J, three of these confined to the 092 matriline. Figure 13.1 traces the births of the golden *M. mulatta* on Cayo Santiago through the matrilines in the population.

Physical examinations were performed on the five golden monkeys on the island as of January 1983. (Golden macaque 489 died in 1975.) All were healthy and normal yearling or juvenile rhesus monkeys except for their golden (D07) or blond (C57, D58, D92, and E27) coat color and light skin pigmentation.

Complete ophthalmological examinations were performed on four golden macaques. The anterior ocular segments were entirely normal by slitlamp biomicroscopy, with normal iris and ciliary body pigmentation, clear lenses, and normal applanation intraocular pressure. Indirect ophthalmoscopy through dilated pupils revealed reduced pigmentation of the retinal pigment epithelium compared to other rhesus monkeys of the same age. The choroidal circulation was much more visible and there was relatively reduced pigmentation in the macula.

DISCUSSION

The paucity of reports on golden rhesus monkeys in the literature, from primatologists in the field, and from major importers of Old World monkeys indicates that this trait is extremely rare. Pickering and van Wagenen (1969) provided the only detailed report on this hereditary anomaly. A juvenile male and three immature female golden rhesus monkeys were obtained from different locations in Asia and raised to maturity in the laboratory. Breeding studies were conducted with the golden monkeys and between golden monkeys and normal rhesus macaques. It was determined that the golden coat color and light skin are probably controlled by an autosomal recessive gene and that golden rhesus monkeys are purely homozygous. Pickering and van Wagenen estimated that the frequency of this phenotype in feral populations of rhesus monkeys is 1:10,000 births or 0.01 percent.

Of the nine golden rhesus monkeys born on Cayo Santiago, eight were closely observed and studied by the authors. Seven of the eight have blond hair while the eighth is distinctly golden. All have light skin. Our observations of these monkeys indicate that they are developmentally and behaviorally normal. Ophthalmological examin-

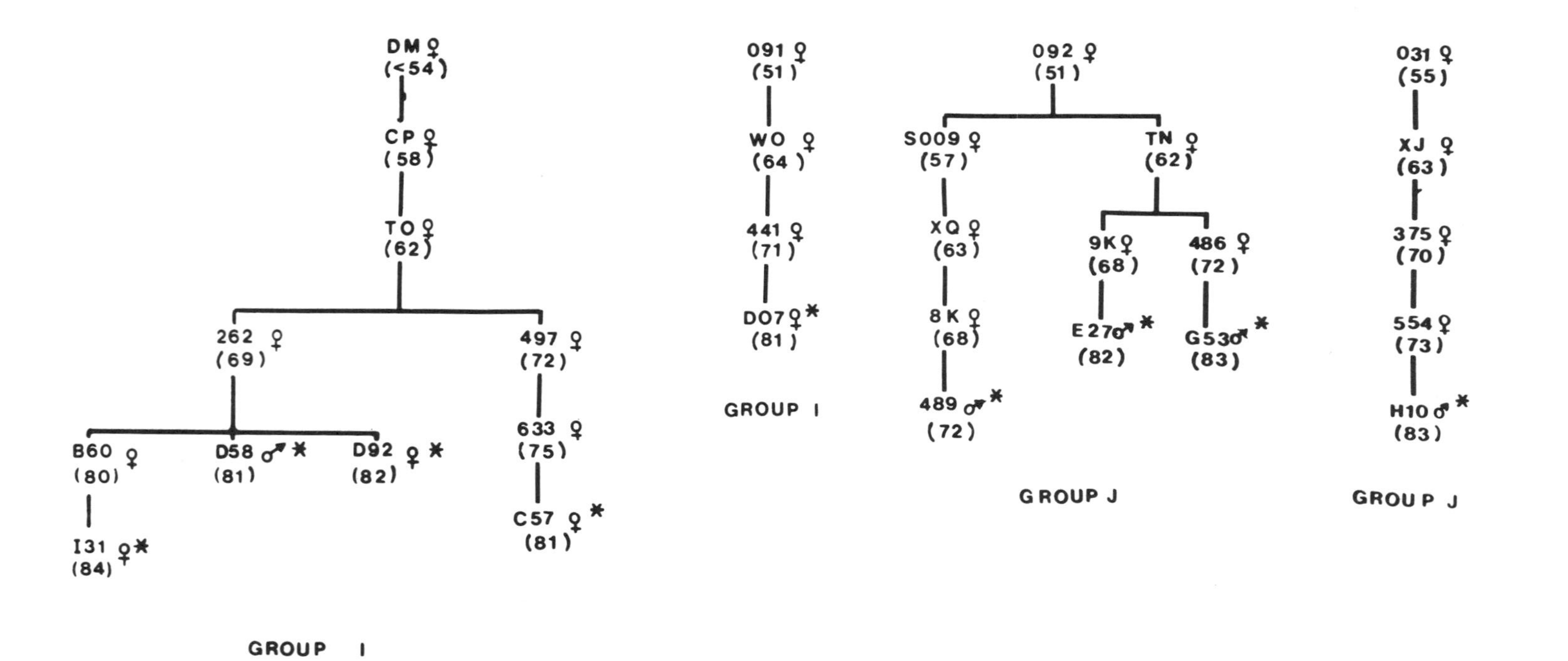

Figure 13.1 Matrilineages of Golden Rhesus Macaques of Cayo Santiago
*indicates golden monkeys
Numbers in parentheses are year of birth
Golden macaque 489 ♂ died in December, 1975; C57 ♀ died in April, 1985.

ations of golden rhesus monkeys have not been conducted previously. The results of these examinations show a reduced pigmentation in the retinal pigment epithelium. This finding is consistent with decreased pigmentation in the skin.

The frequency of the golden rhesus macaque phenotype on Cayo Santiago between 1972 and 1984 was 0.52 percent or 52 times that estimated for feral populations of rhesus monkeys. There have only been nine births of golden macaques on the island since 1966 (A. Figueroa, personal communication, 1984), and as far as the authors can determine from the archives of the Caribbean Primate Research Center, none were born prior to this date. The actual frequency of this phenotype in the population, therefore, is lower than the value derived from only the 1972-84 births. We have estimated the number of births on Cayo Santiago since 1938 at 3850, yielding an overall frequency for the golden macaque trait of 0.23 percent. This figure still far exceeds the estimated rate in the wild.

The eight-fold increase in the frequency of golden macaque births on Cayo Santiago from 1981-1984 compared with the previous nine birth seasons, might indicate an increased distribution of the autosomal recessive allele for golden coat color or, more likely, increased mating among carriers in the colony during those years. Due to the removal of Group J from the island in 1984 (see Chapter 1 this volume), three of the four golden macaque males were taken from the gene pool on Cayo Santiago. To date the one male golden macaque left on Cayo Santiago has not reached sexual maturity. Two females (C57 and D07) have become pregnant. C57 had a stillborn infant in March, 1985 and died in April from complications of the delivery. D07 delivered a phenotypically-normal infant in January 1986. If the remaining golden monkeys on Cayo Santiago, especially the male, survive to mate successfully, then the rate of transmission of this allele and the frequency of the rare and beautiful golden rhesus macaque phenotype should increase markedly on Cayo Santiago in the years to come.

Conclusions

This chapter documented the births and frequency of golden rhesus macaques on Cayo Santiago and traced the nine golden macaques born between 1972 and 1984 through four matrilines in the population. New information was presented on the ocular manifestations of this rare phenotype of rhesus monkeys.

Acknowledgements

The Cayo Santiago Station of the Caribbean Primate Research Center was supported by USPHS contract RR-7-2115 and grant RR-

01293 to the University of Puerto Rico. The ophthalmological examinations were supported by USPHS grants EY-04146, EY-02698 and EY-00137 to the University of Wisconsin. The authors wish to thank the employees of Cayo Santiago, especially Angel Figueroa, for their technical support, and Dr. Bernadette Marriott and Mike Evans for providing the photographs as indicated on the color plate.

REFERENCES

Carpenter, C.R. Breeding colonies of macaques and gibbons on Santiago Island, Puerto Rico, pp. 76-87 in *Breeding Primates.* W. Beveredge, ed. Basel, Switzerland, S. Karger, 1972.

Kessler, M.J.; Rawlins, R.G. Golden rhesus macaques (*Macaca mulatta*) on Cayo Santiago. *American Journal of Primatology* 4:330, 1983.

Pickering, D.E.; van Wagenen, G. The 'golden' mulatta macaque (*Macaca mulatta*): Developmental and reproduction characteristics in a controlled laboratory environment. *Folia Primatologica* 11:161-166, 1969.

Rawlins, R.G. Forty years of rhesus research. *New Scientist* 82:108-110, 1979.

Rawlins, R.G.; Kessler, M.J. Congenital and hereditary anomalies in the rhesus monkeys (*Macaca mulatta*) of Cayo Santiago. *Teratology* 28:169-174, 1983.

Sultana, C.J.; Marriott, B.M. Geophagia and related behavior of rhesus monkeys (*Macaca mulatta*) on Cayo Santiago, Puerto Rico. *International Journal of Primatology* 3:338, 1982.

Windle, W.F. The Cayo Santiago primate colony. *Science* 209:1486-1491, 1980.

CHAPTER FOURTEEN

An Overview of Blood Group Genetic Studies on the Cayo Santiago Rhesus Monkeys

CHRISTINE R. DUGGLEBY, PHILIP A. HASELEY,
RICHARD G. RAWLINS, MATT J. KESSLER

INTRODUCTION

A body of classical population genetic theory bearing on the distribution of genetic variation has been in the literature for some time (Crow and Kimura, 1970; Li, 1976; Wright, 1969). The availability of blood group markers and electrophoretic variants has facilitated investigations of genetic processes in natural populations which were not previously possible. Studies have been carried out in both animal and plant populations to determine the extent and distribution of genetic variation in nature. Data from a wide variety of species have implicated natural selection or density-dependent factors in the distribution of gene and genotype frequencies (Gaines and Whittam, 1980; Tamarin and Krebs, 1969). Attempts have been made to construct more realistic population genetic models, including the potential impact of population structure and demography on genetic variation (Cavilli-Sforza and Bodmer, 1971; Nei, 1975; Wright, 1969). These models have been less extensively tested (but see, Schwartz and Armitage, 1980; Smouse et al., 1981).

With specific reference to primate populations, it is now clear that population size, structure, and demography affect the behavior of individuals and social groups as a whole (Altmann and Altmann, 1979; Dunbar, 1979; Fedigan et al., 1983; Sade, 1972). In turn, social phenomena, such as mating behaviors, dominance relationships, or social group fissioning may affect the distribution of genetic variation, as well as demographic parameters, within and between primate social groups (Altmann and Altmann, 1979; Chapaïs and Schulman, 1980; Dittus, 1979; Fedigan et al., 1983; Sade et al., 1977; Sade, 1980; Schulman and Chapais, 1980). Additionally, of course, gene (or genotype) frequencies may be directly affected by natural selection

Illustration 14.1 An old female with her infant.

through differential fertility or mortality.

Primate population genetic studies attempting to explain the distribution of variation in this milieu have only recently been possible due to: (1) the fairly recent appreciation of the complex interactions briefly outlined above; (2) the difficulties attendant to adequate genetic field studies of long-lived species such as primates; and (3) the recent establishment of simply inherited polymorphisms (genetic markers). Examples of such studies are described.

The history and general features of the Cayo Santiago population of *Macaca mulatta* are described elsewhere (Carpenter, 1972; Sade et al., 1977) and detailed in this volume (see Chapters 1 and 2). Only a few relevant points regarding social organization, social dynamics, and demography will be repeated here. In 1938, 409 rhesus monkeys, captured from a number of localities in India, were released on Cayo Santiago by C.R. Carpenter (Carpenter, 1972). After initial behavioral studies (Carpenter, 1942), the colony was largely ignored, except for the occasional removal of animals, until 1956. At that time, S. Altmann tattooed the monkeys for identification and initiated his ethological studies. His preliminary census indicated a population of about 200 monkeys in two social groups (Altmann, 1962).

Rhesus monkey social groups are defined on the basis of behavioral criteria. Female rhesus monkeys on Cayo Santiago remain in their natal group throughout their lifetime, except when groups undergo fission. Males generally leave their natal group at puberty to join another group or to become solitary (Sade, 1972; 1980). Males tend to change social groups on the average of every four years (Koford, 1965).

A detailed census was started in 1959 (Koford, 1965) and has been maintained continuously since that time (Sade et al., 1977; Chapters 1 and 2, this volume). Maternal genealogies (monkeys tracing descent through the maternal line to a single female alive in 1956) are known for all animals. The social groups on Cayo Santiago had from one to six matrilines each. The importance of these matrilines in terms of behavior, population dynamics, demographics, and genetic architecture cannot be overemphasized. The relevance of matrilines to phenomena associated with genetic studies are discussed below. Phenomena involving matrilines which are not directly related to the genetic aspects of this population are discussed in other chapters of this volume, as well as elsewhere (Chapais and Schulman, 1980; Chepko-Sade, 1974; Missakian, 1972; Sade, 1965; 1972; 1980; Sade et al., 1977; Schulman and Chapais, 1980).

While animals have been removed periodically from the colony, none have been added since 1938 except through births (Chapters 1 and 2, this volume). With provisioning, the population grew to nearly 800 monkeys by 1968. Between 1968 and 1972, several entire social groups were removed or culled, leaving a population of nearly 300 monkeys in four social groups in November 1972 (Sade et al., 1977).

The blood group genetic research conducted on the Cayo Santiago population of *Macaca mulatta* since 1972 has focused on the extent, distribution, and factors affecting genetic variation in the colony. The attributes of this population which have made it particularly suitable for population genetic analysis include:

1. the wealth of ethological and demographic data available,
2. the continuous census records which provide individual identities, group memberships, natality, maternal relationships, mortality, and exact ages of the animals,
3. the depth of maternal genealogies (the maternal relationships of most monkeys are known for more than four generations), and
4. the lack of confounding effects of gene flow from contigous populations.

These factors have allowed us to ask questions relating to the effects of social organization, social dynamics, demography, and episodic events on gene and genotype distributions in the Cayo Santiago population of macaques.

MATERIALS AND METHODS

Monthly census records showing identity, sex, natality, mortality, maternal relationships, social group membership, and age, as well as additional maternal genealogical data, have been provided for the blood group genetic research since its initiation in November 1972, through the courtesy of D.S. Sade, R. Rawlins, M. Kessler and the Caribbean Primate Research Center. In the fall of 1972, the entire population of rhesus monkeys was blood typed at SUNY/Buffalo, using rhesus isoimmune reagents for up to 11 blood group loci (Duggleby and Stone, 1971; Duggleby et al., 1971; Sullivan et al., 1977a, 1977b; Sullivan et al., 1978). Each yearly birth cohort since that time has been blood typed. This chapter summarizes 11 years (1972-1983) of continuous blood group genetic data for this population.

Two specific measures of the degree of genetic differentiation among subdivisions are reported below. Wright (1965) defined F_{ST} as "the correlation between random gametes within subdivisions, relative to gametes of the total population." It is equivalent to the random component of inbreeding, that inbreeding which occurs in a population simply due to finite population size. Nei's (1975) standard genetic distance is a function of the probability of identity of two randomly chosen genes within a subdivision relative to the probability of identity of two genes chosen from different subdivisions. For both measures, higher values indicate greater genetic differentiation among subdivisions. Either measure may be calculated for subdivisions at various levels in a hierarchically subdivided population. Although mathematically not interconvertible, they provide comparable information on the relative genetic differentiation between different sets of loci (protein versus blood group) or between different levels of subdivision (social group versus matriline). Data presented in the next section are given so that comparisons can be made between genetic differentiation among social groups for protein loci and genetic differentiation among social groups for blood group loci; matriline differentiation for the two kinds of loci; and, for either set of loci, the degree of differentiation at the matriline level and the social group level.

Wright's (1965) *F* statistics, specifically F_{IS}, may be used as a measure of inbreeding within subdivisions. F_{IS} is the correlation between uniting gametes relative to those of their own subdivision." (Wright, 1965). It is a measure of the nonrandom component of inbreeding. If related individuals mate more frequently than expected on the basis of chance alone, there will be excess homozygosity. F_{IS} values may be negative, indicating excess heterozygosity and suggesting the subdivision is outbred. Average F_{IS} values are presented below for both social groups and yearly group birth cohorts.

RESULTS AND DISCUSSION

The majority (88 percent) of the rhesus monkeys on Cayo Santiago in November 1972 were derived from a maximum of 15 (presumably unrelated) females alive in 1956 (Duggleby, 1977a). Estimates of effective population size (N_e) for the colony in 1972 suggest that N_e for the 1956 population would have been small, about 70 rhesus monkeys (Duggleby, 1978a). With this small number of progenitors and a small effective population size in 1956 (and probably before), a reduced amount of genetic variation might have been expected. In fact, only one known rhesus blood group allele was absent from the Cayo Santiago population. This was not surprising in view of its rarity in the rhesus population used to elucidate the blood groups (Wisconsin Regional Primate Research Center; Sullivan et al., 1977b). Nei's (1975) formula for genic diversity or average heterozygosity provides an estimate of the amount of variation present in a population. For eight blood group loci, the average heterozygosity of the November 1972 Cayo Santiago population of rhesus was 0.324 (Duggleby, 1980). This may be compared to a value of 0.130 for 57 blood group loci for the human species (Nei, 1975), and is an indication of the large amount of blood group variation remaining after a restriction in population size. This contrasts with the lack of polymorphism for several protein loci known to be polymorphic in other macaque populations (Buettner-Janusch and Sokol, 1977). The lack of polymorphism led the authors to propose that founder effect and drift during population bottlenecks had reduced protein variation in the Cayo Santiago population.

Table 14.1. Differences in Blood Group Phenotype Frequencies Among All Four Social Groups, 1972.

Blood Group System	d.f.	x^2	p
G	6	19.37	<.005
H	4	5.16	>.100
I	9	22.64	<.010
J	3	15.27	<.005
K	3	17.33	<.001
L	3	10.36	<.025
Total	28	90.13	<.001

Table 14.2. Genetic Distances Between Lineages

	1972	1976
Mean within troop distance	.0827	.0507
Mean between troop distance	.0809	.0523

In spite of extensive migration of males among groups, the four social groups present in November 1972 were significantly genetically differentiated for six blood group loci (Table 14.1) (Duggleby, 1978b). Initial reports on the population showed social groups undifferentiated with respect to transferrin types (Buettner-Janusch et al., 1974a) or 6-phosphogluconate dehydrogenase (6-PGD) alleles (Buettner-Janusch et al., 1974b), although in 1976 the groups were significantly differentiated for 6-PGD (Buettner-Janusch and Sokol, 1977). For three protein loci, transferrin, 6-PGD, and carbonic anhydrase II (CA II), Olivier et al. (1981) reported an average F_{ST} value of 0.00992 among the four social groups present in July 1973. For the blood group loci, however, the four social groups present in November 1972 were more highly differentiated, with an average F_{ST} value of 0.0299 (Duggleby, 1977b).

Matrilines were found to be even more differentiated than social groups for eight blood group loci (McMillan and Duggleby, 1981). The average genetic distance between matrilines in November 1972 was 0.0818, while the average genetic distance between social groups was 0.0192 for the same date. Matrilines within the same group were as differentiated from one another as matrilines in different social groups, even controlling for genealogy size (Table 14.2) (McMillan and Duggleby, 1981). Greater genetic differentiation of matrilines compared to social groups was also found for the three protein loci reported above. The average F_{ST} for lineages within groups was found to be 0.0789 and that between social groups, 0.00992 (Olivier et al., 1981).

Using seven blood group loci, the average genetic distance among the four social group birth cohorts for 1973 was 0.1081, demonstrating extensive differentiation of infants born into different groups. These differences were statistically significant (Haseley, 1978).

In an attempt to explain the genetic differentiation of the population found in November 1972, the possibility of inbreeding was explored. Inbreeding within social groups should lead to excess homozygosity within groups and group birth cohorts, and to an overall deficiency of heterozygotes in the population as a whole (Wright, 1969). In fact, while the social groups were highly differentiated, no excess homozygosity was found in the population as a whole, within groups (Duggleby, 1978b) or within group birth cohorts (Haseley, 1980; 1984a). Negative values of F_{IS}, showing approximately a 6 percent excess of heterozygotes, were found for social groups for the *G* and *H* blood group systems for the first few years of this study, indicating outbreeding (Duggleby, 1977b). The migration of most males from their natal groups (Sade, 1972) is apparently sufficient to prevent inbreeding. A more extensive discussion of this and of the factors which may affect zygosity levels over time is in preparation (Duggleby).

Even larger heterozygote excesses were found within group birth cohorts (Haseley, 1980; 1984a). Negative values of F_{IS}, showing for the *H* blood group system within group birth cohorts (nearly 40 percent in 1975) to male migration, Robertson's effect (Robertson, 1965), and viability differences favoring heterozygotes, particularly in the 1975 group birth cohorts. Excess heterozygosity for the *G* blood group system does not appear to be due to viability differences, but instead differential fertility may be implicated (Haseley, 1984a; 1984b).

A variety of explanations for the genetic differences found between matrilines within social groups have been offered, not necessarily mutually exclusive. For the sizeable differences between matrilines for the blood group loci, McMillan (1979, described in McMillan and Duggleby, 1981) proposed lineage-specific mating as a mechanism for the maintenance of genetic differentiation over time (See also Chapter 9, this volume). Random mating of emigrant males within a group should lead to decreased genetic differentiation between matrilines since offspring in different matrilines would come to be paternally related. If individual males mated preferentially with members of a particular matriline, the matrilines would remain differentiated. The effect would be stronger if male relatives mated preferentially with the same matriline. Related males of *M. mulatta* (Drickamer and Vessey, 1973), *M. fuscata* (Enomoto, 1974), and *Cercopithecus aethiops* (Cheney and Seyfarth, 1983) have been known to migrate together. Olivier et al., (1981) ascribed genetic differentiation between matrilines for three protein loci mostly to the pedigree or lineal effect, the propagation of alleles in a matriline through time.

Social group fissioning along genealogical lines has been demonstrated for the rhesus macaques of Cayo Santiago (Chepko-Sade, 1974; Chepko-Sade and Olivier, 1979; Chepko-Sade and Sade, 1979). Insofar as the matrilines within a group are genetically differentiated, fissioning along genealogical lines could lead to large genetic differences between the resultant daughter groups (Buettner-Janusch et al., 1983; Cheverud et al., 1978; Duggleby, 1977a). For human populations, this effect has been called the lineal effect (Neel, 1967; Smouse et al., 1981) to distinguish it from founder effect (Mayr, 1963). Lineal effect exacerbates the founder effect, since members of the same lineage are even less genetically representative of the parental population because they share genes in common.

Two social group fissions occurred in 1973. Group F fissioned into Groups F and M, although not precisely along genealogical lines (Chepko-Sade, 1974). The division of Group J resulted in daughter Groups J and N. The natal portion of each of these daughter groups consisted of a single matriline. Groups J and N rejoined the following year (Sade et al., 1977). Despite the strong lineality of the J-N fission, neither of the two fissions resulted in significant genetic differences between the fission groups for eight blood group loci (Duggleby, 1977a). In 1976, social Group F fissioned into Groups F and O, with the natal segment of Group O consisting of a single matriline (Sade et al., 1977). In this case, however, the daughter groups were significantly differentiated for the eight blood group loci. On the other hand, based on variation for tranferrins, 6-PGD and CA II, Buettner-Janusch et al. (1983) found little differentiation between Groups F and O after fissioning, moderate levels for Groups F and M, but large differences between Groups J and N. Based on the theoretical effects of random versus lineal group fissioning and computer simulations, they concluded that only lineal group fissioning could account for the post-fissioning differentiation found. An examination of three earlier group fission events led Cheverud et al. (1978) to conclude that several factors may all contribute to the differences in transferrin allele frequencies between groups after fissioning. They suggested that combinations of the lineal effects of fission, effects of the migration of new males into groups during a fission (male migration effect), and genetic drift may all function in the differentiation of social groups. The degree of differentiation would depend on the relative strengths of the individual effects, and (in the case of male migration effect and genetic drift) the direction of the effect, increasing or decreasing genetic variation (Cheverud et al., 1978).

That lineal effects of fissioning are not the sole cause of group differentiation is also indicated by the lack of relationship between the recency of group fissions and how differentiated groups are genetically (Duggleby, 1978c). The patterning of genetic distances (Nei, 1975) between groups, based on seven blood group loci, was inconsistent with known timing of fission events. In 1976, immediately after the fission of Group F into Groups O and F, the genetic distance between them was larger than that between any other pair of groups. Even social Groups F and J, which were derived from different groups in 1956 (Sade et al., 1977) were more similar than Groups O and F.

While the overall variation in the Cayo Santiago population has remained high (Duggleby, 1980), the variation between groups has declined since the immediate post-cull period in November 1972 (Duggleby, 1978c; Haseley, 1980). Although some of this decline may be due to the rapid increase in population size that has taken place (the population size, as of January 1984, was nearly 1200 monkeys). a number of our findings suggest that the group differentiation seen in 1972 was partly due to the culling process which preceded our initial genetic work. While 88 percent of the population in 1972 was descended from a maximum of 15 presumably unrelated females alive in the middle 1950s, a sizeable proportion of the adult males, on the other hand, were derived from troops and, therefore, matrilines which had been removed. This meant that the potential fathers were unrelated to, and perhaps quite different genetically from, mothers over the next few years. This may have led to excess heterozygosity (Robertson, 1965) which was found, especially in the earlier years of our studies (Duggleby, 1977b; Haseley, 1980; 1984a; 1984b). Additionally, since these males came from a fairly large number of matrilines (from groups removed between 1968 and 1972), their dispersal into the four groups which remained might have contributed different sets of genes into each of these groups. This might account for some of the group differentiation seen immediately after the culling process. It is worth noting that, as the absolute number and relative proportion of these males declined (representing an ever smaller proportion of all adult males), group differentiation declined as well (Duggleby, 1978c).

CONCLUSIONS

Population genetic studies of nonhuman primates have differed from those carried out with most other groups of organisms by following in the footsteps of extensive ethological studies of the species involved. This has necessitated the inclusion of the potential effects of social behavior and social organization on gene and genotype frequencies. However, the wealth of such data for the Cayo

Santiago population has, at the same time, facilitated these studies.

The major findings, after an analysis of 11 years (1972-1983) worth of blood group population genetic data, may be outlined:

1. Considering the isolation and small size (at least periodically) of the Cayo Santiago population, the amount of variation present is impressive. There is no evidence of inbreeding, either in terms of loss of variation or excess homozygosity. This is true for the population as a whole, for social groups, and for birth cohorts, with the latter two consonant with ethological data on the extent of male migration.
2. The socially important structures and substructures of the Cayo Santiago population of rhesus monkeys have genetic counterparts. Social groups behave as distinct social and genetic units. Matrilines function as social units and exhibit genetic differentiation as well. Social group birth cohorts appear to have a genetic identity and may well serve as socially important groups (see, for example, Cheney and Seyfarth, 1983 on brother and peer migration).
3. Lineal fissioning of social groups can produce high levels of genetic differentiation, but this does not always happen, even when groups fission with complete lineality (with no matriline divided between the daughter groups). The fact that the first two fissions examined in our research did not result in significant genetic differentiation of the daughter groups for eight blood group loci underlines the importance of long-term longitudinal studies. Unique social processes may thus have variable effect. This is undoubtedly one of the reasons for the lack of correspondence between the recency of fissioning and degree of genetic differentiation.
4. Demonstrable fertility and mortality differences have been found between different genotypes despite small effective population size, where the effects of genetic drift might be expected to override those of natural selection.

The difficulties attendant to an adequate genetic field study on a long-lived primate species are largely absent in the free-ranging monkey population of Cayo Santiago. The ethological and demographic data base graphic data base is of exceptional depth, more than four generations, (Sade et al., 1977), and the continuous genetic data covers more than a generation length. This has made it possible to detect long-term genetic differences and to determine the effects of such rare events as group fissions on the population genetics of the colony. In addition, the detailed body of basic information on the colony has enabled us to develop many hypotheses for further testing, such as the effects of the removal of Groups J, M, and O in 1984 and Group P in 1985 (see Chapter 1, this volume) from the island on the genetics of the monkey population remaining on Cayo Santiago.

ACKNOWLEDGEMENTS

This study could not have been accomplished without the encouragement and assistance of Dr. D. S. Sade, who provided blood samples and population data, collected under NSF grant GS-44243X and NIH contract RR-71-2003, until 1976. We gratefully acknowledge the critical comments of Drs. W. F. Duggleby and J. E. Sirianni on this chapter, as well as the many fruitful discussions with Dr. C. Berman. This research was also supported in part by SUNY/Buffalo institutional funds (3S05RR0706607), NSF grants (S039927-X00 and SOC 73-05516 A02), NIH contract RR-71-2115 and grant RR-01293.

REFERENCES

Altmann, S.A., A field study of the sociobiology of rhesus monkeys, *Macaca mulatta*. *Ann. N.Y. Acad. Sci.* 102:338-433, 1962.

Altmann, S.A. and Altmann, J., Demographic constraints on behavior and social organization. In Bernstein, I.S., and Smith, E.O. (eds.), *Primate Ecology and Human Origins: Ecological Influences and Social Organization.* Garland STPM Press, N.Y. pp. 47-63, 1979.

Buettner-Janusch, J., Mason, G.A., Dame, L., Buettner-Janusch, V. and Sade, D.S., Genetic studies of serum transferrins of free-ranging rhesus macaques of Cayo Santiago, *Macaca mulatta* (Zimmerman 1780). *Am. J. Phys. Anthropol* 41:217-232, 1974a.

Buettner-Janusch, J., Dame, L., Mason, G.A. and Sade, D.S., Primate red cell enzymes: Glucose-6-phosphate dehydrogenase and 6-phosphogluconate dehydrogenase. *Am. J. Phys. Anthropol.* 41:7-14, 1974b.

Buettner-Janusch, J., Olivier, T.J., Ober, C.L. and Chepko-Sade, B.D., Models for lineal effects in rhesus group fissions. *Am. J. Phys. Anthropol.* 61:347-353, 1983.

Buettner-Janusch, J. and Sockol, M., Genetic studies of free-ranging macaques of Cayo Santiago. II 6-phosphogluconate dehydrogenase and NADH-methemoglobin reductase (NADH-diaphorase). *Am. J. Phys. Anthropol.* 47:375-379, 1977.

Carpenter, C.R., Sexual behavior of free-ranging rhesus monkeys *(Macaca mulatta) J. Comp. Psychol.* 33:113-162, 1972.

Carpenter, C.R., Breeding colonies of macaques and gibbons on Santiago Island, Puerto Rico. In Beveredge, W.I. (ed.), *Breeding Primates,* Karger, Basel Switzerland pp. 76-87, 1972.

Cavalli-Sforza, L.L. and Bodmer, W.L., *The Genetics of Human Populations.* W.H. Freeman, San Francisco, 1971.

Chapais, B. and Schulman, S.R., An evolutionary model of female dominance relations in primates. *J. Theor. Biol.* 82:47-89, 1980.

Cheney, D.L. and Seyfarth, R.M., Nonrandom dispersal in free-ranging vervet monkeys: social and genetic consequences. *Amer. Nat.* 122:392-412, 1983.

Chepko-Sade, B.D., Division of group F at Cayo Santiago. *Am. J. Phys. Anthropol.* 41:472, 1974.

Chepko-Sade, B.D. and Olivier, T.J., Coefficient of genetic relationship and the probability of intragenealogical fission in *Macaca mulatta*. *Behav. Ecol. Sociobiol.* 5:263-278, 1979.

Chepko-Sade, B.D. and Sade, D.S., Patterns of group splitting within matrilineal kinship groups: A study of social group structure in *Macaca mulatta*. *Behav. Ecol. sociobiol.* 5:67-85, 1979.

Cheverud, J.M., Buettner-Janusch, J. and Sade, D.S., Social group fission and the origin of intergroup genetic differentiation among the rhesus monkeys of Cayo Santiago, Am. J. Phys. Anthropol. 49:449-456, 1978.

Crow, J.L. and Kimura, M., *An Introduction to Population Genetic Theory*. Harper and Row, N.Y., 1970.

Dittus, W.P.J. The evolution of behaviors regulating density and age-specific sex ratios in a primate population. *Behaviour* 69:265-302,1979.

Drickamer, L.C. and Vessey, S.H., Group-changing in free-ranging rhesus monkeys. *Primates* 14:359-368, 1973.

Duggleby, C.R., Blood group antigens and the population genetics of *Macaca mulatta* on Cayo Santiago. II Effects of social group division. *Yrbk. of Phys. Anthropol.* 20:263-271, 1977a.

Duggleby, C.R., Inbreeding in *Macaca mulatta* on Cayo Santiago. Paper delivered at *Ann. Meetings Amer. Soc. Primatol.*, April 1977, 1977b.

Duggleby, C.R., Rhesus of Cayo Santiago: Effective population size and migration rates. In Chivers, P.J. and Herbert, J. (eds.), *Recent Advances in Primatology, Vol. 1: Behavior.* Academic Press, N.Y., pp. 189-191, 1978a.

Duggleby, C.R., Blood group antigens and the population genetics of *Macaca mulatta* on Cayo Santiago. I Genetic differentiation of social groups. *Am. J. Phys. Anthropol.* 48:35-41, 1978b.

Duggleby, C.R., Genetic distance between troops of Cayo Santiago macaques through time. *Am. J. Phys. Anthropol.* 48:390, 1978c.

Duggleby, C.R., Genetic variability and adaptive strategies: Cayo Santiago. Paper delivered at *Ann. Meetings Amer. Soc. Primatol.*, 1980.

Duggleby, C.R., Blystad, C. and Stone, W.H., Immunogenetic studies of rhesus monkeys. II, The H, I, K and L blood group systems. *Vox Sang.* 20:124-136, 1971.

Duggleby, C.R. and Stone, W.H., Immunogenetic studies of rhesus monkeys. I. The G blood group system. *Vox Sang.* 20:109-123, 1971.

Dunbar, R.I.M., Population demography, social organization and mating strategies. In Bernstein, I.S. and Smith, E.O. (eds.), *Primate Ecology and Human Origins: Ecological Influences and Social Organization.* Garland STMP Press, N.Y., pp. 65-88, 1979.

Enomoto, T., The sexual behavior of Japanese monkeys. *J. Hum. Evol.* 3:351-372, 1974.

Fedigan, L.M., Gouzoules, H. and Gouzoules, S., Population dynamics of Arshiyama West Japanese macaques. *Internat. Jour. Primatol.* 4:307-322, 1983.

Gaines, M.S. and Whittam, T.S., Genetic changes in fluctuating role populations: Selective vs. non-selective forces. *Genetics* 96:767-778, 1980.

Haseley, P.A., Genetic variability of birth cohorts of rhesus on Cayo Santiago, *Am. J. Phys. Anthropol.* 48:404, 1978.

Haseley, P.A. Robertson's effect and variability in troop birth cohorts of Cayo Santiago rhesus, *Am. J. Phys. Anthropol.* 52:236, 1980.

Haseley, P.A., Aspects of Genetic Structure and Differential Fertility in Rhesus Macaques of Cayo Santiago. Unpublished doctoral dissertation, State University of New York at Buffalo, 1984a.

Haseley, P.A., Differential fertility by blood group phenotype in Cayo Santiago rhesus. *Am. J. Phys. Anthropol.* 63:169, 1984b.

Kaplan, J.R., Patterns of fight interference in free-ranging rhesus monkeys. *Am. J. Phys. Anthropol.* 47:279-288, 1977.

Koford, C.B., Population dynamics of rhesus monkeys on Cayo Santiago. In DeVore, I. (ed.), *Primate Behavior.* Holt, Rinehart and Winston, N.Y., pp. 160-174, 1965.

Li, C.C. *First Course in Population Genetics.* Boxwood Press, Pacific Grove, 1976.

Mayr., E., *Animal Species and Evolution.* Belknap, Cambridge, 1963.

McMillan, C.A., Genetic differentiation of female lineages in rhesus macaques on Cayo Santiago. *Am. J. Phys. Anthropol.* 50:461-462, 1979.

McMillan, C. and Duggleby, C., Interlineage genetic differentiation among rhesus macaques on Cayo Santiago. *Am. J. Phys. Anthropol.* 56:305-312, 1981.

Missakian, E.A., Genealogical and cross-genealogical dominance relations in a group of free-ranging rhesus monkeys *(Macaca mulatta)* on Cayo Santiago. *Primates* 13:169-180, 1972.

Neel, J.V., The genetic structure of primitive human populations. *Jap. J. Human Genet.* 12:1-16, 1967.

Nei, M., *Molecular Population Genetics and Evolution.* American Elsevier, N.Y., 1975.

Olivier, T.J., Ober, C. Buettner-Janusch, J. and Sade, D.S., Genetic differentiation among matrilines in social groups of rhesus monkeys. *Behav. Ecol. Sociobiol.* 8:279-285, 1981.

Robertson, A., The interpretation of genotypic ratios in domestic animal populations. *Anim. Production* 7:319-324, 1965.

Sade, D.S., Some aspects of parent-offspring relations in a group of rhesus monkeys, with a discussion on grooming. *Amer. J. Phys. Anthropol.* 23:1-17, 1965.

Sade, D.S., A longitudinal study of social behavior of rhesus monkeys. In Tuttle , R. (ed.), *The Functional and Evolutionary Biology of Primates.* Aldine, Chicago, pp. 378-398, 1972.

Sade, D.S. Population biology of free-ranging rhesus monkeys on Cayo Santiago. Puerto Rico. In Cohen, M., Malpas, R.S. and Klein, H.G., (eds.), *Biosocial Mechanisms of Population Regulation.* Yale University Press, New Haven, 1980.

Sade, D.S., Cushing, K., Cushing, P ., Dunaif, J., Figueroa, A., Kaplan, J.R., Lauer, C., Rhodes, D. and Schneider, J., Population dynamics in relation to social structure on Cayo Santiago. *Yrbk. of Phys. Anthropol.* 20:253-262, 1977.

Schulman, S.R. and Chapais, B., Reproductive value and rank relations among macaque sisters. *Am. Nat.* 115:580-593, 1980.

Schwartz, O.A. and Armitage, K.B., Genetic variation in social mammals: The marmot model. *Science* 207:665-667, 1980.

Smouse, P.E., Vitchum, V.J. and Neel, J.V., The impact of random and lineal fission on the genetic divergence of small human groups: A case study among the Yanamamo. *Genetics* 98:179-197, 1981.

Sullivan, P.T., Blystad, C. and Stone, W.H., Immunogenetic studies of rhesus monkeys. IX. The M and N blood group systems. *Immunogenetics* 5:415-421, 1977a.

Sullivan, P.T., Blystad, O. and Stone, W.H., Immunogenetic studies of rhesus monkeys. VIII. A new reagent of the G blood group system and an example of the founder principle. *Anim. Blood Groups Biochem. Genet.* 8:49-53, 1977b.

Sullivan, P.T, Blystad, C. and Stone, W.H., Immunogenetic studies of rhesus monkeys. X. The O, P, Q, R and S blood systems. *Immunogenetics* 7:125-130, 1978.

Tamarin, R.H. and Krebs, C.J., *Microtus* population biology. II. Genetic changes at the transferrin locus in fluctuating populations of voles. *Evolution* 23:183-211, 1969.

Wright, S., The interpretation of population structure by F-statistics with special regard to systems of mating. *Evolution* 19:609-619, 1965.

Wright, S., *Evolution and the Genetics of Populations. Vol. 2, The Theory of Gene Frequencies.* University of Chicago Press, Chicago, 1969.

CHAPTER FIFTEEN

Cayo Santiago Bibliography (1938-1985)

MATT J. KESSLER AND RICHARD G. RAWLINS

The following bibliography is published in response to numerous requests. Publications are listed in chronological order to serve as a historical document of research done over the years at this unique facility.

Locke, C.M. Peopling an island with gibbon monkeys: An ambitious West Indian experiment in biology. *Illust. London News* 13 August: 290-291, 1938.

Anonymous. First American monkey colony starts on Puerto Rico islet. *Life* 6(1):26-27, January 2, 1939.

Mieth, H. Picture of the Week. *Life* 6(3):17, January 16, 1939.

Anonymous. Monkeys—A Research Investment. In *The Economic Review*. Chamber of Commerce, Government of Puerto Rico, San Juan, PR, pp. 61-77 (Summer Qtr.), 1940.

Carpenter, C.R. Rhesus monkeys for American laboratories. *Science* 92:284-286, 1940.

Pomales-Lebron, A., Morales-Otero, P. Hemolytic streptococci from the throat of normal monkeys. *Proc. Soc. Exp. Biol. Med.* 45:509-511, 1940.

Carpenter, C.R., Krakower, C.A. Notes on results of a test for tuberculosis in rhesus monkeys (*Macaca mulatta*). *P.R. J. Publ. Hlth. Trop. Med.* 17:3-13, 1941.

Carpenter, C.R. Sexual behavior of free-ranging monkeys (*Macaca mulatta*). I. Specimens, procedures and behavior characteristics of estrus. *J. Comp. Psychol.* 33:113-142, 1942.

Carpenter, C.R. Sexual behavior of free-ranging monkeys (*Macaca mulatta*). II. Periodicity of estrus, homosexual, auto-erotic and non-conformist behavior. *J. Comp. Psychol.* 33:143-162, 1942.

Poindexter, H.S. A study of the intestinal parasites of the monkeys of the Santiago Island primate colony. *P.R. J. Publ. Hlth. Trop. Med.* 18:175-191, 1942.

Suarez, R.M., Diaz Rivera, R.S., Hernandez Morales, F. Hematological studies in normal rhesus monkeys (*Macaca mulatta*). *P.R. J. Publ. Hlth. Trop. Med.* 18:212-226, 1942.

Illustration 15.1 A yawn displays an adult male's dentition.

Engle, E.T., Krakower, C., Haagensen, C.D. Estrogen administration to aged female monkeys with no resultant tumors. *Cancer Res.* 3:858-866, 1943.

Anonymous. (Advertisement offering Cayo Santiago colony for free.) *Science* 106:32-33, 1947.

Nicholas, W.H., Locke, J. Growing pains beset Puerto Rico. *Nat. Geog.* 99(4):435-436, 1951.

Litter, L. Monkey island. *Conn. State Med. J.* 18:676-679, 1954.

Altmann, S.A. The social play of rhesus monkeys. *Am. Zool.* 1:27, 1958.

Altmann, S.A. How to predict what an animal will do next. *Anat. Rec.* 131:527, 1958.

Frontera, J.G. The Cayo Santiago Primate Colony. In W.F. Windle (ed.), *Neurological and Psychological Deficits of Asphyxia Neonatorum.* Springfield, Il., C.C. Thomas, pp. 246-256, 1958.

Altmann, S.A. A field study of the sociobiology of rhesus monkeys (*Macaca mulatta*). *Ann. N.Y. Acad. Sci.* 102:338-435, 1962.

Altmann, S.A. Social behavior of anthropoid primates: Analysis of recent concepts. In E.L. Bliss (ed.), *Roots of Behavior.* New York, P. Hoeber, pp. 277-285, 1962.

Gavan, J.A. Comparative field investigation of *Macaca mulatta.* (Report to Department of Anatomy, Medical College of South Carolina, Charleston, S.C.), 1963.

Heatwole, H., Sade, D.S., Hildreth, B. Herpetofauna of Cayo Santiago and Cayo Batata. *Carib. J. Sci.* 3:1-5, 1963.

Koford, C.B. Dynamics of an island population of rhesus monkeys. *Proc. XVI Internat. Congress Zool.,* Washington, D.C., Vol. I., p. 263, 1963.

Koford, C.B. Group relations in an island colony of rhesus monkeys. In C.H. Southwick (ed.), *Primate Social Behavior.* Princeton, N.J., Van Nostrand, pp. 136-152, 1963.

Koford, C.B. Rank of mothers and sons in bands of rhesus monkeys. *Science* 141:356-357, 1963.

Carpenter, C.R. Naturalistic Behavior of Nonhuman Primates. University Park, P.A., The Pennsylvania University Press, 1964.

Conaway, C.H., Koford, C.B. Estrous cycles and mating behavior in a free-ranging band of rhesus monkeys. *J. Mammalol.* 45:577-588, 1964.

Sade, D.S. Seasonal cycle in size of testes of free-ranging *Macaca mulatta. Folia Primatol.* 2:171-180, 1964.

Altmann, S.A. Sociobiology of rhesus monkeys. II. Stochastics of social communication. *J. Theoret. Biol.* 8:490-522, 1965.

Conaway, C.H., Sade, D.S. The seasonal spermatogenic cycle in free-ranging rhesus monkeys. *Folia Primatol.* 3:1-12, 1965.

Kaufmann, J.H. A three-year study of mating behavior in a free-ranging band of rhesus monkeys. *Ecology* 46:499-512, 1965.

Koford, C.B. Population dynamics of rhesus monkeys on Cayo Santiago. In I. DeVore (ed.), *Primate Behavior: Field Studies of Monkeys and Apes.* New York, Holt, Rinehart and Winston, pp. 160-174, 1965.

Sade, D.S. Some aspects of parent-offspring and sibling relations in a group of rhesus monkeys, with a discussion of grooming. *Am. J. Phys. Anthropol.* 23:1-18, 1965.

Draper, W.A. Free-ranging rhesus monkeys. Age and sex differences in individual activity patterns. *Science* 151:476-478, 1966.

Farber, P.A. Chromosomes of *Macaca mulatta* leukocytes. *Folia Primatol.* 4:409-415, 1966.
Kaufmann, J.H. Behavior of infant rhesus monkeys and their mothers in a free-ranging band. *Zoologica* 51:17-28, 1966.
Koford, C.B. Population changes in rhesus monkeys: Cayo Santiago, 1960-1964. *Tulane Stud. Zool.* 13:1-7, 1966.
Koford, C.B., Farber, P.A., Windle, W.F. Twins and teratisms in rhesus monkeys. *Folia Primatol.* 4:221-226, 1966.
Morrison, J.A., Menzel, E. W., Jr. Adaptation of a rhesus monkey group to artificial fission and transplantation to a new environment. *Am. Zool.* 6:121, 1966.
United States Public Health Service. The seven tribes of Cayo Santiago. *PHS World* 1:2-7, 1966.
Fisler, G.F. Nonbreeding activities of three adult males in a band of free-ranging rhesus monkeys. *J. Mammalol.* 48:70-78, 1967.
Kaufmann, J.H. Social relations of adult males in a free-ranging band of rhesus monkeys. In S.A. Altmann (ed.), *Social Communication Among Primates.* Chicago, Univ. Chicago Press, pp. 73-98, 1967.
Menzel, E.W., Jr. Naturalistic and experimental research on primates. *Hum. Dev.* 10:170-186, 1967.
National Institute of Neurological Diseases and Blindness. Laboratory of Perinatal Physiology, San Juan, PR, pp. 1-12, 1967.
Quiatt, D. Social dynamics of rhesus monkey groups. *Dissert. Absts.* B28:1325, 1967.
Sade, D.S. Determinants of dominance in a group of free-ranging rhesus monkeys. In S.A. Altmann (ed.), *Social Communication Among Primates.* Chicago, Univ. of Chicago Press, pp. 99-114, 1967.
Sade, D.S. Ontogeny of social relations in a group of free-ranging rhesus monkeys (*Macaca mulatta,* Zimmerman). *Dissert. Absts.* B27:3379, 1967.
Sade, D.S. Ontogeny of social relations in a group of free-ranging rhesus monkeys (*Macaca mulatta,* Zimmerman). *Univ. Microfilms,* Ann Arbor, MI, 67-5156, 1967.
Altmann, S.A. Sociobiology of rhesus monkeys, III. The basic communication network. *Behaviour* 32:17-32, 1968.
Altmann, S.A. Sociobiology of rhesus monkeys, IV. Testing Mason's hypothesis of sex differences in affective behavior. *Behaviour* 32:49-69, 1968.
Sade, D.S. Inhibition of son-mother mating among free-ranging rhesus monkeys. *Sci. Psychoanal.* 12:18-38, 1968.
Wilson, A.P., Vessey, S.H. Behavior of free-ranging castrated rhesus monkeys. *Folia Primatol.* 9:1-14, 1968.
Dicks, D., Myers, R.E., Kling, A. Uncus and amygdala lesions: Effects on social behavior in the free-ranging rhesus monkey. *Science* 165:69-71, 1969.
Meier, G.W. The re-establishment of a rhesus social group. In C.R. Carpenter (ed.), *Behavior.* Basel Switzerland, Karger, pp. 66-71, 1969.
Menzel, E.W., Jr. Naturalistic and experimental approaches to primate behavior. In E. Williams and H. Raush (eds.), *Naturalistic Viewpoints in Psychological Research,* New York, Holt, Rinehart and Winston, 1969.
Missakian, E.A., del Rio, L., Myers, R.E. Reproductive behavior of castrate male rhesus monkeys. *Commun. Behav. Biol.* 4:231-235, 1969.

Morayta, E. Cayo Santiago es el paraiso del mono. In *Apuntos para la Biografia de un Edificio: La Escuela de Medicina Tropical de Puerto Rico.* University of Puerto Rico, School of Medicine, San Juan, PR, pp. 76-81, 1969.

Shah, K.V., Hess, D.M. Presence of antibodies to Simian Virus 40 (SV40) T antigen in rhesus monkeys infected experimentally or naturally with SV40. *Soc. Exp. Biol. Med.* 128:480-485, 1969.

Shah, K.V., Morrison, J.A. Comparison of three rhesus groups for antibody patterns to some viruses: Absence of active simian virus 40 transmission in the free-ranging rhesus of Cayo Santiago. *Am. J. Epidemiol.* 80:308-315, 1969.

Wilson, A.P. Social behavior of free-ranging rhesus monkeys with an emphasis on aggression. *Dissert. Absts.* B29:4494, 1969.

DiGiacomo, R.F., McCann, T.O. Gynecologic pathology in the *Macaca mulatta. Am. J. Obstet. Gynecol.* 108:538-542, 1970.

Loy, J.D. Behavioral responses of free-ranging rhesus monkeys to food shortage. *Am. J. Phys. Anthropol.* 33:263-272, 1970.

Loy, J.D. Perimenstrual sexual behavior among rhesus monkeys. *Folia Primatol.* 13:286-297, 1970.

Vessey, S.H., Morrison, J.A. Molt in free-ranging rhesus monkeys. *J. Mammalol.* 51:89-94, 1970.

Wilson, A.P., Boelkins, R.C. Evidence for seasonal variation in aggressive behaviour by *Macaca mulatta. Anim. Behav.* 18:719-724, 1970.

Conaway, C.H. Ecological adaptation and mammalian reproduction. *Biol. Reprod.* 4:239-247, 1971.

Loy, J.D. Estrous behavior of free-ranging rhesus monkeys. *Primates* 12:1-31, 1971.

Suarez, B., Ackerman, D.R. Social dominance and reproductive behavior in male rhesus monkeys. *Am. J. Phys. Anthropol.* 35:219-222, 1971.

Boelkins, R.C., Wilson, A.P. Intergroup social dynamics of the Cayo Santiago rhesus (*Macaca mulatta*) with special reference to changes in group membership by males. *Primates* 13:125-139, 1972.

Carpenter, C.R. Breeding colonies of macaques and gibbons on Santiago Island, Puerto Rico. In W. Beveredge (ed.), *Breeding Primates.* Basel Switzerland, Karger, pp. 76-87, 1972.

DiGiacomo, R.F., Missakian, E.A. Tetanus in a free-ranging colony of *Macaca mulatta:* A clinical and epizootiological study. *Lab. Animal Sci.* 22:378-383, 1972.

Hausfater, G.H. Intergroup behavior of free-ranging rhesus monkeys (*Macaca mulatta). Folio Primatol.* 18:78-107, 1972.

Missakian, E.A. Genealogical and cross-genealogical dominance relations in a group of free-ranging rhesus monkeys (*Macaca mulatta*) on Cayo Santiago. *Primates* 13:169-180, 1972.

Missakian, E.A. Effects of adult social experience on patterns of reproductive activity of socially deprived male rhesus monkeys (*Macaca mulatta*). *J. Personality Soc. Psychol.* 21:131-134, 1972.

Morrison, J.A., Menzel, E.W., Jr. Adaptation of a free-ranging rhesus monkey group to division and transplantation. *Wildl. Monogr.* 31:6-78, 1972.

Sade, D.S. Sociometrics of *Macaca mulatta.* I. Linkages and cliques in grooming matrices. *Folia Primatol.* 18:196-223, 1972.

Sade, D.S. A longitudinal study of social behavior of rhesus monkeys. In R. Tuttle (ed.), *Functional and Evolutionary Biology of Primates.* Chicago, Aldine-Atherson, pp. 378-398, 1972.

Brueggeman, J.A. Parental care in a group of free-ranging rhesus monkeys (*Macaca mulatta*). *Folia Primatol.* 20:178, 1973.

Gavan, J.A., Hutchinson, T.C. The problem of age estimation: A study using rhesus monkeys (*Macaca mulatta*). *Am. J. Phys. Anthropol.* 38:69-82, 1973.

Marsden, H.M. Aggression within social groups of rhesus monkeys (*Macaca mulatta*). Effect of contact between groups. *Anim. Behav.* 21:247-249, 1973.

Meier, R.J. Considerations of function in macaque dermatoglyphics. *Folia Primate Aggression, Territoriality, and Xenophobia.* New York, Academic

Miller, M.H., Kling, A., Dicks, D. Familiar interactions of male rhesus monkeys in a semi-free-ranging troop. *Am. J. Phys. Anthropol.* 38:605-612, 1973.

Missakian, E.A. Genealogical mating activity in free-ranging groups of rhesus monkeys (*Macaca mulatta*) on Cayo Santiago. *Behaviour* 45:224-241, 1973.

Missakian, E.A. The timing of fission among free-ranging rhesus monkeys. *Am. J. Phys. Anthropol.* 38:621-624, 1973.

Myers, R.E., Sweet, C., Miller, M. Loss of social group affinity following prefrontal lesions in free-ranging macaques. *Brain Res.* 64:257-269, 1973.

Sade, D.S. An ethogram for rhesus monkeys. I. Anti-thetical contrasts in posture and movement. *Am. J. Phys. Anthropol.* 38:537-542, 1973.

Vandenbergh, J.G. Environmental influences on breeding in rhesus monkeys. In C.H. Phoenix (ed.), *Primate Reproductive Behavior.* Basel Switzerland, Karger, pp. 1-19, 1973.

Buettner-Janusch, J., Dame, L., Mason, G.A., Sade, D.S. Primate red cell enzymes: Glucose-6-phosphate dehydrogenase and 6-phosphogluconate dehydrogenase. *Am. J. Phys. Anthropol.* 41:7-14, 1974.

Buettner-Janusch, J., Mason, G.A., Dame, L., Buettner-Janusch, V., Sade, D.S. Genetic studies of serum transferrins of free-ranging macaques of Cayo Santiago, *Macaca mulatta* (Zimmerman, 1780). *Am. J. Phys. Anthropol.* 41:217-232, 1974.

Carpenter, C.R. Aggressive behavioral systems. In R.L. Holloway (ed.), *Primate aggression, Territoriality, and Xenophobia.* New York, Academic Press, pp. 459-496, 1974.

Missakian, E.A. Mother-offspring grooming relations in rhesus monkeys. *Arch. Sex. Behav.* 3:135-141, 1974.

Rawlins, R.G. Metamorphosis of the pubo-ischiadic symphysis in *Macaca mulatta*. *Am. J. Phys. Anthropol.* 41:500, 1974.

Sade, D.S., Schneider, J., Figueroa, A., Conley, J., Correa, O., Cushing, K., Cushing, P., Lauer, C., Serksnys, S. Recent history and characteristics of the Cayo Santiago population. *Am. J. Phys. Anthropol.* 41:501-502, 1974.

Zamboni, L. Fine morphology of the follicle wall and follicle cell-oocyte association. *Biol. Reprod.* 10:125-149, 1974.

Zamboni, L., Conaway, C.H., Van Pelt, L. Seasonal changes in production of semen in free-ranging rhesus monkeys. *Biol. Reprod.* 11:251-267, 1974.

Berman, C.M. Seaside play is a serious business. *New Scientist* 73:671-673, 1975.

Blake, R.N. A free-ranging approach to primate research. *Lab. Anim.* Jul-Aug:28-32, 1975.

Buikstra, J.E. Healed fractures in *Macaca mulatta*, age, sex and symmetry. *Folia Primatol.* 23:140-148, 1975.

Duggleby, C.R. Genetic markers and population structure on Cayo Santiago. *Am. J. Phys. Anthropol.* 42:298, 1975.

Kerber, W.T. The Caribbean Primate Research Center. *Lab. Prim. Newsl.* 14:6-8, 1975.

Lauer, C. The relationship of tooth size and body size in a population of rhesus monkeys (*Macaca mulatta*). *Am. J. Phys. Anthropol.* 43:333-340, 1975.

Lauer, C. Seasonal variation in troop movement patterns of free-ranging rhesus macaques. *Am. J. Phys. Anthropol.* 42:313, 1975.

Loy, J. The descent of dominance in Macaca: Insights into the structure of human societies. In R.H. Tuttle (ed.), *Socioecology and Psychology of Primates.* The Hague, Mouton Publishers, pp. 153-180, 1975.

Rawlins, R.G. Age changes in pubic symphysis of *Macaca mulatta*. *Am. J. Phys. Anthropol.* 42:477-487, 1975.

Sade, D.S. Management of data on social behavior of free-ranging rhesus monkeys. In B. Mittman, L. Borman (eds), *Personalized Data Base Systems.* New York, Wiley, pp. 95-110, 1975.

Sade, D.S., Schneider, J., Figueroa, A., Kaplan, J., Cushing, K., Cushing, P., Dunaif, J., Morse, T., Rhodes, D., Stewart, N.M. Population dynamics in relation to social structure on Cayo Santiago. *Am. J. Phys. Anthropol.* 42:327, 1975.

Rawlins, R.G. Locomotor ontogeny in *Macaca mulatta* I: Behavioral strategies and tactics. *Am. J. Phys. Anthropol.* 44:201, 1976.

Zamboni, L. Seasonal changes in spermatogenesis in free-ranging rhesus macaques. *J. Ultrastruct. Res.* 55:294-295, 1976.

Berman, C.M. Significance of animal play. *Nurs. Mirror* 145:10-13, 1977.

Buettner-Janusch, J., Sockol, M. Genetic studies of free-ranging macaques of Cayo Santiago. I. Description of the population and some nonpolymorphic red cell enzymes. *Am. J. Phys. Anthropol.* 47:371-374, 1977.

Buettner-Janusch, J., Sockol, M. Genetic studies of free-ranging macaques of Cayo Santiago. II. 6-phosphogluconate dehydrogenase and NADH-methemoglobin reductase (NAH-diaphorase). *Am. J. Phys. Anthropol.* 47:375-379, 1977.

DeRousseau, C.J. Variability in normal and pathological bone remodeling at the macaque elbow. *Am. J. Phys. Anthropol.* 47:126, 1977.

DiGiacomo, R.F. Gynecologic pathology in the rhesus monkey (*Macaca mulatta*). II. Findings in laboratory and free-ranging monkeys. *Vet. Pathol.* 14:539-546, 1977.

Duggleby, C.R. Blood group antigens and the population genetics of *Macaca mulatta* on Cayo Santiago. II. Effects of social group division. *Yrbk. Phys. Anthropol.* 20:263-271, 1977.

Kaplan, J.R. Intergroup variation in the social behavior of rhesus monkeys. *Am. J. Phys. Anthropol.* 47:141, 1977.

Kaplan, J.R. Patterns of fight interference in free-ranging rhesus monkeys. *Am. J. Phys. Anthropol.* 47:279-288, 1977.

Rawlins, R.G. Locomotor ontogeny in *Macaca mulatta*. II. Locomotor dimensions of the shoulder girdle and forelimb. *Am. J. Phys. Anthropol.* 47:256, 1977.

Sade, D.S., Cushing, K., Cushing, P., Dunaif, J., Figueroa, A., Kaplan, J.R., Lauer, C., Rhodes, D., Schneider, J. Population dynamics in relation to social structure on Cayo Santiago. *Yrbk. Phys. Anthropol.* 20:253-262, 1977.

Varley, M.A., Vessey, S.H. Effects on geographic transfer on the timing of seasonal breeding of rhesus monkeys. *Folia Primatol.* 28:52-59, 1977.

Berman, C.M. The analysis of mother-infant interaction in groups: Possible influence of yearling siblings. In D. Chivers, J. Herbert (eds.), *Recent Advances in Primatology.* New York, Academic Press, pp. 111-113, 1978.

Breuggeman, J.A. The function of adult play in free-ranging *Macaca mulatta.* In E.O. Smith (ed.), *Social Play in Primates.* New York, Academic Press, pp. 169-192, 1978.

Cheverud, J., Buettner-Janusch, J., Sade, D.S. Social group fission and the origin of intergroup genetic differentiation among the rhesus monkeys of Cayo Santiago. *Am. J. Phys. Anthropol.* 49:449-456, 1978.

Cheverud, J., Buikstra, J. A study of intragroup biological change induced by social group fission in *Macaca mulatta* using discrete cranial traits. *Am. J. Phys. Anthropol.* 48:41-46, 1978.

DeRousseau, C.J. Biomechanical aspects of osteoarthritis in primates, including man. *Am. J. Phys. Anthropol.* 48:389, 1978.

Devor, E.J. Genetic variations in transferrin alleles of rhesus macaques, *Macaca mulatta. Am. J. Phys. Anthropol.* 48:165-169, 1978.

Duggleby, C.R. Blood group antigens and the population genetics of *Macaca mulatta* on Cayo Santiago. I. Genetic differentiation of social groups. *Am. J. Phys. Anthropol.* 48:35-40, 1978.

Duggleby, C.R. Genetic distance between troops on Cayo Santiago macaques through time. *Am. J. Phys. Anthropol.* 48:390, 1978.

Duggleby, C.R. Rhesus on Cayo Santiago: Effective population size and migration rates. In D.J. Chivers, J. Herbert (eds.), *Recent Advances in Primatology Vol. I, Behavior.* New York, Academic Press, pp. 189-191, 1978.

El-Mofty, A., Gouras, P., Eisner, G., Balazs, E.A. Macular degeneration in rhesus monkeys (*Macaca mulatta*). *Exp. Eye Res.* 27:499-502, 1978.

Haseley, P.A. Genetic variability of birth cohorts of rhesus on Cayo Santiago. *Am. J. Phys. Anthropol.* 48:404, 1978.

Kaplan, J.R. Comparison of compound and free-ranging rhesus monkeys. *Am. J. Phys. Anthropol.* 48:410, 1978.

Kaplan, J.R. Fight interference and altruism in rhesus monkeys. *Am. J. Phys. Anthropol.* 49:449-456, 1978.

Olivier, T.J., Ober, C., Buettner-Janusch, J. Genetics of group fission on Cayo Santiago. *Am. J. Phys. Anthropol.* 48:424, 1978.

Rawlins, R.G. Locomotive and manipulative use of the hand in *Macaca mulatta. Am. J. Phys. Anthropol.* 48:427, 1978.

Schulman, S.R. Kin selection, reciprocal altruism, and the principle of maximization: A reply to Sahlins. *Quart. Rev. Biol.* 53:284-286, 1978.

Bito, L.Z., Merritt, S.Q., DeRousseau, C.J. Intraocular pressure of rhesus monkeys (*Macaca mulatta*). I. An initial survey of two free-breeding colonies. *Invest. Ophthal. Vis. Sci.* 18:785-793, 1979.

Chepko-Sade, B.D. Monkey group splits up. *New Scientist* 82:348-350, 1979.

Chepko-Sade, B.D., Sade, D.S. Patterns of group splitting within matrilineal kinship groups: Study of social group structure in *Macaca mulatta* (Cercopithecidae, primates). *Behav. Ecol. Sociobiol.* 5:67-86, 1979.

Chepko-Sade, B.D., Olivier, T.J. Coefficient of genetic relationship and the probability of intragenealogical fission in *Macaca mulatta. Behav. Ecol. Sociobiol.* 5:263-278, 1979.

Cheverud, J., Buikstra, J. Heritability of nonmetric skeletal traits in the rhesus monkeys on Cayo Santiago. *Am. J. Phys. Anthropol.* 50:427, 1979.

Colvin, J. Inter-group transfer in semi-free-ranging male rhesus macaques. I. Relationships of immature males with adult females. *Am. J. Phys. Anthropol.* 50:429, 1979.

DeRousseau, C.J. Aging in *Macaca mulatta*. I. Passive movement as an indicator of joint function. *Am. J. Phys. Anthropol.* 50:432, 1979.

Dunaif, J.C. The effects of estrus on the social network of female rhesus monkeys (*Macaca mulatta*). *Am. J. Phys. Anthropol.* 50:434, 1979.

Lauer, C. Population density and intergroup relations in free-ranging rhesus monkeys, *Macaca mulatta. Am. J. Phys. Anthropol.* 50:456, 1979.

McMillan, C. Genetic differentiation of female lineages in rhesus macaques at Cayo Santiago. *Am. J. Phys. Anthropol.* 50:461-462, 1979.

Ober, C.L., Buettner-Janusch, J., Olivier, T.J. Genetic differentiation between matrilines in the Cayo Santiago macaque groups. *Am. J. Phys. Anthropol.* 50:468, 1979.

Quiatt, D. Aunts and mothers: Adaptive implications of allomaternal behavior of nonhuman primates. *Am. Anthropologist* 81:310-319, 1979.

Rawlins, R.G. Forty years of rhesus research. *New Scientist* 82:108-110, 1979.

Rawlins, R.G. Parturient and post-partum behavior in a free-ranging rhesus monkey (*Macaca mulatta*). *J. Mammalol.* 60:432-433, 1979.

Rawlins, R.G., Decarie, D. Sampling positional behavior of Cayo Santiago macaques (*Macaca mulatta*). *Am. J. Phys. Anthropol.* 50:473, 1979.

Baulu, J., Redmond, D.E., Jr. Social and nonsocial behaviours of sex- and age-matched enclosed and free-ranging rhesus monkeys (*Macaca mulatta*). *Folia Primatol.* 34:239-258, 1980.

Berman, C.M. Early agonistic experience and rank acquisition among free-among free-ranging infant rhesus monkeys. *Am. J. Phys. Anthropol.* 52:204, 1980.

Berman, C.M. Mother-infant relationships among free-ranging rhesus monkeys on Cayo Santiago: A comparison with captive pairs. *Anim. Behav.* 28:860-873, 1980.

Berman, C.M. Social relationships among free-ranging infant rhesus monkeys. *Dissert. Absts. Internat.* B40:3614, 1980.

Berman, C.M. Early Agonistic experience and rank acquisition among free-ranging infant rhesus monkeys. *Internat. J. Primatol.* 1:153-170, 1980.

Chapais, B., Schulman, S.R. An evolutionary model of female dominance relations in primates. *J. Theoret. Biol.* 82:47-89, 1980.

Chapais, B., Schulman, S.R. Alarm responses to raptors by rhesus monkeys at Cayo Santiago. *J. Mammalol.* 61:739-741, 1980.

Cheverud, J. Phenotypic, genetic and environmental correlations among cranial measurements in the rhesus monkeys of Cayo Santiago. *Am. J. Phys. Anthropol.* 52:513, 1980.

Denlinger, J.L., Eisner, G., Balazs, E.A. Age-related changes in the vitreus and lens of rhesus monkeys (*Macaca mulatta*). *Exp. Eye Res.* 31:67-79, 1980.
El-Mofty, A.A.M., Eisner, G. Balazs, E.A., Denlinger, J.L., Gouras, P. Retinal degeneration in rhesus monkeys, *Macaca mulatta*. Survey of three seminatural free-breeding colonies. *Exp. Eye Res.* 31:147-166, 1980.
Haseley, P.A. Robertson's effect and the variability in troop birth cohorts of Cayo Santiago rhesus. *Am. J. Phys. Anthropol.* 52:236, 1980.
Lauer, C. Seasonal variability in spatial defense by free-ranging rhesus monkeys (*Macaca mulatta*). *Anim. Behav.* 28:476-482, 1980.
McMillan, C.A. A lineage-specific mating model for the maintenance of inter-lineage differentiation among *Macaca mulatta* on Cayo Santiago. *Am. J. Phys. Anthropol.* 52:253-254, 1980.
McMillan, C.A. The possibility of selection as a factor in the patterning of genetic distances between matrilineages of rhesus monkeys on Cayo Santiago. *Am. J. Phys. Anthropol.* 52:251-252, 1980.
Meier, G.W. The Caribbean Primate Research Center: A facility for life-cycle research. *N. Y. Med. Qtr.* 2:10-13, 1980.
Ober, C., Olivier, T.J., Buettner-Janusch, J. Genetic aspects of migration in a rhesus monkey population. *J. Human Evol.* 9:197-203, 1980.
Rhodes, R.L. Patterning of mating in a free-ranging group of rhesus monkeys. *Am. J. Phys. Anthropol.* 52:270, 1980.
Sade, D.S. Population biology of free-ranging rhesus monkeys on Cayo Santiago, Puerto Rico. In M.N. Cohen, R.S. Malpass, H.G. Klein (eds.), *Biosocial Mechanisms of Population Regulation*. New Haven, Yale U. Press, pp. 171-187, 1980.
Schulman, S.R., Chapais, B. Reproductive value and rank relations among macaque sisters. *Am. Natural.* 115:580-593, 1980.
Schulman, S.R. Intragroup spacing and multiple social networks in *Macaca mulatta*. *Dissert. Absts. Internat.* B41:1633, 1980.
Windle, W.F. The Cayo Santiago primate colony. *Science* 209:1486-1491, 1980.
Brereton, A.R. Intergroup consorting by a free-ranging female rhesus monkey. *Primates* 22:417-423, 1981.
Cheverud, J. Ontogenic and phenotypic, genetic and environmental static allometry in the primate cranium. *Am. J. Phys. Anthropol.* 54:208, 1981.
Cheverud J.M. Variations in highly and lowly heritable morphological traits among social groups of rhesus macaques (*Macaca mulatta*) on Cayo Santiago. *Evol.* 35:75-83, 1981.
Cheverud, J.M., Buikstra, J.E. Quantitative genetics of skeletal nonmetric traits in the rhesus macaques on Cayo Santiago. I. Single trait heritabilities. *Am. J. Phys. Anthropol.* 54:43-49, 1981.
Cheverud, J.M., Buikstra, J.E. Quantitative genetics of skeletal nonmetric traits in the rhesus macaques on Cayo Santiago. II. Phenotypic, genetic and environmental correlations between traits. *Am. J. Phys. Anthropol.* 54:51-58, 1981.
Cheverud, J.M. Epiphyseal union and dental eruption in *Macaca mulatta*. *Am. J. Phys. Anthropol.* 56:157-167, 1981.
DeRousseau, C.J., Bito, L.Z. Intraocular pressure of rhesus monkeys (*Macaca mulatta*). II. Juvenile ocular hypertension and its apparent relationship to ocular growth. *Exp. Eye Res.* 32:407-417, 1981.

DeRousseau, C.J., Bito, J.W., Kaufman, P.L. Development of presbyopia, degenerative disk disease and other predictable age dependent afflictions in rhesus monkeys. *Am. J. Phys. Anthropol.* 54:214, 1981.

Duggleby, C.R. The use of red cell antigen markers in paternity exclusions. *Am. J. Primatol.* 1:332-333, 1981.

McMillan, C.A. Synchrony of estrus in macaque matrilines at Cayo Santiago. *Am. J. Phys. Anthropol.* 54:251, 1981.

McMillan, C.A., Duggleby, C.R. Interlineage genetic differentiation among rhesus macaques on Cayo Santiago. *Am. J. Phys. Anthropol.* 56:305-312, 1981.

Olivier, T.J., Ober, C., Buettner-Janusch, J., Sade, D.S. Genetic differentiation among matrilines in social groups of rhesus monkeys. *Behav. Ecol. Sociobiol.* 8:279-285, 1981.

Berman, C.M. The roles of maternal age and the presence of close kin on mother-infant interaction among free-ranging rhesus monkeys on Cayo Santiago. *Internat. J. Primatol.* 3:261, 1982.

Berman, C.M. The ontogeny of social relationships with group companions among free-ranging infant rhesus monkeys. I. Social networks and differentiation. *Anim. Behav.* 30:149-162, 1982.

Berman, C.M. The ontogeny of social relationships with group companions among free-ranging infant rhesus monkeys. II. Differentiation and attractiveness. *Anim. Behav.* 30:163-170, 1982.

Berman, C.M. The social development of an orphaned rhesus infant on Cayo Santiago: Male care, foster mother-orphan interaction and peer interaction. *Am. J. Primatol.* 3:131-141, 1982.

Cheverud, J.M. Relationships among ontogenetic, static and evolutionary allometry. *Am. J. Phys. Anthropol.* 59:139-149, 1982.

Cheverud, J.M. Phenotypic, genetic and environmental morphological integration in the cranium. *Evol.* 36:499-516, 1982.

Cheverud, J.M., Buikstra, J.E. Quantitative genetics of skeletal nonmetric traits in the rhesus macaques of Cayo Santiago. III. Relative heritability of skeletal nonmetric and metric traits. *Am. J. Phys. Anthropol.* 59:151-155, 1982.

Colvin, J.D. Proximate causes of male emigration in rhesus monkeys. *Internat. J. Primatol.* 3:277, 1982.

Datta, S. The role of alliances in the acquisition of intra-family rank in the rhesus macaque (*Macaca mulatta*). In M.L. Roonwal (ed.), *Abstracts of the International Symposium on Primates.* Jodhpur, India, p. 39, 1982.

DeRousseau, C.J. Patterns of skeletal aging in rhesus monkeys. *Am. J. Phys. Anthropol.* 57:180, 1982.

Gouzoules, H., Gouzoules, S., Marler, P. Meaning, social context and intentionality in macaques' vocalizations. *Internat. J. Primatol.* 3:249, 1982.

Kessler, M.J. Nasal and cutaneous anatrichosomiasis in the free-ranging rhesus monkeys (*Macaca mulatta*) of Cayo Santiago. *Am. J. Primatol.* 3:55-60, 1982.

Kessler, M.J., Rawlins, R.G., London, W.T. Tetanus in monkeys: Symptomatology, epizootiology, treatment and prophylaxis, pp. 37-39 In *Proc: 1982 Ann. Meeting Am. Assoc. Zoo Vets.*, M.E. Fowler (ed.), 1982.

Marriott, B.M., Smith, J.C., Jr., Jones, A.O., Rawlins, R.G., Kessler, M.J. Hair mineral content in rhesus monkeys. *Fed. Proc.* 41:770, 1982.

McMillan, C.A. Dominance, adult-subadult status and male mating success in rhesus macaques. *Am. J. Phys. Anthropol.* 57:207, 1982.

McMillan, C.A. Male age and mating success among rhesus macaques. *Internat. J. Primatol.* 3:312, 1982.

Rawlins, R.G., Kessler, M.J. A five-year study of tetanus in the Cayo Santiago rhesus monkey colony: Behavioral description and epizootiology. *Am. J. Primatol.* 3:23-39, 1982.

Rawlins, R.G., Kessler, M.J. Behavioral description and epidemiology of tetanus in the rhesus monkeys of Cayo Santiago. *Internat. J. Primatol.* 3:326, 1982.

Roemer, J., Marriott, B.M. Feeding patterns of rhesus monkeys (*Macaca mulatta*) on Cayo Santiago island, Puerto Rico. *Internat. J. Primatol.* 3:327, 1982.

Sultana, C.J., Marriott, B.M. Geophagia and related behavior of rhesus monkeys (*Macaca mulatta*) on Cayo Santiago island, Puerto Rico. *Internat. J. Primatol.* 3:338, 1982.

Berman, C. Infant social developments: Studies on Cayo Santiago, 1974-present. *Am. J. Primatol.* 4:327, 1983.

Berman, C.M. Effects of being orphaned: A detailed case study of an infant rhesus. In R.A. Hinde (ed.), *Primate Social Relationships: An Integrated Approach.* Oxford, Blackwells Scientific Publ., pp. 79-81, 1983.

Berman, C.M. Differentiation of relationships among rhesus infants. In R.A. Hinde (ed.), *Primate Social Relationships: An Integrated Approach.* Oxford, Blackwells Scientific Publ., pp. 89-93, 1983.

Berman, C.M. Matriline differences in infant development. In R.A. Hinde (ed.), *Primate Social Relationships: An Integrated Approach.* Oxford, Blackwells Scientific Publ., pp. 132-134, 1983.

Berman, C.M. Early differences in relationships between infants and other group members based on mother's rank. In R.A. Hinde (ed.), *Primate Social Relationships: An Integrated Approach.* Oxford, Blackwells Scientific Publ., pp. 154-156, 1983.

Berman, C.M. Influence of close female relatives on peer-peer rank acquisition. In R.A. Hinde (ed.), *Primate Social Relationships: An Integrated Approach.* Oxford, Blackwells Scientific Publ., pp. 157-159, 1983.

Buettner-Janusch, J., Olivier, T.J., Ober, C.L., Chepko-Sade, B.D. Models for lineal effects in rhesus group fissions. *Am. J. Phys. Anthropol.* 61:347-353, 1983.

Chapais, B. Reproductive activity in relation to male dominance and the likelihood of ovulation in rhesus monkeys. *Behav. Ecol. Sociobiol.* 12:215-228, 1983.

Chapais, B. Mate selection among the Cayo Santiago rhesus monkeys. *Am. J. Primatol.* 4:328, 1983.

Chapais, B. Matriline membership and male rhesus reaching high ranks in their natal troops. In R.A. Hinde (ed.), *Primate Social Relationships: An Integrated Approach.* Oxford, Blackwells Scientific Publ., pp. 171-175, 1983.

Chapais, B. Structure of the birth season relationship among adult male and female rhesus monkeys. In R.A. Hinde (ed.), *Primate Social Relationships: An Integrated Approach.* Oxford, Blackwells Scientific Publ., pp. 200-208, 1983.

Chapais, B. Dominance, relatedness and the structure of female relationships in rhesus monkeys. In R.A. Hinde (ed.), *Primate Social Relationships: An Integrated Approach.* Oxford, Blackwells Scientific Publ., pp. 209-219, 1983.

Chapais, B. Autonomous, bisexual subgroups in a troop of rhesus monkeys. In R.A. Hinde (ed.), *Primate Social Relationships: An Integrated Approach.* Oxford, Blackwells Scientific Publ., pp. 220-221, 1983.

Chapais, B. Male dominance and reproductive activity in rhesus monkeys. In R.A. Hinde (ed.), *Primate Social Relationships: An Integrated Approach.* Oxford, Blackwells Scientific Publ., pp. 267-271, 1983.

Chapais, B. Adaptive aspects of social relationships among adult rhesus monkeys. In R.A. Hinde (ed.), *Primate Social Relationships: An Integrated Approach.* Oxford, Blackwells Scientific Publ., pp. 286-289, 1983.

Chapais, B., Schulman, S.R. Fitness and female dominance relationships. In R.A. Hinde (ed.), *Primate Social Relationships: An Integrated Approach.* Oxford, Blackwells Scientific Publ., pp. 271-278, 1983.

Chepko-Sade, B.D. Role of males in group fission in *Macaca mulatta. Dissert. Absts. Internat.* A43(10):3363, 1983.

Colvin, J.D. Proximate causes of male emigration at puberty in rhesus monkeys. *Am. J. Primatol.* 4:328, 1983.

Colvin, J. Description of sibling and peer relationships among immature male rhesus monkeys. In R.A. Hinde (ed.), *Primate Social Relationships: An Integrated Approach.* Oxford, Blackwells Scientific Publ., pp. 20-27, 1983.

Colvin, J. Rank influences rhesus male peer relationships. In R.A. Hinde (ed.), *Primate Social Relationships: An Integrated Approach.* Oxford, Blackwells Scientific Publ., pp. 57-64, 1983.

Colvin, J. Influences of the social situation on male emigration. In R.A. Hinde (ed.), *Primate Social Relationships: An Integrated Approach.* Oxford, Blackwells Scientific Publ., pp. 160-171, 1983.

Colvin, J. Familiarity, rank and the structure of rhesus male peer networks. In R.A. Hinde (ed.), *Primate Social Relationships: An Integrated Approach.* Oxford, Blackwells Scientific Publ., pp. 190-200, 1983.

Datta, S.B. Relative power and the acquisition of rank. In R.A. Hinde (ed.), *Primate Social Relationships: An Integrated Approach.* Oxford, Blackwells Scientific Publ., pp. 93-103, 1983.

Datta, S.B. Relative power and the maintenance of dominance. In R.A. Hinde (ed.), *Primate Social Relationships: An Integrated Approach.* Oxford, Blackwells Scientific Publ., pp. 103-112, 1983.

Datta, S.B. Patterns of agonistic interference. In R.A. Hinde (ed.), *Primate Social Relationships: An Integrated Approach.* Oxford, Blackwells Scientific Publ., pp. 289-297, 1983.

DeRousseau, C.J., Bito, L.Z., Kaufman, P.L. Age-dependent impairment of the rhesus visual and musculoskeletal systems and possible behavioral consequences. *Am. J. Primatol.* 4:329, 1983.

DeRousseau, C.J., Rawlins, R.G., Denlinger, J.L. Aging in the musculoskeletal system of rhesus monkeys: I. Passive joint excursion. *Am. J. Phys. Anthropol.* 61:483-494, 1983.

Duggleby, C., Haseley, P., Sade, D., Rawlins, R., Kessler, M. Findings of blood group genetic studies on the colony 1972 to present. *Am. J. Primatol.* 4:330, 1983.

Gouzoules, S., Gouzoules, H. Mothers' and others' responses to immature rhesus monkey screams. Am. J. Primatol. 4:327, 1983.

Hill, D.A. Seasonal differences in the social relationships of adult male rhesus macaques. *Am. J. Primatol.* 4:328, 1983.

Kessler, M.J. Cayo Santiago, home of the rhesus monkey. *Lab. Prim. Newsl.* 22:9-10, 1983.

Kessler, M.J., Rawlins, R.G. Age- and pregnancy-related changes in serum total cholesterol and triglyceride levels in the Cayo Santiago rhesus macaques. *Exp. Gerontol.* 18:1-4, 1983.

Kessler, M.J., Rawlins, R.G. The hemogram, serum biochemistry and electrolyte profile of the free-ranging Cayo Santiago rhesus macaques (*Macaca mulatta*). *Am. J. Primatol.* 4:107-116, 1983.

Kessler, M.J., Rawlins, R.G. Golden rhesus macaques (*Macaca mulatta*) on Cayo Santiago. *Am. J. Primatol.* 4:330, 1983.

Kessler, M.J., Rawlins, R.G. Animal Models of Human Disease: Tetanus. *Comp. Pathol. Bull.* 15(4):3, 1983.

Marriott, B., Smith, J.C., Jr., Jacobs, R.M., Lee Jones, A.O. Hair mineral content as an indicator of mineral nutriture in rhesus monkeys. *Am. J. Primatol.* 4:329, 1983.

McGrath, J.W., Cheverud, J.M., Buikstra, J.E. Genetic correlations between sides and heritability of asymmetry for nonmetric traits in rhesus macaques on Cayo Santiago. *Am. J. Phys. Anthropol.* 60:224, 1983.

McMillan, C.A. Lineage-specific mating, does it exist? *Am. J. Primatol.* 4:329, 1983.

McMillan, C.A. Factors affecting mating success among rhesus macaque males on Cayo Santiago. *Dissert. Absts. Internat.* A43(9):3050, 1983.

Pearl, M.C., Schulman, S.R. Techniques for the analysis of social structure in animal societies. *Adv. Study Behav.* 13:107-146, 1983.

Rawlins, R.G., Kessler, M.J. Demography of the free-ranging Cayo Santiago macaques (*M. mulatta*): 1976-1982. *Am. J. Primatol.* 4:331, 1983.

Rawlins, R.G., Kessler, M.J. Congenital and hereditary anomalies in the rhesus monkeys (*Macaca mulatta*) of Cayo Santiago. *Teratology* 28:169-174, 1983.

Richtmeier, J.T., Cheverud, J.M., Buikstra, J.E. Etiology of cranial nonmetric traits in the rhesus macaques of Cayo Santiago. *Am. J. Phys. Anthropol.* 60:244-245, 1983.

Sade, D.S., Chepko-Sade, B.D., Loy, J., Kaplan, J.R. Early infant mortality among free-ranging rhesus monkeys. *Am. J. Primatol.* 4:361, 1983.

Scanlon, C.E. Social development in a congenitally blind infant rhesus monkey. *Am. J. Primatol.* 4:327, 1983.

Schulman, S.R. Analysis of social structure. In R.A. Hinde (ed.), *Primate Social Relationships: An Integrated Approach.* Oxford, Blackwells Scientific Publ., pp. 221-225, 1983.

Turnquist, J.E. Age and joint mobility in free-ranging rhesus monkeys (*Macaca mulatta*). *Am. J. Primatol.* 4:330, 1983.

Berman, C.M. Variation in mother-infant relationships: Traditional and nontraditional factors. In M. Small (ed.), *Female Primates: Studies by Women Primatologists.* New York, Alan R. Liss, Inc., pp. 17-36, 1984.

Common, L. The monkey's puzzle: Researchers seek clues to degenerative arthritis in "Monkey Heaven." *Arthritis News* (Canada) 2(3):35-36, 1984.

DeRousseau, C.J. Variation in patterns of growth in rhesus monkeys. *Am. J. Phys. Anthropol.* 63:151, 1984.

Duggleby, C.R., Sade, D.S., Rawlins, R.G., Kessler, M.J. Kin-structured migration and genetic differentiation. *Am. J. Phys. Anthropol.* 63:153, 1984.

Falk, D. The Caribbean Primate Research Center. *Internat. J. Primatol.* 5:337, 1984.

Gouzoules, S., Gouzoules, H., Marler, P. Rhesus monkey *(Macaca mulatta)* screams: representational signaling in the recruitment of agonistic aid. *Anim. Behav.* 32:182-193, 1984.

Haseley, P.A. Differential fertility by blood group phenotype in Cayo Santiago rhesus. *Am. J. Phys. Anthropol.* 63:169, 1984.

Kessler, M., Figueroa, A., Kapsalis, E., Berard, J., Davila, E., Martinez, H., Gonzalez, J. Trapping, removal and translocation of a group of free-ranging rhesus monkeys (*Macaca mulatta*) from Cayo Santiago to a semi-natural hill enclosure. *Am. J. Primatol.* 6:409, 1984.

Kessler, M., Rawlins, R. Absence of naturally-acquired tetanus immunity in the free-ranging Cayo Santiago rhesus monkeys (*Macaca mulatta*). *Am. J. Primatol.* 6:410, 1984.

Kessler, M.J., Rawlins, R.G. Absence of naturally-acquired tetanus antitoxin in the free-ranging Cayo Santiago rhesus monkeys (*Macaca mulatta*). *J. Med. Primatol.* 13:57-66, 1984.

Kessler, M.J., Yarbrough, B., Rawlins, R.G., Berard, J. Intestinal parasites of the free-ranging Cayo Santiago rhesus monkeys (*Macaca mulatta*). *J. Med. primatol.* 13:57-66, 1984.

Loy, J.D., Loy, G., Keifer, G., Conaway, C. *The Behavior of Gonadectomized Rhesus Monkeys.* Basel Switzerland, Karger, 1984.

McGrath, J.W., Cheverud, J.M., Buikstra, J.E. Genetic conditions between sides and heritability of asymmetry for nonmetric traits in rhesus macaques on Cayo Santiago, *Am. J. Phys. Anthropol.* 64:401-411, 1984.

McMillan, C.A. Mating length and visibility in relation to male dominance rank. *Am. J. Primatol.* 6:414-415, 1984.

Ober, C., Olivier, T.J., Sade, D.S., Schneider, J.M., Cheverud, J., Buettner-Janusch, J. Demographic components of gene frequency change in free-ranging macaques on Cayo Santiago. *Am. J. Phys. Anthropol.* 64:223-231, 1984.

Quiatt, D. Juvenile/adolescent role functions in a rhesus monkey troop: An application of household analysis to nonhuman primate social organization. *Internat. J. Primatol.* 5:374, 1984.

Rawlins, R.G., Kessler, M.J. Serum total cholesterol levels and social dominance in adult male rhesus monkeys at Cayo Santiago. *Am. J. Primatol.* 6:420, 1984.

Rawlins, R.G., Kessler, M.J., Turnquist, J.E. Reproductive performance, population dynamics and anthropometrics of the free-ranging Cayo Santiago rhesus macaques. *J. Med. Primatol.* 13:247-259, 1984.

Richtsmeier, J.T., Cheverud, J.M., Buikstra, J.E. The relationship between cranial metric and nonmetric traits in the rhesus macaques from Cayo Santiago. *Am. J. Phys. Anthropol.* 64:213-222, 1984.

Sade, D.S., Dow, M.M., Chepko-Sade, B.D. Multi-state life tables for free-ranging male rhesus monkeys. *Colegium Anthropologicum* 8:237-247, 1984.

Sade, D.S., Rhodes, D.L., Loy, J., Haufsater, G., Breuggeman, J.A., Kaplan, J.R., Chepko-Sade, B.D., Cushing, Kaplan K. New findings on incest among free-ranging rhesus monkeys. *Am. J. Phys. Anthropol.* 63:212, 1984.

Scanlon, C.E. Play and the development of social skills in infant rhesus macaques. *Internat. J. Primatol.* 5:377, 1984.

Turnquist, J.E. Joint mobility and body proportions: a comparison between free-ranging rhesus and patas monkeys. *Am. J. Primatol.* 6:424-425, 1984.

Berard, J. Female transfer in free-ranging rhesus monkeys. *Am. J. Primatol.* 8:331-332, 1985.

Berman, C.M., Rawlins, R.G. Maternal dominance, sex ratio and fecundity in one social group on Cayo Santiago. *Am. J. Primatol.* 8:332, 1985.

Busse, C.D. Post-ejaculatory behavior of free-ranging rhesus monkeys. *Am. J. Primatol.* 8:334, 1985.

Cheverud, J.M., Dow, M.M. An autocorrelation analysis of genetic variation due to lineal fission in a social group of rhesus macaques. *Am. J. Phys. Anthropol.* 67:113-121, 1985.

Colvin, J.D. Breeding-season relationships of immature male rhesus monkeys with females. I. Individual differences and constraints on partner choice. *Internat. J. Primatol.* 6:261-287, 1985.

Colvin, J., Tissier, G. Affiliation and reciprocity in sibling and peer relationship among free-ranging immature male monkeys. *Anim. Behav.* 33:959-977, 1985.

DeRousseau, C.J. Aging in the musculoskeletal system of rhesus monkeys. II. Degenerative joint disease. *Am. J. Phys. Anthropol.* 67:177-184, 1985.

DeRousseau, C.J. Aging in the musculoskeletal system of rhesus monkeys. III. Bone loss. *Am. J. Phys. Anthropol.* 68:157-167, 1985.

Estape-Wainwright, E., Cangiano, J.L., Rodriguez-Sargent, C., Martinez-Maldonado, M. Variability of indirect blood pressure in anesthetized rhesus monkeys: Dependence on blood pressure level. *Fed. Proc.* 44:628, 1985.

Grynpas, M.D., Pritzker, K.P.H., Kessler, M.J. Cortical bone aging in free ranging rhesus monkeys. *Am J. Primatol.* 8:341, 1985.

Haseley, P.A. Genetic structural analysis of Cayo Santiago birth cohort, 1973-1980. *Am. J. Primatol.* 8:342, 1985.

Howard Jr., C., Kessler, M., Schwartz, S. Impaired glucose tolerance (IGT), gestational diabetes, and noninsulin diabetes mellitus (NIDDM) in Cayo Santiago *Macaca mulatta. Am. J. Primatol.* 8:344, 1985.

Howard, C.F., Jr., Kessler, M.J., Schwartz, S. Spontaneous NIDDM and impaired glucose tolerance (IGT) in rhesus monkeys (Macaca mulatta). *Diabetes* 34:191A, 1985.

Kapsalis, E. The status of a translocated troop from Cayo Santiago to a hill corral after one year. *Am. J. Primatol.* 8:346, 1985.

Kessler, M.J., Howard Jr., C.F., London, W.T. Gestational diabetes mellitus and impaired glucose tolerance in an aged *Macaca mulatta. J. Med. Primatol.* 14:237-244, 1985.

Kessler, M.J., Rawlins, R.G. Congenital cataracts in a free-ranging rhesus monkey, *J. Med. Primatol.* 14:225-228, 1985.

Knezevich, M., DeRousseau, C.J. Patterns of epiphyseal fusion in caged and free-ranging rhesus macaques (*Macaca mulatta*). *Am. J. Primatol.* 8:348, 1985.

Marler, P. Representational vocal signals of primates. In B. Holldobler and M. Lindauer (eds.), *Experimental Behavioral Ecology and Sociobiology.* Sunderland, Massachusetts, Sinauer Assoc., Inc., 1985, pp. 211-221.

Opava-Stitizer, S., Rodriguez-Sargent, C. Naturally occuring high blood pressure in rhesus monkeys. *Fed. Proc.* 44:628, 1985.

Pritzker, K.P.H. Cheng, P.-T., Hunter, G.K., Grynpas, M.D., Kessler, M.J., Renlund, R.C. In vitro and in vivo models of calcium pyrophosphate crystal formation. In W.T. Butler (ed.) *The Chemistry and Biology of Mihealited Tissues,* EBSCO Media, Birmingham, Al., pp. 381-384, 1985.

Pritzker, K.P.H., Kessler, M.J., Renlund, R.C., Turnquist, J.E., Tepperman, P.F. Degenerative arthritis in aging rhesus macaques. *Am. J. Primatol* 8:358, 1985.

Pritzker, K.P.H., Kessler, M.J., Turnquist, J., Renlund, R.C. Osteoarthritis and CPPD deposition disease in free ranging rhesus monkeys: Preliminary studies. *Clin. Res.* 33:780A, 1985.

Rawlins, R.G., Kessler, M.J. Environmental factors and seasonal reproduction in the Cayo Santiago macaques. *Am. J. Primatol.* 8:359, 1985.

Rawlins, R.G., Kessler, M.J., Berman, C.M. Sex ratio adjustment in the Cayo Santiago macaques. *Am. J. Primatol.* 8:359, 1985.

Rawlins, R.G., Kessler, M.J. Climate and seasonal reproduction in the Cayo Santiago macaques. *Am. J. Primatol.* 87-99, 1985.

Reichs, K.J., DeRousseau, C.J. Secular trend toward increasing body size in rhesus monkeys. *Am. J. Phys. Anthropol.* 66:217, 1985.

Renlund, R.C., Pritzker, K.P.H., Kessler, M.J. Rhesus primates (*Macaca mulatta*) as a model for calcium pyrophosphate dihydrate crystal deposition disease. *Lab. Invest.* 52:55a, 1985.

MOTION PICTURES

Behavioral Characteristics of the Rhesus Monkey. C.R. Carpenter, producer. Sponsors: The Motion Picture Studio of The Pennsylvania State College (now University), New York Zoological Society; The Viking Fund Inc., and the Department of Anatomy of Columbia University's College of Physicians and Surgeons. Released 1947.

Social Behavior of Rhesus Monkeys. C.R. Carpenter, producer. Sponsors: The Motion Picture Studio of The Pennsylvania State College (now University), New York Zoological Society, The Viking Fund Inc., and the Department of Anatomy of Columbia University's College of Physicians and Surgeons. Released 1947.

The Rhesus Monkeys of Santiago Island. Laboratory of Perinatal Physiology, National Institute of Neurological Diseases and Blindness, National Institutes of Health, San Juan, and the University of Puerto Rico. C.W. Schwartz, E.R. Schwartz, C.B. Koford. Released 1966.

Behavior of Free Ranging Rhesus Monkeys on Cayo Santiago: Mother-Offspring Interaction. M.H. Miller and A. Kling, Brain Research Laboratory, Department of Psychiatry, University of Illinois, College of Medicine and R.E. Meyers, Laboratory of Perinatal Physiology, National Institute of Neurological Diseases and Blindness, National Institute of Health, San Juan. Released 1971.

Island of Monkeys. "The Nature of Things" television series. Nancy Archibald producer, R.G. Rawlins technical consultant Canadian Broadcasting Corporation, Toronto. Released 1979.

The Discovery of Animal Behavior Part III. "Nature" television series. Dr. John Sparks, producer. British Broadcasting Corporation, Natural History Unit, Bristol. Released 1982.

What Animals Have Been Saying Behind Our Backs. "Discover: The World of Science" television series. Peter Graves, producer. Sponsor: Atari, a Warner Communications Company. (Rhesus monkey communications study by Drs. H. and S. Gouzoules.) Released 1983.

POSTSCRIPT

Cayo Santiago

MATT J. KESSLER

Trapped in the wilderness of India
Assembled on the loading docks of Calcutta
After sailing 14 K with Carpenter
They arrived in San Juan bay in November.

Consumption was the scourge of humanity
So the monkeys were tested with PPD
Some succumbed during the long journey
And many more died in captivity.

Four hundred and nine finally made it to freedom
A lush little island with Tomilin to feed them
Although there were problems of disease and socialization
The colony survived thanks to the Markle Foundation.

The war years were lean and so were the *chavos*
German U-boats sank the SS Coamo
With severe food shortages and gasoline rationing
Donations from the townspeople was what was happening.

The monkeys bred so well that some had to be sold
To universities and the army we are told
But all wasn't good and in 1947
They were offered free in Science to anyone who would have them.

But no one wanted Cayo's monkeys then, not even for free
(Isn't it ironic how foolish some people can be?)
So now after 48 years of breeding and observations
Let's not forget to thank all of the monkeys, people and organizations.

Index